THE LAW OF THE RIVER

THE LAW OF THE RIVER

Transboundary River Basin Management and Multi-Level Approaches to Water Quantity Management

Cathy SUYKENS

intersentia
Cambridge – Antwerp – Portland

Intersentia Ltd
Sheraton House | Castle Park
Cambridge | CB3 0AX | United Kingdom
Tel.: +44 1223 370 170 | Fax: +44 1223 370 169
Email: mail@intersentia.co.uk
www.intersentia.com | www.intersentia.co.uk

Distribution for the UK and Ireland:
NBN International
Airport Business Centre, 10 Thornbury Road
Plymouth, PL6 7 PP
United Kingdom
Tel.: +44 1752 202 301 | Fax: +44 1752 202 331
Email: orders@nbninternational.com

Distribution for Europe and all other countries:
Intersentia Publishing nv
Groenstraat 31
2640 Mortsel
Belgium
Tel.: +32 3 680 15 50 | Fax: +32 3 658 71 21
Email: mail@intersentia.be

Distribution for the USA and Canada:
International Specialized Book Services
920 NE 58th Ave. Suite 300
Portland, OR 97213
USA
Tel.: +1 800 944 6190 (toll free) | Fax: +1 503 280 8832
Email: info@isbs.com

The Law of the River. Transboundary River Basin Management and Multi-Level Approaches to Water Quantity Management

ISBN 978-1-78068-631-8 (paperback)
ISBN 978-1-78068-660-8 (PDF)
D/2018/7849/28
NUR 828

GPRC Guaranteed Peer Reviewed Content

British Library Cataloguing in Publication Data. A catalogue record for this book is available from the British Library.

ACKNOWLEDGMENTS

This book is based on the author's PhD research, which has partly been written in the framework of the European Union's Seventh Programme for Research, Technological Development, and Demonstration within the STAR-FLOOD project (grant agreement no. 308364).

This book had been printed with the financial support of the Stichting Schilthuisfonds in The Hague. The author would like to thank the Flemish Research Foundation (FWO) for its support in enabling a research stay at Vermont Law School (Environmental Law Center) in the United States in the summer of 2014 and the faculty of the Environmental Law Center for accommodating my research stay, which has been very inspiring. Warmest thanks is extended to Pascale Van Houtte at Intersentia for her support in publishing this book.

The author's sincere thanks is extended to the members of the jury for their invaluable feedback and support and in particular, prof. dr. Kurt Deketelaere and prof. dr. Marleen van Rijswick for their continuous support throughout the course of this PhD and book. Thank you professors Michael Faure, Annelies Freriks, Geert Van Calster, Simone Borg and Ellen Hey for all your help. The author would like to thank all the water management professionals who agreed to be interviewed as part of this research. A special thank you to Marina, Titia and Klaartje for your support in finalising this PhD and in the run up to the defence.

Warmest thanks to my parents and brother for their unconditional love and support, which have been instrumental attributes to the successful finalisation of the PhD preceding this book. To my friends and colleagues: thank you for all our shared laughter and conversations. A special thank you to Olivier and Claire, for making daily life joyful, warm and exciting.

This work is dedicated to every person that has been affected by natural disasters and aims to contribute to the efforts of rendering our society more resilient.

Cathy Suykens

CONTENTS

Chapter V
Recommendations ... 297

FIGURES AND TABLES

ACRONYMS

BGTC	Benelux Grouping for Territorial Cooperation
CCA	Dutch Calamities Compensation Act
CCR	French Central Reinsurance Fund
CAT-NAT	French National Catastrophes Insurance Mechanism
CIW	Coordination Committee Integrated Water Management
CSF	Critical Success Factor
CWA	Clean Water Act
DG02	Department Navigable Watercourses Walloon Region
DG03	Department Non-Navigable Watercourses Walloon Region
DIWP	Decree Integrated Water Policy
DRBC	Delaware River Basin Commission
ECM	Enhanced Cooperation Model
EGTC	European Grouping for Territorial Cooperation
EIA	Environmental Impact Assessment
EPA	Environmental Protection Agency
EPCI	Territorial Cooperation Instrument for Municipalities
EPTB	Public Basin Organisation
EPTC	Inter-Communal Territorial Organisation
FD	Floods Directive
FRMP	Flood Risk Management Plan
FFMP	Flexible Flow Management Plan
GALA	Dutch General Administrative Law Act
GAO	United States Government Accountability Office
GEIP	General Emergency Intervention Plan
INCODEL	Interstate Commission on the Delaware River Basin
ISC	International Scheldt Commission
KPI	Key Performance Indicator
LCTC	Local Cooperation Mechanism for Transboundary Cooperation
MLWS	Multi-Layer Water Safety
MOW	Department of Mobility and Public Works
MS	EU Member States
NEPA	National Environmental Policy Act
NFIP	National Flood Insurance Program
NPS	National Park Service

PAPI	Action Plan for Flood Prevention
PB	Planetary Boundary
PB-Water	Planetary Boundary for freshwater use
PLU	Local Urbanistic Plan
PPRI	Flood Risk Prevention Plan
PPRN	Natural Disasters Risk Prevention Plan
PGRI	Flood Risk Management Plan
RBD	River Basin District
RBMP	River Basin Management Plan
SAGE	Local Water Management Plan
SCOT	Plan for Territorial Cohesion
SDAGE	Water Management Master Plan (River Basin Management Plan for France)
SDG	Sustainable Development Goal
SEA	Strategic Environmental Assessment
SEIP	Specific Emergency Intervention Plan
SNGRI	National Flood Risk Management Plan
SPW	Service Public de Wallonie
SWCD	Soil and Water Conservation District
TDML	Total Maximum Daily Load
TVA	Tennessee Valley Authority
UDC	Upper Delaware Council
USACE	United States Army Corps of Engineers
USGS	United States Geological Survey
VMM	Flemish Environment Agency
WFD	Water Framework Directive
WIIN	Water Infrastructure for Improvements to the Nation
WRPA	Water Resources Planning Act

CHAPTER I

INTRODUCTION*

1. SUBJECT AND PROBLEM DEFINITION

1. This book deals with transboundary river basin management in the European Union, primarily from the perspective of water quantity management.

2. Since the entry into force of the Water Framework Directive (WFD) in 2000, the EU has attempted to shift the governance of water resources away from the traditional administrative boundaries toward hydrological boundaries through the river basin approach.[1] The "sister Directive" of the WFD, the 2007 Floods Directive (FD), has also adopted this approach.[2] The focal point of the WFD and FD is the River Basin District (RBD), which is defined as "the area of land and sea constituted by one or more river basins, including their associated groundwaters and coastal waters".[3] This focus is a reflection of the "principle of holistic water management at catchment level"[4]: the management of water from source to sea, including both surface waters and associated groundwaters. For transboundary waters, the Directives require Member States to designate International River Basin Districts (IRBDs).[5] In these IRBDs, States have a duty to cooperate through joint institutional mechanisms to attain the objectives of the Directives. As the IRBD is the natural hydrological unit, it is important that there is coherence in policies and measures within this unit. Indeed, measures promulgated upstream in one jurisdiction have an impact on the state of the water system in another jurisdiction further downstream and vice versa. The river basin approach therefore entails several legal challenges and questions. How should Member States cooperate in these IRBDs and which tools should they use?

---

* The research for this book has been concluded in July 2017.

1 Directive (EC) 2000/60 establishing a framework for Community action in the field of water [2000] OJ L327 (Water Framework Directive or WFD).

2 Directive (EC) 2007/60 of 23 October 2007 on the assessment and management of flood risks [2007] OJ L 288 (Floods Directive or FD).

3 River basins are defined as: "the area of land from which all surface run-off flows through a sequence of streams, rivers and, possibly lakes into the sea at a single river mouth, estuary or delta". See, respectively, Article 2(13) WFD and Article 2(15) WFD.

4 Commission, 'Report from the Commission to the European Parliament and the Council on the Implementation of the Water Framework Directive (2000/60/EC) River Basin Management Plans', COM(2012) 670 final, 9.

5 This will be discussed in detail in Chapter II of this book.

The first assumption of this book is that the EU legal framework is inadequately equipped to foster effective transboundary river basin management, as will be demonstrated by selected examples in the following paragraphs. The WFD stipulates, for IRBDs falling entirely within the EU, that Member States must ensure coordination with the aim of producing a single International River Basin Management Plan (RBMP).[6] However, if Member States do not succeed in agreeing on such a single RBMP, they may adopt RBMPs covering at least the parts of the IRBD located on their territory.[7] The same logic applies to the drafting of Flood Risk Management Plans (or FRMPs) for IRBDs.[8] This means that the failure to jointly produce the RBMPs and/or FRMPs does not have any legal repercussions for the Member States concerned in terms of compliance and punitive measures.

The WFD and FD further provide that Member States in IRBDs may cooperate through shared goals and planning processes.[9] However, the WFD and the FD do not provide instruments to safeguard cooperation in these basins. On the contrary, Member States are to rely on existing international treaties. These treaties have been adopted pursuant to the applicable international legal framework, namely on the basis of the UNECE Water Convention[10] and the UN Watercourses Convention.[11] Cooperation pursuant to the existing treaties between riparian States is often weak from a legal perspective as enforcement mechanisms regarding shared responsibilities are lacking.[12] In addition, the level of formalisation and the success factors of these traditional treaties differ widely between the various International River Basin Districts.

3. In these IRBDs, the question also arises which aspects of water management are covered by the regime. Water management encompasses several aspects, broadly categorised into water quality management, i.e. point-source and diffuse pollution, and water quantity management i.e. scarcity, droughts and flood risk management. This book takes the perspective of water quantity management.

---

6 Ellen Hey, 'Multi-Dimensional Public Governance Arrangements for the Protection of the Transboundary Aquatic Environment in the European Union: The Changing Interplay between European and Public International Law' (2009) 6 International Organizations Law Review 191, 199. Article 13(2) WFD. For IRBDs extending beyond the territory of EU Member States, states should "endeavour" to produce a single river basin management plan. See Article 13(3) WFD.

7 Article 8(2) WFD.

8 Article 8(2) FD.

9 Andrea Keessen and Marleen van Rijswick, 'Adaptation to Climate Change in European Water Law and Policy' (2012) 8 Utrecht Law Review 38.

10 Convention on the Protection and Use of Transboundary Watercourses and International Lakes (adopted 17 March 1992, entered into force 6 October 1996) 31 ILM 1992.

11 Convention on the Law of the Non-Navigational Uses of International Watercourses (adopted 21 May 1997, entered into force 17 August 2014) 36 ILM 700.

12 Ellen Hey and Marleen van Rijswick, 'Transnational Water Management' in Oswald Jansen and Bettina Schöndorf-Haubold (eds), *The European Composite Administration* (Intersentia 2010). Andrea M Keessen and Marleen Van Rijswick (n 9) 42. Also, see Marleen van Rijswick and Herman Havekes, *European and Dutch Water Law* (Europa Law Publishing 2012) 252.

Although flood risk management can also be considered as "water safety management", it is captured by the umbrella notion "water quantity management" in this book. The reasons for choosing water quantity management as the lens through which the legal framework for river basin management is viewed relate to the second assumption of this book, namely that this aspect of water management has been underdeveloped in the EU legal framework with the dichotomy between water quantity management and water quality management in EU primary law constituting a major barrier to integrated river basin management, the importance of which will be established in Section 2. Not only in the EU legal framework has the quantitative aspect of water management received less attention in comparison to the qualitative aspect, most legal research and literature with respect to EU river basin management has also focused on water quality management.[13]

The reasoning behind this second assumption is as follows. As an exception to the ordinary legislative procedure which applies to measures issued on the basis of the EU's Environment Title, measures with regard to "the quantitative management of water resources" are subject to the unanimity requirement pursuant to Article 192 (2) TFEU.[14] However, it still is not entirely clear what constitutes "quantitative management of water resources". For example, the WFD and FD are both based on Article 192 (1) TFEU, circumventing the unanimity requirement, even though both Directives clearly deal with quantitative water management. As to the WFD, the "centre of gravity" is water quality, and issues related to water quantity are accessory. As the link between quantitative and qualitative water management is becoming increasingly clear, it should be evaluated whether the unanimity requirement for measures related to the quantitative management of water resources will indeed constitute a barrier to further legislative initiatives related to IRDBs.

2. THE IMPORTANCE OF INTEGRATED RIVER BASIN MANAGEMENT

4. Three important considerations lie at the basis of the necessity to adequately address the topic of this book.

(a) First, freshwater resources are under ever-increasing *pressure*;

13 For example, the analysis of the legal nature of the environmental objectives of the WFD, see Jasper Van Kempen, 'Countering the Obscurity of Obligations in European Environmental Law: An Analysis of Article 4 of the European Water Framework Directive' (2012) 24 Journal of Environmental Law 499. Of course, water quantity-focused research in the EU is not non-existent, see e.g. Marleen van Rijswick, 'Mechanisms for Water Allocation and Water Rights in Europe and the Netherlands: Lessons from a General Public Law Perspective' (2015) 24 Journal of Water Law 141.

14 See, for an explanation of the Environment Title and the applicable decision-making procedures: Geert Van Calster and Leonie Reins, *EU Environmental Law* (Edward Elgar Publishing 2017) part I, section 3.

(b) Second, the idea of approaching water management in an *integrated* manner is more and more becoming part of the general discourse whereas water quantity management in the EU has only been addressed as accessory to achieving good water quality;
(c) Third, water management is an inherently *transboundary* issue and cannot be dealt with effectively from an exclusively national perspective.

5. These three considerations are discussed below.

2.1. INCREASING PRESSURE ON WATER RESOURCES: THE HOLOCENE HANGOVER

6. The importance of a healthy water sector has been identified as a crucial contributing factor in reaching and maintaining a healthy society, from an environmental, human health and economic point of view.[15] However, freshwater resources are subject to increasing pressure. This is illustrated through the Planetary Boundaries research.

In 2009, Rockström and others introduced the framework of Planetary Boundaries (PBs), which designate the safe operating space for humanity to continue to thrive in the earth system.[16] The underlying reason for defining this safe operating space has to do with the consideration that human pressure on nature is influencing the systems that enable the maintaining of the Holocene state, which represents an epoch of stability during which Earth's regulatory capacity allows for the continuation of human development.[17] Since the Industrial Revolution, the Holocene epoch has been tipping over toward the Anthropocene, which is characterised by the impact of human pressure on the planet and represents a shift away from the stability of the Holocene, also dubbed the "Holocene hangover".[18] The Planetary Boundaries framework thus aims to maintain the Holocene state by identifying the boundaries of Earth-system processes, beyond which irreversible and abrupt change is expected to occur.[19]

15 David Gray and Claudia W Sadoff, 'Water for Growth and Development – A Theme Document of the 4th World Water Forum Water for Growth and Development' (2006). E.g. Water security has been identified as one of the United Nation's Sustainable Development Goals and the European Investment Bank has provided financing for approximately 4 billion euros in water related projects in 2015.

16 Johan Rockström and others, 'A Safe Operating Space for Humanity' (2009) 461 Nature 472.

17 I.e. for the last 11,700 years. See Giacomo Certini and others, 'Holocene as Anthropocene Geological Evidence for the Anthropocene' (2015) 349 Science 246.

18 Ibid. For an Article in popular press, see Fredrik Albritton Jonsson, 'The Holocene hangover: it is time for humanity to make fundamental changes' (The Guardian, 7 December 2016) <www.theguardian.com/books/2016/dec/07/the-holocene-hangover-it-is-time-for-humanity-to-make-fundamental-changes> accessed 31 March 2017.

19 Johan Rockström and others (n 16) 472.

Nine planetary systems have been identified, one of which is global freshwater use. At the time of Rockström's research in 2009, the boundaries of three such systems had been exceeded.[20]

In 2015, Steffen and others. updated the Planetary Boundaries research and added a sub-global dimension.[21] Freshwater use has an important sub-global dimension of course, with river basins spanning international, regional, national, provincial, and municipal scales. The Planetary Boundary for freshwater (PB-Water) respects the basin-scale boundary and represents the assessment of "the maximum rate of blue water withdrawal along rivers, based on the amount of water required in the river system to avoid regime shifts in the functioning of flow-dependent ecosystems".[22] It sets certain thresholds related to the amount of freshwater that humans can appropriate. Exceeding these thresholds would trigger processes that would have a detrimental impact on societies, e.g. shifts in moisture feedback.[23] One of the concepts that is paramount in this book and will be used throughout is the concept of environmental flow, which relates to the amount of flow and water in the river needed for the river to thrive, i.e. maintain a good ecosystem state.[24] The environmental flow constitutes an important variable in assessing the PB-Water. In the updated version of the PB research, the PB-Water was amended. Initially, the PB-Water was provisionally set on 4000 km3 per year, based on global consumption of blue water.[25] The revised framework also takes into account other determinants such as environmental flow requirements, which has led to a new estimate of 2800 km^3 per year.[26] The main lesson learned from the PB-research is that society is moving closer toward reaching and exceeding the PB-Water than originally thought in the 2009 research paper. Moreover, importantly, a substantial volume of environmental flows has already been

---

20 The three earth-system processes for which the safe operating space has been exceeded relate to climate change, rate of biodiversity loss and the nitrogen cycle. See Johan Rockström and others (n 16) 374.

21 Indeed, the original study only focused on global boundaries, whereas changes in variables at the sub-global level clearly have an impact on the global-level boundaries. See Will Steffen and others, 'Planetary Boundaries: Guiding Human Development on a Changing Planet' (2015) 347 Science. Local and regional environmental issues are therefore addressed in the new study. The four boundaries that were exceeded are climate change, loss of biosphere integrity, land-system change, and altered biogeochemical cycles.

22 Steffen and others (n 21) 5.

23 Dieter Gerten and others, 'Towards a Revised Planetary Boundary for Consumptive Freshwater Use: Role of Environmental Flow Requirements' (2013) 5 Current Opinion in Environmental Sustainability 551.

24 Steffen and others use the term " Environmental Water Flows " and refer to the maintenance of a fair-to-good ecosystem state. Steffen and others (n 21) 5. The term will be further explained below.

25 With a range of 4000-6000 km^3/year. This was based on estimates e.g. related to water availability and stress. Gerten and others (n 23) 551.

26 With the lower value of the range being 1100 km3 per year and the upper value 4500 km^3 per year. Gerten and others (n 23) 551.

appropriated.[27] Of course, as the PB-Water range already suggests, the question whether or not basins are flirting with the boundaries strongly depends on the regions concerned.

The Planetary Boundaries framework is a good example of the current pressure on freshwater. Targeted data for the specific pressures exist in abundance. For example, the 2030 Water Resources Group predicts that there will be a 40% freshwater gap between demand and availability.[28] Climate change is expected to have a substantial impact on the hydrological cycle and freshwater resources.[29] The risk of flooding in many areas is likely to increase as a result of intensifying precipitation patterns.[30] Combined with among other things, higher water temperatures, these phenomena are also expected to exacerbate water pollution and to influence water quality in general.[31] The vulnerabilities related to freshwater resources lie in the combination of physical pressures and human developments or decisions, such as increased population, economic development, urbanisation, (in)sufficient funding and planning, ageing infrastructure, and so forth.[32]

An integrated approach is required to cope with these pressures, and the various aspects of water management cannot be viewed in isolation. This leads us to the second consideration necessitating the analysis presented in this book.

2.2. QUANTITY MANAGEMENT IN RELATION TO QUALITY MANAGEMENT

7. Whereas the EU legal and policy framework for water management has primarily focused on water quality[33], a combination of factors calls for a reconsideration of this approach toward a more thoroughly developed integrated approach.

It is becoming increasingly apparent that good water quality cannot be achieved without proper water quantity management and vice versa; these two aspects of water management are indeed inextricably linked. In its "Blueprint to Safeguard Europe's Water Resources", the European Commission emphasises the intimate

27 Ibid 554.

28 2030 Water Resources Group, 'Charting Our Water Future: Economic Frameworks to Inform Decision-Making' (2009) <www.mckinsey.com/~/media/mckinsey/dotcom/client_service/sustainability/pdfs/charting%20our%20water%20future/charting_our_water_future_full_report_.ashx> accessed 28 March 2017.

29 Stephen McCaffrey, *The Law of International Watercourses* (Oxford University Press 2007). Agenda 21 also stressed the impact of climate change on freshwater resources.

30 Bryson Bates and others, 'Climate Change and Water: IPCC Technical Paper of the Intergovernmental Panel on Climate Change' (2008).

31 Ibid 43.

32 Julie C Padowski and others, 'Measuring Global Water Security towards Sustainable Development Goals' (2016) 11 Environmental Research Letters.

33 This will be discussed further in this chapter and in more detail in Chapter II.

connection between water quality and water quantity in the context of achieving "good status" and acknowledges the need to provide a stronger foundation for quantitative water management.[34] Several legal tools and instruments exist to institutionalise this intricate link. The concepts that are important for this book and that will make recurring appearances, will be concisely discussed in this section.[35] For example, one of the concepts that enables tackling all aspects of water management in a comprehensive manner, is water security. Indeed, water security relates both to the sustainable access to good quantity and quality freshwater and to the mitigation of the risks associated with water.[36] Captured in the framework for water security is the integrated water resources paradigm, which has been the main discourse in water management policy in the past ten years and implies that water should be approached in an integrated manner and from the perspective of its functions, i.e. quality, quantity, the relevant actors, rules, and so forth.[37] Integrated water resources management also links to and helps achieve nature-based solutions to water security, e.g. by making use of natural infrastructure such as floodplains in the context of flood risk management.[38] Another concept which will be used throughout this book is "environmental flow", which has been described above in relation to the Planetary Boundaries research and which abundantly clarifies the link between water quality and water quantity.[39] The concept embodies the line of reasoning that there should be sufficient water to safeguard environmental, social and economic benefits.[40]

The 'rise' of these integration-enabling concepts thus indicates that scholars and policy makers have acknowledged the intricate connection between water

---

34 Commission, 'A Blueprint to Safeguard Europe's Water Resources' (Communication) COM (2012) 673 final, 6. Good status and related provisions will be discussed in Chapter II.

35 And further elaborated in the context of the theoretical framework presented in this Chapter, as well as in Chapters V and VI.

36 Bjorn-Oliver Magsig, International Water Law and the Quest for Common Security (Routledge 2015).

37 Hannelore Mees, Cathy Suykens and Ann Crabbé, 'Evaluating Conditions for Integrated Water Resource Management at Sub-Basin Scale. A Comparison of the Flemish Sub-Basin Boards and Walloon River Contracts' (2017) 27 Environmental Policy and Governance 2. Global Water Partnership, 'Integrated Water Resources Management' (2000) TAC Background Papers 4/2000 <www.gwp.org/Global/GWP-CACENA_Files/en/pdf/tec04.pdf> accessed 20 February 2017.

38 E Cohen-Shacham and others, Nature-Based Solutions to Address Societal Challenges (IUCN 2016).

39 The environmental flow concept has been defined above as the amount of flow the river needs to maintain a good ecosystem state. The concept has been defined elsewhere as "The water regime provided within a river, wetland or coastal zone to maintain ecosystems and their benefits where there are competing water uses and where flows are regulated" M Dyson, G Bergkamp and J Scanlon, *Flow: The Essentials of Environmental Flows* (IUCN 2003) 2.

40 Other tools have emerged to link the several facets of water management, such as "peak ecological water", which have been described by Gleick and Palaniappan as the point beyond which the ecological harm exceeds the benefits from human use. Peter H Gleick and Meena Palaniappan, 'Peak Water Limits to Freshwater Withdrawal and Use' (2010) 107 Proceedings of the National Academy of Sciences.

quantity and water quality. However, as will be discussed in Chapter II, this has not yet thoroughly seeped through to the EU primary and secondary law level, where water quality overshadows water quantity issues.[41] This is problematic, not only for reasons of the physical connection between these aspects of water management, but also the reality that water quantity related issues are increasingly apparent in the EU, which in turn is connected to the first consideration in the section above. For example, over-abstraction of water has been identified as the second most common pressure on the ecological status in the EU.[42]

2.3. TRANSBOUNDARY, TRANSBOUNDARY, TRANSBOUNDARY

8. The third consideration is that water management, whether tackling point-source and diffuse pollution, flood risks or scarcity or droughts, is an inherently transboundary issue. This cannot be over-emphasised. As has been stressed time and time again by legislators, policy makers and academics throughout the world, these issues cannot be effectively addressed in a comprehensive manner from the national perspective without taking into account the water resources and their management upstream or downstream across the border. In the European Union, approximately 60% of EU surface water is located in a district that crosses a border, and there are transboundary waters in all Member States with the exception of Cyprus and Malta.[43] Actions upstream (in one jurisdiction) influence the quality and quantity of water downstream (in the part of the river basin located in a neighbouring jurisdiction) and vice versa.[44] Reference can be made to the floods that occurred in central Europe in May and June 2013.[45] After days of heavy rain, several Central European countries were hit by deadly floods, resulting in overall losses of approximately EUR 12bn.[46] These 2013 floods demonstrate the transboundary nature of the phenomena, as the floods emerged in one jurisdiction

---

41 For several reasons, which will be discussed in-depth in Chapter II.

42 Commission, 'A Blueprint to Safeguard Europe's Water Resources' (Communication) COM (2012) 673 final, 6.

43 Commission, 'Water Notes on the Implementation of the Water Framework Directive: Joining Forces for Europe's Shared Waters: Coordination in international river basin districts' (2008). Although Cyprus and Malta do not have transboundary waters running through their territories, it is important to mention that EU legislation, rightly so, regulates water bodies in EU Member States, regardless of whether these are transboundary or not.

44 D Grey and C Sadoff, 'Beyond the River: The Benefits of Cooperation on International Rivers' (2002) 4 Water Policy 389.

45 Some of the headlines of the worldwide media coverage: Tony Paterson, 'Central Europe under Water: Floods Wreak Havoc in Germany and Hungary' *The Independent* (2013); Kirsten Grieshaber, 'Central Europe Floods Are Most Dramatic In A Decade' *The Huffington Post* (2013).

46 Michael Able, 'Floods Dominate Natural Catastrophe Statistics in First Half of 2013' *Munich RE* (2013) 4.

and proliferated in other jurisdictions. In the aftermath of these floods in Central Europe, the European Parliament issued a resolution stressing the importance of coordinated approaches in cross-border flood risk management strategies.[47]

At the other end of the spectrum, the scarcity of water, numerous examples of transboundary externalities can be mentioned. Failure to take into account environmental flows in the decision-making mechanism related to the use of water resources in the transboundary Rio Grande River Basin shared between the United States and Mexico frequently results in the failure of the river to reach the Gulf of Mexico because the flows have been over-allocated upstream.[48]

9. In addition to the physical reality of freshwaters and the above-mentioned climatic development driving cooperation[49], the river basin approach mentioned above in Section 1 increasingly necessitates strong and legally enshrined, enforceable mechanisms of cooperation between (EU Member) States sharing these river basins.[50] This is connected to one of the major challenges of the governance spectrum in this regard: the multi-level governance challenge and the enormous degree of fragmentation. In broad terms, there are five levels of governance that need to be taken into account:

(a) The international level, governed by the UN Watercourses Convention and the UNECE Water Convention;
(b) The level of the European Union, through the Water Framework Directive and its sister and daughter Directives;
(c) The sub-regional level[51]: the structures formed on the basis of these Conventions for specific rivers, such as the Rhine;
(d) The national level of the riparian Member States: the level at which the instruments issued at the level of the European Union is implemented in the national orders;
(e) The sub-national level of the riparian Member States: the local authorities for sub-basins and sub-sub-basins entering into consultations with their cross-border counterparts.

---

47 EP Resolution on the floods in Europe, 2013.

48 2030 Water Resources Group, 'Charting Our Water Future: Economic Frameworks to Inform Decision-Making' (2009) <www.mckinsey.com/~/media/mckinsey/dotcom/client_service/sustainability/pdfs/charting%20our%20water%20future/charting_our_water_future_full_report_.ashx> accessed 28 March 2017, 114.

49 Increasing the need for cooperation between States sharing river basins. McCaffrey (n 29) 405. UN-Water, *Transboundary Waters: Sharing Benefits, Sharing Responsibilities* (United Nations 2008) 2 <www.unwater.org/app/uploads/2017/05/UNW_TRANSBOUNDARY.pdf> accessed 16 July 2017.

50 In other words, cooperation in these shared basins is pivotal, regardless of the details of the various climate change prognoses.

51 Monika Ambrus, Herman Kasper Gilissen and Jasper Van Kempen, 'Public Values in Water Law: A Case of Substantive Fragmentation?' (2014) 10 Utrecht Law Review 8, 18.

When developing frameworks for the governance of transboundary waters for a river basin subject to the EU legal framework, each of these five levels needs to be taken into account, as well as the interaction taking place between these levels. At each of these levels, there is a plethora of competent authorities that operate in parallel with overlapping competences. Moreover, there is interaction between each of these levels.

2.4. SUMMING UP: ADDED VALUE

10. This book aims to fill gaps in the existing body of literature in a twofold manner. First, the functioning of the designated IRDBs is in full development, on the one hand driven by the recent legislative efforts such as the WFD and the FD. On the other hand, there is increasing pressure on the environment, intensifying phenomena such as floods, droughts and scarcity. The former element encourages cooperation in these IRDBs, whereas the latter element creates an unparalleled necessity for cooperation in these districts. In light of these two elements – the driver in the form of the legal incentives on the one hand and the climatic development on the other – it is clear that a blueprint for cooperation in these IRBDs has been established, but still requires substantial development and fine-tuning.

3. METHODS

3.1. COMPARATIVE ANALYSIS OF THE RIVER BASINS

3.1.1. Selection of river basins

11. This book covers external comparative legal research regarding two transboundary river basins situated in the European Union and the United States: the Scheldt River and the Delaware River.

The Scheldt is a river located entirely within the European Union, and is therefore fully subject to EU law. The river is shared between Belgium – i.e. the Flemish Region, the Walloon Region and the Brussels Capital-Region[52] – France and the Netherlands. These countries are obviously different in their legal and institutional DNA and in how hydrological scale governance has been addressed. In addition to Belgium, the other two countries studied in the context of the Scheldt case study are the Netherlands and France, both of which were involved in the STAR-FLOOD project. The researcher therefore gained an affinity with

52 As competences for water management are within the province of the three Regions in Belgium.

flood risk management in these countries during the course of the four-year project.

The Scheldt basin district is a tidal, precipitation-driven river, susceptible to phenomena such as transboundary floods, where its vulnerability is amplified by factors such as its high population density and degree of urbanisation. In one city in one region in one country of the basin, no fewer than eight water managers are active.[53] The multi-level governance challenges with respect to the Scheldt are therefore rather daunting. It represents a good testing case for the robustness of the EU legal framework, and the vulnerability assessment carried out in Chapter II.

The Delaware River Basin and its water resources have been subject to conflicts, landmark litigation, and various attempts at formal joint governance structures since the emergence of the United States as an independent country. The Delaware River is shared between the States of New York, Pennsylvania, New Jersey and Delaware.[54] Stakes in the Delaware River urge reconciliation of 19 federal agencies, 14 interstate agencies and 43 state agencies that are involved in the management of water resources in the basin. The multi-level governance challenge is therefore significant. In this basin, a centralised body with regulatory authority over the whole of the basin has been established. Whereas the scope of many interstate compacts is restricted to water allocation, the Delaware Compact[55] tackles all aspects of water resources management, ranging from water quality to flood risk management to water supply to watershed management. For this reason, the manner in which this model has been formed and operates in practice provides valuable insights into the governance of transboundary waters in a federalist setting. Studies have shown that the Delaware Compact, due to its adaptive nature, which allows tapping into new and unexpected developments, is an appropriate tool, among other things, to deal with future climatic pressure.[56] This mechanism has been referred to as a best practice in river basin management, and has served as a model for other water-sharing agreements.[57]

---

53 I.e. in the city of Antwerp. See Hannelore Mees, Cathy Suykens and Ann Crabbé, 'Evaluating Conditions for Integrated Water Resource Management at Sub-Basin Scale. A Comparison of the Flemish Sub-Basin Boards and Walloon River Contracts' (2017) 27 Environmental Policy and Governance 2.

54 Lynn A Mandarano, Jeffrey P Featherstone and Kurt Paulsen, 'Institutions for Interstate Water Resources Management' (2008) 44 Journal of the American Water Resources Association 136, 364.

55 The Delaware Compact is the agreement governing the Delaware River, which will be explained in Chapter IV.

56 Watermark Initiative, 'US Water Stewardship: A Critical Assessment of Interstate Watershed Agreements' (2009) <http://watermarkinitiative.com/wp-content/uploads/2013/10/WMI_Report_09-13.pdf> accessed 14 July 2017.

57 Emily Jeffers, 'Creating Flexibility in Interstate Compacts' (2009) 36 Ecology Law Quarterly 209, 227.

3.1.2. *Validity of the comparison*[58]

12. In the United States, the term that has been used to refer to the body of rules applicable to a river[59] is the "Law of the River", which encompasses the applicable international treaty, the specific agreements governing the transboundary river, the relevant US Supreme Court decisions and the various federal statutes and regulations.[60] Transposing this term to the situation in the European Union, the "Law of the River" in the EU would encompass the applicable international treaty and international customary law, the specific agreements governing the specific river, the relevant CJEU judgments and EU primary and secondary law in the form of especially the Water Framework Directive and the Floods Directive, plus the national legislative frameworks in the respective riparian Member States.

13. Comparing these two river basins, their respective "laws of the river", and how multi-level governance with regard to transboundary waters is approached in the respective governance frameworks will yield relevant lessons for the EU legal framework.

14. The comparative research will adopt a functionalist approach, i.e. starting from a particular issue (cooperation)[61] and aiming at the identification of best practices and lessons learnt, keeping in mind the five pillars of transboundary water cooperation. For the validity of any legal comparative exercise it is important to find the common denominator for comparison. This common denominator is often referred to as the "tertium comparationis", which entails a "common quality that two things, which are being compared, share".[62] Here, from a territorial perspective, the object of comparison is the transboundary water body, and from a material perspective, the manner in which said water body is governed. In a broader sense, the transboundary river basins studied in the context of this book are situated in the European Union and the United States respectively. Therefore, the comparative exercise is executed within the setting of the legal and institutional differences between these continents. The EU has been granted competences in the context of water management on the basis of primary EU law, and thus multi-level governance challenges in the relationship

---

58 A detailed account of the validity of the comparison is provided in Chapters III and IV.

59 Specifically for the Colorado River.

60 Jason Robison and Douglas Kennedy, 'Equity and the Colorado River Compact' (2012) 42 Envtl. L. 1157.

61 Linda Kestemont and Paul Schoukens, *Rechtswetenschappelijk Schrijven*, Leuven, Acco, 2012, 61. Wouter Devroe, *Rechtsvergelijking in een Context van Europeanisering en Globalisering*, Leuven, Acco, 2010, 39. Mathias Siems, *Comparative Law* (Cambridge University Press 2014) 26; Michele Graziadei, 'The Functionalist Heritage' in Pierre Legrand and Roderick Munday (eds), *Comparative Legal Studies: Traditions and Transitions* (Cambridge University Press 2003) 561.

62 Jaakko Husa, *A New Introduction to Comparative Law* (Bloomsbury 2015) 148.

between the Union and its Member States present themselves. The United States encounters similar challenges posed by multi-level governance models related to transboundary freshwaters and is equally faced with the increasing necessity to cooperate resulting from, among other things, climate change.[63] The validity of the comparison between river basins in the two continents is further supported by the fact that the focus with respect to water resources management in the United States lies with the States. Issues related to the delicacy of the transfer of competences and sovereignty from the state levels to a transboundary water body and the "federal level" are therefore clearly apparent both in the EU and in the US.

15. When reviewing water resources management in the US with the aim of drawing lessons for the EU legal framework, differences in competences of the relevant institutions in general and particularly with regard to water management in the US and EU legal and institutional settings will be taken into account.[64]

From a constitutional point of view, it is relevant to mention that the dynamics strongly differ in the EU and the US. The US Constitution is from 1789, and 27 amendments have since been included in its scope.[65] The last amendment to the Constitution dates from 1992. In the EU, since the promulgation of the Treaty of Rome in 1957, the Treaty has regularly been subject to reforms decided at intergovernmental conferences.[66]

The EU is a *sui generis* legal order with EU Member States having separate legal personality under international law and can enter into international treaties.[67] This in contrast to US federal states. The EU as an economic and political union is clearly far less developed in comparison to the US. The federal institutions in the US have more power and clout than those existing in the EU. The United States Congress has more extensive powers to legislate.[68] It is important in this regard that the institutions in the US have greater competences to follow up on implementation of legal instruments and enforce should need be. In the US, the executive agencies have direct administrative authorities with regard to the implementation of legislation, as well as oversight authority.[69] Depending on the

63 Catherine Valcke, 'Reflections on Comparative Law Methodology', in Maurice Adams and Jacco Bomhoff (eds), *Practice and Theory in Comparative Law* (Cambridge University Press 2012) 33.

64 Geert De Baere and Kathleen Gutman, 'Federalism and International Relations in the European Union and the United States: A Comparative Outlook' in Elke Cloots, Geert De Baere and Stefan Sottiaux (eds), *Federalism in the European Union* (Hart Publishing 2012) 135.

65 It has been amended 33 times in total.

66 I.e. around 1986, 1993, 1996, 1999, 2003 and 2009.

67 Nikolaos Lavranos, 'The MOX Plant and IJzeren Rijn Disputes: Which Court is the Supreme Arbiter' (2006) 19 Leiden Journal of International Law 221, 233.

68 Clíona JM Kimber, 'A Comparison of Environmental Federalism in the United States and the European Union' (1995) 54 Maryland Law Review 1658, 1686. E.g. with regard to federal lands.

69 John Hoornbeek, Policy-Making Institutions and Water Policy Outputs in the European Union and the United States: A Comparative Analysis, (2004) 11 Journal of European Public Policy 461, 465.

scope of the national legal instrument in question, the national administrative authority has the competence to implement the law in the jurisdictions of the State.[70]

The effect of Acts adopted is different, as legislation in the US is directly binding from the perspective of its population, whereas in the EU, with respect to the instruments most commonly used in the context of environmental and water law, i.e. Directives, the "direct effect" principle applies.[71] The EU Directives need to be implemented in the national legal frameworks of the Member States, a process in which the objectives set by the Directives need to be attained and the Member States have flexibility in how these objectives are attained. Regulations, on the other hand, are legal acts that are binding in their entirety and directly applicable in the Member States.

In the context of water resources management, as mentioned, the dominant instrument is the Directive, i.e. the Water Framework and Floods Directives.[72] Once the Directive has been implemented in national legislation, the Member States are responsible for enforcing their national laws. This is a translation of Article 4 (3) of the Treaty on the European Union, which states that "the Member States shall take any appropriate measure, general or particular, to ensure fulfilment of the obligations arising out of the Treaties or resulting from the acts of the institutions of the Union".[73] Among other things, this emphasis on national implementation is a reflection of the subsidiarity principle included in Article 5 TEU. The Member States should also provide adequate remedies to ensure effective legal protection of EU law.[74] The European Commission, in addition to having the authority to introduce legislative initiatives, works as a watchdog in this regard, as it may bring infringement proceedings before the Court of Justice of the European Union for a breach of EU law within the territory of a Member State.[75] In the context of environmental law, a breach may be triggered by lack of or late transposition of the EU legal instrument in question, or by cases where the EU provisions are applied incorrectly.[76]

Whereas the US Supreme Court has binding law of precedent, this is not the case for the EU Court of Justice. Moreover, the Supreme Court has the power to

---

70 Ibid 465.

71 This applies when the relevant provisions are sufficiently clear and precise and unconditional, in combination with the situation that the EU country has not yet transposed the instrument. Case C-26/62 *Van Gend en Loos* [1963] ECLI:EU:C:1963:1.

72 As will be further elaborated upon in Chapter II.

73 This also flows from Article 291 TFEU, which states that: "Member States shall adopt all measures of national law necessary to implement legally binding Union acts". For a full account of enforcement of EU environmental law, see Martin Hedemann-Robinson, *Enforcement of European Union Environmental Law: Legal Issues and Challenges* (2007). There is also a role for independent agencies in the EU.

74 Article 19 TEU.

75 This competence also covers Title XX of the TFEU with regard to the environment.

76 Martin Hedemann-robinson, Enforcement of European Union Environmental Law: Legal Issues and Challenges (2007) 35.

apportion waters when proceedings are brought before it and Congress, on the basis of the Commerce Clause, can allocate interstate waters.

Also, stemming from the federal competences in water management in the US, representatives from the federal level are involved in the agreements concluded between the States sharing the water bodies in question. Although the EU Commission is also party or observer to several transboundary water agreements in the EU, it is not involved as closely compared to the US federal level involvement.[77]

Finally, one of the important differences with regard to water management in the EU versus the US is that water in the US has a more private character than in the EU, where water is very much seen as a public domain and public responsibility. In the US, private water rights exist through different doctrines: in the East, the riparian doctrine is dominant, and in the West, the prior appropriation doctrine is prominent. However, this difference in legal characterisation of water resources does not form an impediment to a valid EU-US comparison for the governance of rivers. Whereas these water-rights doctrines are crucial in determining the rights and obligations of private parties within the States, they are not quintessential in interstate relationships and disputes.[78] This was repeatedly confirmed by the Supreme Court, for example in *New Jersey v New York*, where the Court held that it was not bound by a strict application of the private riparian rights doctrine in interstate water disputes, as usually applied within the State in question.[79] This line of reasoning, namely that one doctrine should not govern interstate disputes before the Supreme Court holds true especially in disputes between States where in one of the States the prior appropriation doctrine is enshrined in its constitution, and in the other, the riparian doctrine, for example in *Kansas v Colorado*.[80] For the comparison between specific river basins in the two continents, the doctrines that States apply within their State are not decisive. As the Court ruled in *Kansas v Colorado*, States may stipulate rules on the manner in which water is allocated within the state boundaries, but these rules may not be extrapolated to interstate water management.[81] On the other hand, if both States have adopted the same doctrine, this doctrine should prevail in the Supreme Court's allocation method, as was the case in *Wyoming v Colorado*.[82]

16. This analysis of the US legal framework with regard to river basin management as applied to the Delaware River Basin and the subsequent

---

77 E.g. the Rhine. The US federal government is involved in terms of financial contributions, voting, and so forth.

78 Josh Clemons, 'Interstate Water Disputes: A Road Map for States' (2004) 12 Southeastern Environmental Law Journal 115.

79 *New Jersey v New York* 283 US 336 (1931).

80 This is because the riparian doctrine was dominant in Kansas and the appropriation doctrine in Colorado. *Kansas v Colorado* 185 US 146 (1902).

81 Ibid 95.

82 This is because both States have adopted the prior appropriation method. *Wyoming v Colorado* 259 US 419 (1922).

comparison with the legal landscape in the EU does not aim to transplant the legal mechanisms existing in the US into the EU legal framework. This would not be possible due to the legal and institutional differences between the United States and the European Union. This book does not have the pretence to argue that a "one size fits all" solution is feasible in transboundary water resources management. Indeed, even within the US, the Eastern type of interstate water Compact is different from its Western equivalent, due to differences from a geographical and historical perspective.[83]

3.2. METHOD OF EVALUATION

17. In addition to comparative legal research, this book also includes an evaluation exercise targeted at river basin management. Resilience and adaptive management theory has been used to identify benchmarks guiding the evaluation. The figure below illustrates the evaluation framework, which will be explained in the following Section 5.

18. Before teasing out the individual evaluation benchmarks and their relevance to this book, it should be mentioned that the evaluation framework will be applied to the two case studies as hand, i.e. the Scheldt and the Delaware. This will in turn generate a set of conclusions and recommendations. Indeed, based on the application of the evaluation benchmarks on "the laws of the river" of these two transboundary river basins in Chapters III and IV and an analysis of relevant jurisprudence and doctrine, it is the intention of this book to set forth Critical Success Factors (CSFs) for achieving resilient transboundary river basin management.

A Critical Success Factor is a management concept that is used to indicate elements required to ensure the success of a certain business. These are the few key areas *where things must go right* in order for the business to thrive, and these therefore require the necessary continuous attention from the executives of the company.[84] The concept was presented for the first time in 1979 to define the information needs of a CEO, and, since then, has been used in the context of governance as well.[85]

Chapter V thus aims to determine the elements that have been conducive to the success of the river basins studied, as well as associated with their failures, by

---

83 Zachary Lorne McCormick, The Use of Interstate Compacts to Resolve Transboundary Water Allocation Issues' (Oklahoma State University 1973) 433.

84 John Rockart, 'Chief Executives Define Their Own Data Needs' (1979) 3 Harvard Business Review 81.

85 Ibid. For the governance aspects, see e.g. Zyad Alreemy and others, 'Critical Success Factors (CSFs) for Information Technology Governance (ITG)' (2016) 36 International Journal of Information Management 907 <http://eprints.soton.ac.uk/397195/1/csf_itg_ijim_accepted.pdf>.

using the resilience benchmarks as applied to their respective legal frameworks. This is thus the connection between the resilience evaluative criteria developed in this Chapter and the Critical Success Factors developed in Chapter V. As the Critical Success Factors are drawn from an analysis of river basin management regimes, in combination with doctrine and jurisprudence review, these can mainly be considered as a specification of the first resilience criterion "integrated river basin management at the hydrological scale and nested governance". The relationship between each individual CSFs and the resilience benchmarks will be explained in more detail in Chapter V.

Finally, Critical Success Factors should be distinguished from Key Performance Indicators (KPI). Both these tools can be used to guide and measure "good governance". But whereas Key Performance Indicators are a quantitative measurement tool, i.e. have the form of a ratio or percentage to assess the effects of certain actions, Critical Success Factors are qualitative, and pinpoint the elements that drive the success of a business, governance mechanism or action. In order to comprehensively evaluate the performance of a river basin management mechanism and its underlying legal framework, Key Performance Indicators should be drawn up as well. However, this is beyond the scope of this book.

19. The evaluation framework is visualised in figure 1.

Figure 1. Evaluation framework

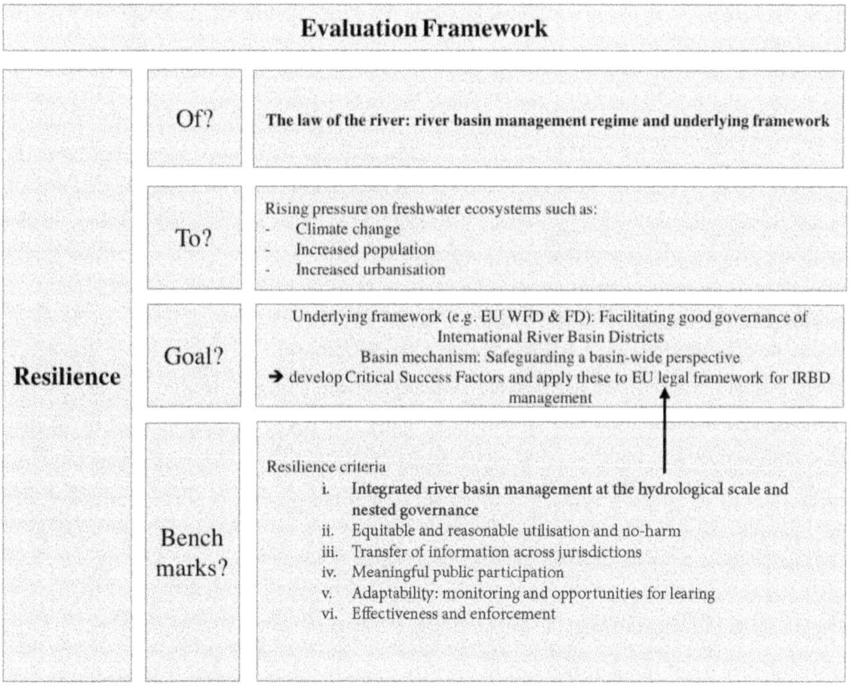

4. THEORETICAL FRAMEWORK

4.1. INTRODUCTION

20. The overarching theoretical framework that functions as a common thread throughout this book is based on social-ecological resilience theory as applied to transboundary river basin management. Several theories can be used to tackle the "common action" problem inherent to the management of common-pool resources such as transboundary waters and the associated externalities.[86] This thesis has opted for resilience theory because it fits well with the problems at hand, i.e. it aims to accommodate the changes and uncertainties that are inherent to water management. Dynamics in the physical context requires dynamism in the legal and governance response. Resilience thinking revolves around grasping the manner in which a social-ecological system can continue to thrive in the face of uncertainties and change. The essential logic underpinning this line of thinking entails that a certain system, such as a transboundary water governance mechanism, should govern for flexibility instead of maintaining stability. A practical translation of this logic in terms of water management is that rigid water use allocation schemes provided for in agreements between riparian States do not bode well with changes in water supply caused by events such as climate change.

4.2. SUBJECT OF EVALUATION: WHAT SHOULD BE RESILIENT?

21. The evaluation framework revolves around the legal and governance framework for transboundary river basin management, as embodied by the Scheldt River in the EU and the Delaware River in the US The locus of evaluation is therefore "the law of the river" for these two river basins. Evaluating these river basins also implies that the underlying legal framework is included in the evaluation, i.e. an evaluation of the law of the Scheldt River also includes an evaluation of the EU level to the extent that it affects the law of the river. Indeed, the WFD and FD are implemented by the Member States sharing the Scheldt River, and therefore form an integral part of the governance mechanism of the basin. It is noteworthy that this research applies the resilience theory to the "law of the river", in other words, the legal system needs to be resilient. The locus of evaluation thus does not *a priori* relate to the ecological resilience of the transboundary water system, but rather to the institutional resilience of the river basin management mechanism and its underlying legal framework.[87]

---

86 E.g. Eyal Benvenisti, 'Collection Action in the Utilisation of Shared Freshwater: The Challenges of International Water Resources Law' (1996) 90 The American Journal of International Law 384.

87 Olivia Green and Charles Perrings, 'Institutionalized cooperation and resilience in international water law' in Craig Allen and Ahjond Garmestani (eds), *Social-Ecological*

4.3. UNCERTAINTIES IN TRANSBOUNDARY WATER MANAGEMENT: RESILIENT REGARDING WHICH PHENOMENA?

22. Unpredictability and uncertainty are inherent features of the governance of natural resources. The most apparent reason for uncertainty that comes to mind relates to climate change. For example, there is scientific consensus that extreme events will intensify in future, but it is difficult to pinpoint exactly when and to which degree such events will take place in specific basins.

23. In general, moving beyond the aspect of climate change, several types of uncertainties can be discerned when assessing a certain water body.[88]

24. A first type of uncertainty relates to the hydrological cycle and climatic events, and more specifically, the question of the amount of precipitation in a given period of time, the flow of the river, evaporation rates, and so forth. This uncertainty includes climate input.[89] An example of how the provisions of a transboundary agreement may potentially clash with changes in the water system as a result of this type of uncertainty relates to a rigid and inflexible water quantity allocation mechanism of the use of water by basin states. Such allocation mechanism may quickly go out of date when the quantity of water in the basin changes. The second uncertainty that has been identified related to the performance of hydraulic infrastructure in place in the basin in question.[90] Reference can be made to the failure of the Oroville Dam, which caused catastrophic flooding in the State of California in the United States in February 2017.[91] A third type relates to political instability and economic uncertainty, which could influence relations between States. A practical example of this type of uncertainty relates to the funding of the joint entity appointed to coordinate the transboundary water body by the States

---

Resilience and Law (Columbia University Press 2013) 176. Olivia O Green, Barbara A Cosens and Ahjond S Garmestani, 'Resilience in Transboundary Water Governance: The Okavango River' (2013) 18 Ecology and Society 2.

88 Alain Bensoussan and Nadir Farhi, 'Uncertainties and Risks in Water Resources Management' in Jean-Michel Lasry, Delphine Lautier and Damien Fessler (eds), *The Economics of Sustainable Development* (Economica 2010).

89 Newsha Ajami, George Hornberger and David Sunding, 'Sustainable Water Resource Management under Hydrological Uncertainty' (2008) 44 Water Resources Research 1.

90 Alain Bensoussan and Nadir Farhi, 'Uncertainties and Risks in Water Resources Management' in Jean-Michel Lasry, Delphine Lautier and Damien Fessler (eds), *The Economics of Sustainable Development* (Economica 2010).

91 Benjamin Oreskes and Shelby Grad, 'The Government Failure at the Heart of the Oroville Dam Crisis' *LA Times* (Los Angeles, 20 February 2017) <www.latimes.com/local/lanow/la-me-ln-essential-california-20170220-story.html> accessed 10 July 2017. Jenny Rowland and Kevin DeGood, 'What the Oroville Dam Disaster Says About America's Aging Infrastructure' (Fortune, 18 February 2017) <http://fortune.com/2017/02/18/oroville-dam-california-flood/> accessed 21 February 2017.

or, for example in the case of the United States, the federal government. Even with a decreased flow of financial resources from these governments, the joint entity should be able to carry out its activities. Foreign policies of riparian States may influence joint governance structures for shared basins. Finally, other factors resulting in pressure on the water systems relate to demographic developments, urbanisation and so forth. Projections are that the population of EU-28 will grow by approximately 2.6% in the period 2014-2080.[92] A cross-cutting issue, which influences all categories of uncertainties referred to above, is the topic of climate-induced human displacement. This issue represents an ever-increasing challenge and the extent of the problem remains a question mark for policy makers all over the world. The specific numbers in this regard vary greatly: certain studies refer to 150 million climate refugees by 2050.[93] Globally, floods were considered to be the main trigger for relocations in 2015.[94]

25. A certainty in this realm of uncertainties is that these various phenomena will increase the pressure on legal and governance mechanisms related to transboundary water management.[95] Robust transboundary water management agreements and institutional mechanisms that are (i) able to weather the various types of pressures and are (ii) flexible enough to appropriately respond to the changes inevitably brought by these uncertain factors, are therefore quintessential.

4.4. FOUNDATIONS OF RESILIENCE AND ADAPTIVE CAPACITY

26. Resilience and adaptive capacity are two notions that are closely linked and that aim to provide a systematic and comprehensive manner for a legal and governance mechanism to deal with change and uncertainty inherent to natural resources management. There are many working definitions of the concept of resilience, which have evolved over time. Resilience in its application to ecological systems was first discussed by C.S. Holling in 1973. Holling described resilience as

---

92 Eurostat, 'People in the EU – Who are we and how do we live?' (Eurostat 2015) <http://ec.europa.eu/eurostat/statistics-explained/index.php/People_in_the_EU_%E2%80%93_who_are_we_and_how_do_we_live%3F> accessed 21 February 2017.

93 Environmental Justice Foundation, 'No Place Like Home: Where next for Climate Change Refugees?' (Environmental Justice Foundation 2009) <http://ejfoundation.org/sites/default/files/public/no%20place%20like%20home.pdf> accessed 21 February 2017.

94 IDMC, 'Global Report on Internal Displacement' (IDMC 2016) 21 <www.internal-displacement.org/assets/publications/2016/2016-global-report-internal-displacement-IDMC.pdf> accessed 21 February 2017.

95 Lukas Rüttinger and others, 'A New Climate for Peace: Taking Action on Climate and Fragility Risks', (Adelphi, International Alert, Woodrow Wilson International Center for Scholars, European Union Institute for Security Studies 2015) <http://newsroom.unfccc.int/media/252731/newclimateforpeace.pdf> accessed 17 July 2017.

"the capacity of a system to absorb disturbance and reorganize while undergoing change so as to still retain essentially the same function, structure, identity, and feedback".[96] Over time, this definition developed into an interpretation of the concept which did not focus as much on the capacity of the system to maintain the same elements, but shifted toward the capacity of the system to evolve in the face of change.[97] The following definition provides an encompassing view of the concept, by defining resilience as "the ability of a system, community or society exposed to hazards to resist, absorb, accommodate, adapt to, transform and recover from the effects of a hazard in a timely and efficient manner, including through the preservation and restoration of its essential basic structures and functions through risk management".[98] In turn, adaptive capacity refers to the ability of such a system, community or society to adjust to changes and new conditions in an appropriate manner.[99]

27. This means that resilience and adaptive capacity are two sides of the same coin. Resilience theory has been applied outside the ecology field since the promulgation of Holling's work, e.g. in the context of cultural theory, anthropology and human geography.[100] Resilience and adaptive management thinking can also be applied in the context of legal research. According to Ebbesson and Hey, the correlation between law and resilience revolves around two aspects: the manner in which law can influences the resilience of social-ecological systems and the question of the impact of change on the law itself.[101] Key questions ensuing from the nexus of law and resilience are, for example, to what extent the law can allow flexibility without compromising legal certainty, and so forth.

28. The question then arises what the specific translation is of the resilience and adaptive management thinking to transboundary water resources. Before addressing the benchmarks ensuing from this line of thinking, it is worth mentioning that adaptive management thinking favours multiple,

---

96 CS Holling, 'Resilience and Stability of Ecological Systems' (1973) 4 Annu.Rev.Ecol.Syst. 1.

97 Carl Folke and others, 'Adaptive Governance of Social-Ecological Systems' (2005) 30 Annual Review of Environment and Resources 441, 457. Brian Chaffin, Hannah Gosnell and Barbara Cosens, 'A Decade of Adaptive Governance Scholarship: Synthesis and Future Directions' (2014) 19 Ecology & Society 56.

98 United Nations International Strategy for Disaster Reduction, 'UNISDR Terminology on Disaster Risk Reduction' (2009) <www.unisdr.org/files/7817_UNISDRTerminologyEnglish.pdf> accessed 17 July 2017.

99 Adaptive governance was coined as a notion by Dietz and others. in 2003: Thomas Dietz, Elinor Ostrom and Paul Stern, 'Struggle to Govern the Commons' (2003) 302 Science 1907. Robin Kundis Craig, 'Stationarity Is Dead. Long Live Transformation: Five Principles for Climate Change Adaptation Law.' (2010) 34 Harvard Environmental Law Review 9, 28.

100 Carl Folke, 'Resilience: The Emergence of a Perspective for Social-Ecological Systems Analyses' (2006) 16 Global Environmental Change 253, 255.

101 Jonas Ebbesson and Ellen Hey, 'Introduction: Where in Law Is Social-Ecological Resilience?' (2013) 18 Ecology and Society 1.

even overlapping, levels of jurisdiction over a strictly divided, hierarchical and centralised mechanism, which does not rule out the existence of strong authorities, coordination mechanisms and binding decision-making powers. This line of thinking is also apparent in the widely supported polycentric governance approach to governing "wicked problems", i.e. issues that are drenched in a great degree of complexity, uncertainty and multi-dimensionality, as is typical of transboundary water management.[102] Indeed, polycentric governance essentially is about embracing multiple levels of governance, which fits well with the multi-level reality of water management at the national and international scale.[103] Local capacity-building is a key aspect of this approach, for example of the local water manager operating at the sub-sub-basin scale.[104] This is linked to another important aspect of adaptive management: communication and information exchange across the nested scales.[105] Information gathered at the local scale, e.g. by the local water manager, should find its way to the higher-level, international basin scale.[106] Such successful inter-scalar links require strong coordination across scales.[107]

29. It should be noted that it is rather difficult to precisely measure the level of resilience of a legal or governance framework. This would require the occurrence of a shock event such as a large-scale transboundary flood with substantial

102 In legal research terms, polycentric governance has also been referred to as legal pluralism. Thomas Dietz, Elinor Ostrom and Paul Stern, 'The Struggle to Govern the Commons'; Barbara Cosens, 'Transboundary River Governance in the Face of Uncertainty: Resilience Theory and the Columbia River Treaty' (2010) 30 J. Land Resources & Envtl. L.; Elinor Ostrom, *Governing the Commons: The Evolution of Institutions for Collective Action* (Cambridge University Press 1990). Folke and others. Catrien Termeer and others, 'The Regional Governance of Climate Adaptation: A Framework for Developing Legitimate, Effective, and Resilient Governance Arrangements' (2011) 2 Climate Law 159, 164.

103 Ostrom's polycentric governance approach as such is not used as an evaluation framework. However, it is worth mentioning that its basic ideas and notions, at times, resemble adaptive management thinking.

104 Barbara Cosens, 'Transboundary River Governance in the Face of Uncertainty: Resilience Theory and the Columbia River Treaty' (2010) 30 J. Land Resources & Envtl. L. 240.

105 Olivia O Green and others, 'EU Water Governance: Striking the Right Balance between Regulatory Flexibility and Enforcement?' (2013) 18 Ecology and Society. Ahjond S Garmestani and others, 'The Integration of Social-Ecological Resilience and Law' (2014); Hannelore Mees, Cathy Suykens and Ann Crabbé (n 37); Sally J Priest and others, 'The European Union Approach to Flood Risk Management and Improving Societal Resilience: Lessons from the Implementation of the Floods Directive in Six European Countries' (2016) 21 Ecology and Society. Kofi Akamani, 'Adaptive Water Governance: Integrating the Human Dimensions into Water Resource Governance' (2016) Journal of Contemporary Water Research & Education 2.

106 This is one of the benchmarks guiding the evaluation exercise, i.e. information exchange, which will be discussed in the following sections. See Olivia O Green, Barbara A Cosens and Ahjond S Garmestani, 'Resilience in Transboundary Water Governance: The Okavango River' (2013) 18 Ecology and Society 2. Tanya Heikkila, Edella Schlager and Mark W Davis, 'The Role of Cross-Scale Institutional Linkages in Common Pool Resource Management: Assessing Interstate River Compacts' (2011) 39 Policy Studies Journal 121.

107 Olivia O Green and others (n 105).

damage in the different countries sharing the basin, and the subsequent inventory of the manner in which the framework has responded to the shock event and which elements of the framework were conducive to its ability to "resist, absorb, accommodate, adapt to, transform and recover" from the effects of the event. However, thorough literature review shows that when a certain governance framework includes certain elements, it is likely to be considered as resilient. These criteria are explained in the section below.

4.5. BENCHMARKS OF EVALUATION

4.5.1. Schematic overview of benchmarks

30. Based on a combination of legal and policy literature, several benchmarks[108] have been identified to guide the evaluation exercise carried out in this book. These benchmarks will be explained in the following section. An overview of the literature inspiring these evaluation criteria is provided in the table below.

Table 1. Literature at the Heart of the Benchmarks

Benchmarks	Meaning	Literature
Equitable and Reasonable Utilisation and the No-Harm Rule	There is a mechanism and understanding of the equitable and reasonable utilisation standard and the no-harm rule.	Freriks (1994); Benvenisti (1996); Wouters and others. (2005); McCaffrey (2007); McIntyre (2007); Van Kempen (2012); Leb (2013); UNECE (2013); Tanzi and others. (2015); Van Rijswick (2015); Dai, Van Rijswick an Schmidt (2017); McIntyre (2017)[109]

[108] In this book, the words 'benchmarks' and 'evaluation criteria' are used interchangeably.

[109] Annelies Freriks, *Juridische Bescherming van Beeksystemen* (WEJ Tjeenk Willink 1994). Patricia K Wouters and others, 'Sharing Transboundary Waters: An Integrated Assessment of Equitable Entitlement: The Legal Assessment Model' (International Hydrological Programme 2005) IHP-VI Technical Document in Hydrology N°74; McCaffrey (n 29); Owen McIntyre, *Environmental Protection of International Watercourses Under International Law* (Ashgate 2007); Jasper Van Kempen, *Europees Waterbeheer: Eerlijk Zullen We Alles Delen?* (Boom Juridische Uitgevers 2012); Christina Leb, *Cooperation in the Law of Transboundary Water Resources* (Cambridge University Press 2013); Unece, 'Guide to Implementing the Water Convention' (2013) <www.unece.org/fileadmin/DAM/env/water/publications/WAT_Guide_to_implementing_Convention/ECE_MP.WAT_39_Guide_to_implementing_water_convention_small_size_ENG.pdf> accessed 17 July 2017; Attila Tanzi and others (eds), *The UNECE Convention on the Protection and Use of Transboundary Watercourses and International Lakes* (Brill Nijhoff 2015); Owen McIntyre, 'Substantive Rules of International Water Law', *Routledge Handbook of Water Law and Policy* (Routledge 2017). Marleen van Rijswick, 'Mechanisms for Water Allocation and Water Rights in Europe and the Netherlands: Lessons from a General Public Law Perspective' (2015) 24 Journal of Water Law 141. Liping Dai, Marleen van Rijswick and Bram Schmidt, *Towards a Sustainable, Balanced and Equitable Allocation of Water Use Rights* (In press, 2017).

Benchmarks	Meaning	Literature
Integrated River Basin Management at the hydrological scale and nested governance	The legal framework provides the necessary tools to induce governance at the hydrological scale through the river basin approach, including surface waters and groundwaters and water quality and quantity management.	Dietz and others. (2003); Jaspers (2003); Cosens (2010); Ebbesson (2010); Keessen and Van Rijswick (2012); Van Kempen (2012); Cosens (2013; Green and others. (2013a); Green and others. (2013b); Ebbesson and Hey (2013); Van Rijswick and others (2014); Suykens (2015); Mees, Suykens and Crabbé 2017); Priest and others (2017)[110]
Horizontal and vertical transfer of information	Information must be shared, both from the local level to the basin level and all the appropriate chains, and between the various jurisdictions involved, and this implies that there is coordination in decision-making.	Verwijmeren and Wiering (2007); Cosens (2010); Cosens (2013); Green and others. (2013), Gilissen and others. (2016)[111]
Meaningful public participation and co-production	Two-way flow of information, where input is used by governments.	Cosens (2010); Van Rijswick and Salet (2012); Green and others. (2013b); Ebbesson and Hey (2013), Ek and others (2016), Mees and others (2016), Mees and others (2017) Pettersson and others (2017)[112]

110 Green, Cosens and Garmestani (n 106). Folke and others. Robin Kundis Craig, 'Stationarity Is Dead. Long Live Transformation: Five Principles for Climate Change Adaptation Law.' (2010) 34 Harvard Environmental Law Review 9. Barbara Cosens, 'Transboundary River Governance in the Face of Uncertainty: Resilience Theory and the Columbia River Treaty' (2010) 30 Journal of Land Resources and Environmental Law 229; Jonas Ebbesson, 'The Rule of Law in Governance of Complex Socio-Ecological Changes' (2010) 20 Global Environmental Change 414; Andrea M Keessen and Marleen Van Rijswick (n 9) 38; Olivia O Green and others, 'EU Water Governance: Striking the Right Balance between Regulatory Flexibility and Enforcement?' (2013a) 18 Ecology and Society; Jonas Ebbesson and Ellen Hey, 'Introduction: Where in Law Is Social-Ecological Resilience?' (2013) 18 Ecology and Society 1; Marleen van Rijswick and others, 'Ten Building Blocks for Sustainable Water Governance: An Integrated Method to Assess the Governance of Water' (2014) 39 Water International 725; Cathy Suykens, 'EU Water Quantity Management in International River Basin: Crystal Clear?' (2015) European Energy and Environmental Law Review 134; Sally J Priest and others, 'The European Union Approach to Flood Risk Management and Improving Societal Resilience: Lessons from the Implementation of the Floods Directive in Six European Countries' (2017) 21 Ecology and Society.

111 Joris Verwijmeren and Mark Wiering, Many Rivers to Cross: Cross-Border Cooperation in River Management (Eburon Delft 2007) 173; Herman-Kasper Gilissen and others, 'Bridges over Troubled Waters: An Interdisciplinary Framework for Evaluating the Interconnectedness within Fragmented Domestic Flood Risk Management Systems' (2016) 25 Journal of Water Law 12.

112 Pettersson, M, Van Rijswick, M., Suykens, C., Alexander, M., Priest, S., Beyers, J., Ek, K. 'How Legitimate is Flood Risk Governance in Europe? Insights from Intra-Country Assessments'. In review with Water International; Hannelore Mees and others, 'Coproducing Flood Risk Management through Citizen Involvement: Insights from Cross-Country Comparison in Europe' (2016) 21 Ecology and Society; Kristina Ek and others, 'Strengthening and Redesigning European Flood Risk Practices: Towards Appropriate and Resilient Flood Risk Governance Arrangements Design Principles for Resilient, Efficient and Legitimate Flood Risk Governance; Lessons from Cross-Country Comparisons' (STAR-FLOOD Consortium 2016).

Benchmarks	Meaning	Literature
Monitoring and opportunities for learning	The legal framework and institutional mechanism should provide for the ability to respond appropriately to changing circumstances.	Keessen and Van Rijswick (2012); Green and others (2013a); Clarvis (2014); McDonald and Styles (2014); Van Rijswick and others (2014); Mees, Suykens and Crabbé (207); Priest and others (2017)[113] Benvenisti (1996); Craig (2010)
Effectiveness and Enforcement	The legal framework should have the tools to achieve the goals set, and the means to enforce instruments.	Ostrom (2005); Buijze (2009), Dinar and others (2010); Keessen and Van Rijswick (2012); Wouters and others (2012); Green and others (2013a); Faure (2014), Van Rijswick and other. (2014); Priest and others (2017); Mee, Crabbé and Suykens (2017)[114]

4.5.2. Discussion of benchmarks

4.5.2.1. Equitable and reasonable utilisation and the no harm-rule

31. The equitable and reasonable utilisation principle and the no-harm rule are part of the body of international customary law applicable to transboundary water management, as codified by the UN Watercourses Convention.[115] Furthermore, especially relevant for the river basin situated in the EU, these have been coined as the normative cornerstones of the UNECE Water Convention, to which the European Union is party. The preambles of the Water Framework Directive and the Floods Directive refer to this Convention, stating that these Directives are meant to contribute to the achievement by the Union and its Member States of the obligations deriving from the Convention.[116] The equitable and reasonable utilisation standard has been said to have found its origin in the United States' Supreme Court rulings.[117] Equitable apportionment is one of the key principles of interstate water management, as will be discussed in Chapter IV of this book.

113 Jan McDonald and Megan C Styles, 'Legal Strategies for Adaptive Management under Climate Change' (2014) 26 Journal of Environmental Law 25. Margot Hill Clarvis, Andrew Allan and David M Hannah, 'Water, Resilience and the Law: From General Concepts and Governance Design Principles to Actionable Mechanisms' (2014) 43 Environmental Science and Policy 98.

114 Michael G Faure, 'The Complementary Roles of Liability, Regulation and Insurance in Safety Management: Theory and Practice' (2014) 17 Journal of Risk Research 689. Hannelore Mees, Ann Crabbe and Cathy Suykens, 'Flood risk governance in Belgium. Towards a resilient, efficient and legitimate arrangement?' in Corinne Larue and others, *Facing hydro-meteorological extreme events in Europe: a governance issue* (Wiley In Press 2018).

115 UN Watercourses Convention (n 11).

116 Recital 21 WFD and Recital 6 of the FD.

117 Apportionment in the US has often occurred via Supreme Court judgments, see Chapter IV. Joseph W Dellapenna, 'The Customary International Law of Transboundary Fresh Waters' (2001) 1 International Journal of Global Environmental Issues 264, 276.

32. It is important to explain why these principles are important to EU water management, and are used as an evaluation benchmark both for the functioning of the river basin commission and for the underlying EU legal framework in the form of the WFD and the FD.[118] As mentioned, these principles together with the duty to cooperate are considered to constitute the normative triangle of the Water Convention.[119] Because of the ratification by the European Union of the Water Convention and on the basis of Article 216.2 TFEU, the Water Convention is an integral part of the legal regime of the EU.[120] This also means that the EU cannot adopt secondary EU law that violates the provisions of the Water Convention, as the latter will automatically prevail.[121] Another important consequence of the EU's ratification of the Convention is that the Convention can be used to interpret EU water law, which is coherent with the harmonious interpretation approach.[122] This technique of harmonious interpretation also applies to international customary law, which is entirely relevant in the context of transboundary water management.[123]

It is noteworthy that the Water Convention specifically focusses on transboundary aspects of water management, whereas the EU's WFD and FD aim to provide a legal framework for all water bodies in the EU.[124] The territorial scope

---

118 As these are part of international customary law and are used in the US as well, they constitute a relevant benchmark for the Delaware River as well. On the application of international customary law, see Simone Borg, 'The Conservation of Marine Living Resources under International Law: The 1982 United Nations Convention on the Law of the Sea and Beyond' in Algosia Fitzmaurice, and Norman A. Martinez Gutierrez (eds), The IMLI Manual on International Maritime Law – Volume I: The Law of the Sea (Oxford University Press 2014) ch 12.

119 Christina Leb, *Cooperation in the Law of Transboundary Water Resources* (Cambridge University Press 2013) 85. The duty to cooperate is also a normative requirement stemming from the Water Convention, but acts as a catalyst for achieving compliance with the above-mentioned principles, and is therefore considered a prerequisite present throughout this book.

120 See Council Decision 95/308/EC of 24 July 1995 on the conclusion, on behalf of the Community, of the Convention on the protection and use of transboundary watercourses and international lakes. See, generally, Ramses Wessel, 'Reconsidering the Relationship between International and EU Law: Towards a Content-Based Approach?' in Enzo Cannizzaro, Paolo Palchetti and Ramses Wessel (eds), *International Law as Law of the European Union* (Koninklijke Brill NV 2012).

121 Whereas the international agreements need to conform with primary EU law. Gabor Baranyai, 'The Water Convention and the European Union: The Benefits of the Convention for EU Member States' in Attila Tanzi and others (eds), *The UNECE Convention on the Protection and Use of Transboundary Watercourses and International Lakes* (Brill Nijhoff 2015) 94.

122 Pieter Jan Kuijper, "It Shall Contribute to … the Strict Observance and Development of International Law …'" in Court of Justice of the European Union (ed), *The Court of Justice and the Construction of Europe: Analyses and Perspectives on Sixty Years of Case-law* (The Asser Press 2013) 601.

123 See below on the customary international law status of the principles discussed.

124 For an overview of the differences between the UN Watercourses Convention, see the UNECE Water Convention and the EU Directives, Alistair S Rieu-clarke and Flavia Loures, 'The Role and Relevance of the UN Convention on the Law of the Non-Navigational Uses of International Watercourses to the EU and Its Member States' (IHP-HELP Centre for Water Law, Policy and Science, 2010). See, for the interplay between international and EU law with regard to the river basin approach: Hey and van Rijswick (n 12).

especially of the Water Framework Directive is wider than the Water Convention. This may explain why the Water Convention, in a way, is more developed on the topic of transboundary cooperation. Moreover, whereas the WFD mainly revolves around water quality issues, with water quantity considered as accessory to achieving water quality, the important principles underpinning the Water Convention relate to all activities resulting in transboundary impact, both in the context of water quality and of water quantity.[125] As this book focusses specifically on transboundary cooperation using water quantity issues as examples, the examination of water quantity management in International River Basin Districts in light of the key components of cooperation as defined by the Convention is conducive to determining the resilience of this aspect of the EU legal regime.

33. Before addressing the specifics of the no harm-rule and the principle of equitable and reasonable utilisation, it is relevant to provide the context in which these principles exist.

(A) Background: International theories governing shared waters

34. Over the years, several doctrines governing non-navigational uses of international watercourses have been promulgated. These theories are often conflicting in their attempt to reconcile the shared character of the watercourses with the principle of sovereignty of States under international law.[126] Two theories are at opposite ends of the spectrum: the theory of absolute territorial sovereignty, which entails that States have the right to unimpeded use of shared waters, e.g. the right to divert water as seen fit[127], and the theory of absolute territorial integrity, which grants the downstream State the right to demand the full unaltered flow of water from the upstream States.[128] From a practical perspective, the latter theory would necessitate the downstream States' prior consent for every action of the upstream State influencing the flow of the river.[129] The question of prior consent has been the subject of the landmark *Lac Lanoux* case, in the context of which Spain opposed a project initiated by the French which entailed the diversion

---

125 Although the Convention includes more provisions on environmental protection than on water quantity issues. However, the no-harm and equitable and reasonable utilisation principles apply to water quality, quantity and floods. Moreover, these principles are part of customary international law, and in that context, are also geared toward quantity issues. Gabor Baranyai, 'The Water Convention and the European Union: The Benefits of the Convention for EU Member States' in Attila Tanzi and others (eds), *The UNECE Convention on the Protection and Use of Transboundary Watercourses and International Lakes* (Brill Nijhoff 2015) 95.

126 Owen McIntyre, *Environmental Protection of International Watercourses Under International Law* (Ashgate 2007) 13.

127 This theory coincides with the well-known Harmon Doctrine, taken after the US Advocate-General Harmon who emphasised the absolute sovereignty of the United States in diverting the Rio Grande river, to the detriment of Mexico.

128 McCaffrey (n 29) 126.

129 Owen McIntyre (n 126) 17.

of water from the shared Carol river into the Ariège river and the return of an equal amount of water into the river.[130] Spain argued that the natural flow of the water would be altered and the return of the water would be based on human intervention.[131] In 1957, the Arbitral Tribunal found that the rule that requires prior agreement of the downstream State before an upstream State can utilise an international watercourse could not be seen as a custom or a general principle of law but could only result from a Treaty.[132] The Tribunal did emphasise the duty to negotiate in interstate waters.

35. Both the theory of absolute territorial sovereignty and absolute territorial integrity have been rejected by doctrine and courts as irrelevant in light of mutual dependence of States in shared waters and increasing pressure on water resources.[133] Since *Lac Lanoux*, with the development of international environmental law and international water law, the cooperative approach in interstate waters further developed and the realisation that the rights of a state with respect to transboundary resources are limited by the other states' rights over those resources.[134]

36. The third theory represents the middle ground between these two theories and is that of limited territorial sovereignty, which coincides with the principle of equitable and reasonable utilisation.[135] It entails that States' interests in shared water resources must be balanced. This principle is deeply rooted in state practice, case law and doctrine and has been acknowledged as the governing principle and cornerstone of international water law.[136] It has been codified in various international instruments, such as the 1966 Helsinki Rules on the Uses of Waters of International Rivers drawn up by the International Law Association (ILA), the ILA's 2004 Berlin Rules on Water Resources and the 1997 UN Watercourses Convention.[137] The wording of the UN Watercourses Convention is clear:

---

130 *Lake Lanoux Arbitration* (France v Spain) [1957] 12 RIAA 281; 24 ILR 101.

131 McCaffrey (n 29) 130.

132 *Lake Lanoux Arbitration* (France v Spain) [1957] 12 RIAA 281; 24 ILR 101 para 13.

133 McCaffrey (n 29) 133.

134 Reference can specifically be made to the UN Watercourses Convention and the UNECE Water Convention, in the context of which the duty to cooperate in the form of notification and consultation has been strongly anchored. Christina Leb, *Cooperation in the Law of Transboundary Water Resources* (Cambridge University Press 2013) 68. McIntyre (n 150) 343.

135 This principle is explained in-depth in section b) below.

136 *Gabčikovo-Nagymaros Project (Hungary-Slovakia)* [1997] ICJ Rep 7; McIntyre (n 126) 26; McCaffrey (n 29) 136.

137 The International Law Association is an NGO composed of legal scholars and founded in 1873 – these scholars drafted the 1966 Helsinki Rules, which for the first time codified the law of transboundary waters. The 2004 Berlin Rules on Water Resources, drafted by the International Law Association, supersede the 1966 Helsinki Rules. The UN Watercourses Convention is based on draft articles prepared by the International Law Commission, which is a United Nations body that codifies international customary law.

"Watercourse States *shall* cooperate on the basis of sovereign equality, territorial equality, territorial integrity, mutual benefit and good faith in order to attain optimal utilisation and adequate protection of an international watercourse".[138] This duty to cooperate has gained further traction with the entry into force of the UN Watercourses Convention on 17 August 2014, following the 35th ratification of the Convention by Vietnam.[139]

37. The fourth approach is referred to as the community of interest or "common management". The community of interest as the "basis of a common legal right" has been put forth by the PCIJ in the *Case Relating to the Territorial Jurisdiction of the International Commission of the River Oder.*[140] This approach implies that the watercourse is subject to joint governance mechanisms with the drainage basin constituting the vantage point. This theory most profoundly recognises the physical unity of shared water bodies irrespectively of legal borders and is most strongly in line with the logic of managing common-pool resources.[141] However, this principle has not yet been accepted as a general principle of international law, having the same normative power as the limited territorial sovereignty principle. Indeed, there have been discussions on the question whether the community of interest approach can be categorised as a distinct legal theory.[142] In the *Pulp Mills on the River Uruguay* case, where the ICJ acknowledged the importance of Environmental Impact Assessments in the context of transboundary notifications in shared waters, the ICJ referred to the specific Treaty in its consideration of the question whether there is a community of interest.[143] It has been stated that international law can restrict unilateral actions by States, but does not have the same ability to superimpose joint governance structures. However, many watercourses throughout the world are governed by joint institutions and the approach is gaining traction in international water law.[144] For the European Union it is relevant that the Water Convention clearly requires parties to enter

---

138 Article 8 UN Watercourses Convention. Own emphasis.

139 Christina Leb, 'The Significance of the Duty to Cooperate for Transboundary Water Resources Management under International Water Law' in Alistair Rieu-Clarke, Andrew Allan and Sarah Hendry (eds), *Routledge Handbook of Water Law and Policy* (Routledge 2017).

140 *Case Relating to the Territorial Jurisdiction of the International Commission of the River Oder*, PCIJ (1929) Ser. A.23.

141 See on common-pool resources: Eyal Benvenisti, 'Collection Action in the Utilisation of Shared Freshwater: The Challenges of International Water Resources Law' (1996) 90 The American Journal of International Law 403. Thomas Dietz, Elinor Ostrom and Paul C Stern, 'The Struggle to Govern the Commons' (2008) 302 Urban Ecology: An International Perspective on the Interaction Between Humans and Nature 611.

142 Christina Leb, *Cooperation in the Law of Transboundary Water Resources* (Cambridge University Press 2013) 52.

143 *Pulp Mills on the River Uruguay* (Argentina v Uruguay) [2006] ICJ Rep 113, Para 204. Owen McIntyre, 'The Proceduralisation and Growing Maturity of International Water Law'(2010) 22 Journal of Environmental Law 475, 482. Christina Leb, *Cooperation in the Law of Transboundary Water Resources* (Cambridge University Press 2013) 54.

144 Ibid 56.

into legal arrangements for the establishment of joint bodies.[145] This in contrast to the UN Watercourses Convention, which states that parties may consider the establishment of joint mechanisms. It can therefore be argued that the community of interest approach reigns over International River Basin Districts in the EU, considering the obligation stipulated in Article 9 (1) and (2) of the Water Convention and the fact that this Convention forms an integral part of the EU legal order following its ratification.

(b) The principle of equitable and reasonable utilisation

38. The validity of the principle of equitable and reasonable utilisation as a pinnacle of the governance of transboundary watercourses and the related duty to cooperate have been widely recognised in international law.[146] It is the general rule on the basis of which States should determine their respective rights and obligations in shared waters.[147] For example, the ICJ stated that Slovakia deprived Hungary of its equitable and reasonable utilisation of the shared natural resources of the Danube river and violated the proportionality principle required by international law by unilaterally diverting 90% of the flow of the river.[148]

39. The principle of equitable and reasonable utilisation constitutes a reflection of the limited territorial sovereignty theory and is found in the middle of the spectrum of the two extreme and opposing theories benefiting the upstream State and the downstream State respectively.[149] The principle boils down to the consideration that States sharing water resources each have the right to an equitable and reasonable utilisation of the transboundary waters.[150] "Equitable" or "equity" are not synonymous to "equal" in this context.[151] Indeed, watercourse States have a right to use the waters in an equitable manner, but do not have the right to an equal share thereof. The notion "equitable" implies that States have the

145 Article 9(1) and 9(2) of the UNECE Water Convention.

146 See Commentary to the Draft Articles on the Law of the Non-Navigational Uses of International Watercourses, in Report of the International Law Commission on the work of its 46th session, United Nations GAOR, DOC. A/49/10 (1994). Unece (n 109).

147 Commentary to the Draft Articles on the Law of the Non-Navigational Uses of International Watercourses, in Report of the International Law Commission on the work of its 46th session, United Nations GAOR, DOC. A/49/10 (1994).

148 *Gabčikovo-Nagymaros Project* (Hungary-Slovakia) [1997] ICJ Rep 7, at 85.

149 Owen McIntyre, 'Substantive Rules of International Water Law', *Routledge Handbook of Water Law and Policy* (Routledge 2017) 236.

150 The precise wording in the Water Convention is as follows: "To ensure that transboundary waters are used in a reasonable and equitable way, taking into particular account their transboundary character, in the case of activities which cause or are likely to cause transboundary impact". See Article 2(2)(c) of the Water Convention.

151 Commentary to the Draft Articles on the Law of the Non-Navigational Uses of International Watercourses, in Report of the International Law Commission on the work of its 46th session, United Nations GAOR, DOC. A/49/10 (1994).

obligation to pay reasonable regard to the legal rights of other States through good faith negotiations, as confirmed, inter alia, in the *Fisheries Jurisdiction* cases.[152]

40. The need for reasonable and equitable utilisation applies both to the water quantity and to water quality issues, this despite the focus of the Water Convention on water quality issues.[153] The principle of sustainable development is considered as being part of the reasonable and equitable utilisation principle. This is apparent in the manner in which the relevant provisions are formulated in the Water Convention and the UN Watercourses Convention, the latter stipulating that "an international watercourse shall be used and developed by watercourse States with a view to attaining optimal and *sustainable utilisation* thereof and benefits therefrom, taking into account the interests of the watercourse States concerned, consistent with *adequate protection* of the watercourse".[154]

41. It is important to note that this normative principle, as is true for the no-harm rule discussed below and in line of the framework nature of the Water Convention, in each basin is to be translated on a case-by-case basis.[155] The UN Watercourses Convention does provide guidance by including a non-exhaustive list containing several criteria that are helpful in determining what is meant with the notion "equitable and reasonable".[156] These criteria largely result from two types of categories, i.e. natural characteristics of the basin on the one hand and social and economic characteristics on the other hand.[157] These are further fleshed out in criteria related to natural factors, e.g. hydrographic, climatic and ecological factors; the social and economic needs of the States in question; the demographic reality, i.e. the number of people relying on the water body; the impact of the use in one watercourse State on the other; existing and potential uses of the watercourse; the conservation, protection, development and economy of use of the water resources of the watercourse; and finally, the available alternatives to a planned or existing

152 *Fisheries Jurisdiction Cases* (UK v Iceland; FRG v Iceland) [1974] ICJ 205.

153 See Owen McIntyre, 'The Principle of Equitable and Reasonable Utilisation' in Attila Tanzi and others (eds), *The UNECE Convention on the Protection and Use of Transboundary Watercourses and International Lakes* (Brill Nijhoff 2015) 151. For a discussion of the water quality aspects of the principle, see Jasper Van Kempen, *Europees Waterbeheer: Eerlijk Zullen We Alles Delen?* (Boom Juridische Uitgevers 2012).

154 Own emphasis. See Article 2.2(c) and Article. 2.5(c) of the UN Watercourses Convention. See Unece, 'Guide to Implementing the Water Convention' (2013) para 102 <www.unece.org/fileadmin/DAM/env/water/publications/WAT_Guide_to_implementing_Convention/ECE_MP.WAT_39_Guide_to_implementing_water_convention_small_size_ENG.pdf> accessed 17 July 2017.

155 Ibid 237.

156 The Guide to implementing the Water Convention also refers to these factors set forth by the UN Watercourses Convention to explain the principle in the Convention (n 154).

157 Patricia K Wouters and others, 'Sharing Transboundary Waters: An Integrated Assessment of Equitable Entitlement: The Legal Assessment Model' (International Hydrological Programme 2005) IHP-VI Technical Document in Hydrology N°74, 21.

use.[158] The equitable and reasonable utilisation standard is not a clear rule, but rather a vague standard, which, according to Benvevisti, invites watercourse States to negotiate in order to determine how it should be translated in their specific basin.[159]

Although there is no hierarchy in these factors, special attention should be given to securing "vital human needs".[160] This implies that sufficient water, and flow, should be provided to sustain human life, which relates both to water required for the production of food and drinking water.[161] This provision also emphasises the human rights dimension of (transboundary) water management, which offers useful guidance in allocating water use shares.[162] Essentially, States should find an agreement which bolsters maximum benefits for all watercourse States but also safeguards protection of the resources and the sustainability of the watercourse in the long term, and taking into account the criteria together.[163] Competing interests of States must be weighed.[164] Algorithmic and other models have been developed to further specify the Article 6 criteria, which may help States in drafting water allocation provisions as part of their transboundary water management agreements.[165] The principle of equitable and reasonable utilisation does not entail an absolute prohibition of States to carry out an activity that causes harm to another watercourse State.[166] The question whether such harm has occurred or will occur or to what extent is to be weighed against the other factors.

158 See Article 6 of the UN Watercourses Convention.

159 Eyal Benvenisti, 'Collection Action in the Utilisation of Shared Freshwater: The Challenges of International Water Resources Law' (1996) 90 The American Journal of International Law 403.

160 See Article 10(2) of the UN Watercourses Convention, which provides that: "In the event of a conflict between uses of an international watercourse, it shall be resolved with reference to Articles 5 to 7, with special regard being given to the requirements of vital human needs".

161 See statements of understanding accompanying the Draft Articles on the Law of the Non-Navigational Uses of International Watercourses. See Article 3(8) (b) of the Protocol on Shared Watercourses in the Southern African Development Community (SADC), Adopted 7 August 2000 <www.sadc.int/files/3413/6698/6218/Revised_Protocol_on_Shared_Watercourses_-_2000_-_English.pdf> accessed 28 February 2017. Wouters and others (n 109).

162 It is remarkable however, that the human rights standard is included in the broader category of social and economic reality as part of the equitable and reasonable utilisation principle, i.e. not identified as an autonomous parameter. Equally, the human right to water has not been explicitly acknowledged in EU law, see below, e.g. in Chapter III. Eyal Benvenisti, 'Collection Action in the Utilisation of Shared Freshwater: The Challenges of International Water Resources Law' (1996) 90 The American Journal of International Law, 407. This will be relevant especially to Chapter II on the EU legal framework and Chapter V related to normative recommendations.

163 See the Commentary on the draft article 5 of the UN Watercourses Convention: ILC, 'Report of the International Law Commission on the work of its 46th session' (2 May-22 July 1994) UN Doc A/49/10.

164 Alistair Rieu-Clarke and Christopher J Spray, 'Ecosystem Services and International Water Law: Towards a More Effective Determination and Implementation of Equity?' (2013) 16 Potchefstroom Electronic Law Journal 12, 19.

165 Bruce Lankford, 'Does Article 6 (Factors Relevant to Equitable and Reasonable Utilisation) in the UN Watercourses Convention Misdirect Riparian Countries?' (2013) 38 Water International 130, 135.

166 See Patricia K Wouters and others, 'Sharing Transboundary Waters: An Integrated Assessment of Equitable Entitlement: The Legal Assessment Model' (International Hydrological Programme 2005) IHP-VI Technical Document in Hydrology N°74, 21.

42. Finally, the application of the equitable and reasonable utilisation principle is inherently flexible, and therefore coherent with adaptive management principles.[167] Indeed, it is flexible because the determination of the equitable apportionment of the use of water may easily change over time, as also emphasised by the United States Supreme Court: "it is obvious that if the depletion of the waters of the river by Colorado continues to increase, there will come a time when Kansas may justly say that there is no longer an equitable division of benefits, and may rightfully call for relief against the action of Colorado, its corporations and citizens, in appropriating the waters of the Arkansas for irrigation purposes".[168]

(c) The "no-harm" rule

43. The "no-harm" rule (also referred to as the *sic utero duo* principle) is also deeply entrenched in international customary water law and embraces a due diligence approach to ensure that activities of watercourse States do not cause significant harm to other watercourse States, reflecting a good neighbourliness requirement.[169] The due diligence approach is translated in the wording "must take all appropriate measures to prevent, control and reduce any transboundary impact".[170] Specifically, the due diligence obligation implies adopting and implementing appropriate laws and regulations to keep other States from suffering any harm. Article 7 of the UN Watercourses Convention includes similar wording by stipulating that watercourse States must take all appropriate measures so as not to cause significant harm to other watercourse States. The first version of the no-harm provision as drafted by the International Law Commission did not entail a due diligence obligation but an obligation of result to refrain from causing significant harm to other watercourse States.[171]

If the affected State must show that it has been harmed, the State whose activity has caused the significant harm must show that it has applied due diligence. However, the State still needs to demonstrate that its use of the shared transboundary waterbody has been equitable and reasonable.[172]

---

167 The adaptive management principles will be explained in the resilience sections below.

168 *Kansas v Colorado*, 206 US 46 (1907).

169 Joanna Kulesza, *Due Diligence in International Law* (Brill Nijhoff 2016) 94.

170 Article 2.1 Water Convention.

171 Because of the controversy surrounding the more absolute character of the no-harm rule in the draft version. See Joseph W Dellapenna, 'The Customary International Law of Transboundary Fresh Waters' (2001) 1 International Journal of Global Environmental Issues 264, 281. The exact wording of the initial version of the provision was as follows: "Watercourse States shall utilize an international watercourse in such a way as not to cause appreciable harm to other watercourse States". See Article 7 of the draft articles: ILC, 'Report of the International Law Commission on the work of its 46th session' (2 May-22 July 1994) UN Doc A/49/10.

172 McCaffrey (n 29) 441.

44. The no-harm rule has been used abundantly in seminal case law, such as *Trail Smelter*, where the Court stated that States do not have the right to cause serious harm to another State in the context of transboundary air pollution, and in *Corfu Channel*, where the ICJ held that it is "every State's obligation not to allow knowingly its territory to be used for acts contrary to the rights of other states".[173] The principle has also been endorsed in international instruments such as Principle 21 of the Stockholm Declaration, which provides: "States have, in accordance with the Charter of the United Nations and the principles of international law, the sovereign right to exploit their own resources pursuant to their own environmental policies, and the responsibility to ensure that activities within their jurisdiction or control do not cause damage to the environment of other states or of areas beyond the limits of national jurisdiction".[174] The rule necessitates a two-way street of good neighbourliness pursuant to the "de minimis" rule: States have the duty to refrain from acts causing significant harm, but also to tolerate minor inconveniences on their territories caused by other States.[175] Key in understanding when the no-harm rule has been breached in a specific river basin is the word "significant", which goes beyond a minor inconvenience, but is less than substantial.[176] There has to be "real impairment of a significant use of the water body or of its environment by a riparian".[177] The scope of activities triggering the no-harm rule is not restricted to those linked to the use of the watercourse but to any activity having a transboundary impact, regardless of where it takes place in the originating State.[178]

(D) Conclusive remarks on the application of the principles

45. The no-harm rule and the equitable and reasonable utilisation principle are to be seen as complementary in their application to transboundary watercourses.[179] The link with the equitable and reasonable utilisation principle lies in the fact that harm, even when it may be qualified as significant, can be justifiable if it

173 *Corfu Channel Case* (UK v Albania) (Merits) [1949] ICJ Rep 22.
174 *Trail Smelter* [1941] 35 AJIL 716. Declaration of the United Nations Conference on the Human Environment (16 June 1972) UN Doc A/Conf.48/14/Rev. 1(1973); 11 ILM 1416 (1972).
175 McCaffrey (n 29) 410.
176 Commentary to the Draft Articles on the Law of the Non-Navigational Uses of International Watercourses, in ILC, 'Report of the International Law Commission on the work of its 46th session' (2 May-22 July 1994) UN Doc A/49/10, 94 para 15.
177 UNECE, 'Guide to Implementing the Water Convention' (UNECE 2013) 15, para 80.
178 Attila Tanzi and Alexandros Kolliopoulos, 'The No-Harm Rule' in Attila Tanzi and others (eds), *The UNECE Convention on the Protection and Use of Transboundary Watercourses and International Lakes* (Brill Nijhoff 2015) 135.
179 See e.g. Jasper Van Kempen, *Europees Waterbeheer: Eerlijk Zullen We Alles Delen?* (Boom Juridische Uitgevers 2012) 274; Attila Tanzi and Alexandros Kolliopoulos, 'The No-Harm Rule' in Attila Tanzi and others (eds), *The UNECE Convention on the Protection and Use of Transboundary Watercourses and International Lakes* (Brill Nijhoff 2015) 139.

can be seen as equitable and reasonable under the specific circumstances.[180] The difference between the principle of equitable and reasonable utilisation and the no harm-rule may also be explained through the difference between violation of a legal right and factual harm. The equitable and reasonable utilisation is based on the vantage point that each riparian State has a legal stake and interest in an equitable part of the shared watercourse that should be protected.[181]

46. This question of significant effects is rather basin-specific, as it depends on the natural characteristics of the shared watercourse, for example. The same goes for the question "what is equitable and reasonable utilisation", which should be further detailed in the legal and institutional mechanisms governing the transboundary watercourse in question. However, and importantly, on the basis of the UN Watercourses Convention and the Water Convention, States are expected to provide for a mechanism detailing the translation of the equitable and reasonable utilisation principle in their shared waters. Chapter III will further address the practical implications of such translation to the EU legal framework for transboundary waters.[182]

47. As mentioned above, cooperation between watercourse States is the catalyst in the achievement of the principle of equitable and reasonable utilisation and the no-harm rule. For example, in order to fully grasp what the equitable and reasonable share of the use of the water would be, up-to-date information for the entire basin is quintessential. Moreover, a watercourse State cannot adequately determine its equitable and reasonable share unilaterally, as it needs information related to the various criteria from the other watercourse States.[183] Reference can be made, for example, to the social and economic needs of the respective States and potential uses of the watercourse. This presupposes, firstly, that information is gathered through monitoring throughout the entire basin and, secondly, that this information is shared by the States in question. The UN Watercourses Convention and the Water Convention therefore provide a whole range of substantive and procedural obligations underpinning the compliance with these two normative principles of international water law. These roughly overlap with the principles stemming from resilience thinking as outlined in the sections below: for example, the emphasis on continuous monitoring so that changing circumstances can be taken into account, notification requirements and information exchange, and dispute settlement.[184] These resilience-specific benchmarks will be discussed below.

---

180 Wouters and others.

181 McCaffrey (n 29) 389.

182 Considering the scope of this book, the translation of the principles will focus on water quantity and flood risk related issues.

183 McCaffrey (n 29) 401.

184 E.g. Article 9 of the Water Convention, Part III of the Water Convention and Article 33 of the Water Convention on dispute settlement. These provisions will also be referred to in Chapter III of this book related to the EU legal framework.

4.5.2.2. Integrated river basin management at the hydrological scale

48. One of the major issues in the context of (transboundary) water management relates to the discrepancy between the scale of decision-making and coordination on the one hand and the scale of ecological processes and problems on the other.[185] The legal framework should therefore remedy this issue by matching these respective scales. Needless to say, water has an utter disregard for administrative boundaries between municipalities, provinces, regions and nations. This means that to govern water purely from the perspective of legal boundaries, i.e. within the administrative layers of governance referred to in the preceding sentences, is counterintuitive to the logic of water flows.[186] Indeed, individual States cannot take measures in isolation without being affected by upstream or downstream activities and, vice versa, affecting these actions in turn, and this relates to both the material scope of water management, i.e. water quality and quantity, and to the hydrological scope, i.e. surface water and groundwater.[187]

49. In the context of (transboundary) water management, the appropriate ecological scale is the river basin scale.[188] As stated in the Dublin Principles, "the most appropriate geographical entity for the planning and management of water resources is the river basin, including surface and groundwater".[189] The shift from the focus on political boundaries toward the river basin scale has been widely applauded at the international and multilateral level for a long time.[190] For example, the ILA's 1966 Helsinki Rules and subsequent 2004 Berlin Rules[191], relevant for international watercourses, presented the concept of "drainage basin", which is defined as "an area determined by the geographic limits of a system of interconnected waters, the surface waters of which normally share a common terminus".[192] By analogy, the SADC Protocol of Shared Watercourse

---

185 Alistair Rieu-Clarke and Christopher J Spray, 'Ecosystem Services and International Water Law: Towards a More Effective Determination and Implementation of Equity?' (2013) 16 Potchefstroom Electronic Law Journal 12, 43.

186 Basin boundaries can exist both within national borders and beyond national borders.

187 See e.g. Marleen Van Rijswick, Herman Kasper Gilissen and Jasper van Kempen, 'The Need for International and Regional Transboundary Cooperation in European River Basin Management as a Result of New Approaches in EC Water Law' (2010) 11 ERA Forum 129. Cathy Suykens, 'EU Water Quantity Management in International River Basin: Crystal Clear?' (2015) European Energy and Environmental Law Review 134.

188 Dave Huitema and others, 'Adaptive Water Governance: Assessing the Institutional Prescriptions of Adaptive (Co-) Management from a Governance Perspective and Defining a Research Agenda' (2009) 14 Ecology and Society.

189 The International Conference on Water and the Environment: The Dublin Statement on Water and Sustainable Development, 31 January 1992.

190 Although not necessarily effectively practised and implemented.

191 See the above sections on the equitable and reasonable utilisation and no-harm principles for an explanation of the ILA and the rules.

192 International Law Association Report of the Fifty-Second Conference (International Law Association, Helsinki 1967). International Law Association Report of the Seventy-First

Systems of 1995 places the concept of "drainage basins" front and centre.[193] The UNECE Water Convention uses the following definition of scope with respect to transboundary waters: "any surface or groundwaters which mark, cross or are located on boundaries between two or more States".[194] In the United States, reference to governance through ecological delineation, dates from as early as 1879, when John Wesley Powell suggested drawing state boundaries based on watershed areas.[195] Throughout the country, States have concluded legal agreements to govern joint bodies from the basin-wide perspective.[196]

50. In the context of river basin management, an integrated approach to such management has widespread support.[197] Integrated river basin management has been defined as "the management of all surface and subsurface water resources of the river basin in its entirety with due attention to water quality, water quantity and environmental integrity".[198] This is strongly related to the concept of integrated water resources management, which promotes "the coordinated development and management of water, land and related resources in order to maximise the resultant economic and social welfare in an equitable manner without compromising the sustainability of vital ecosystems".[199] This integrative approach therefore relates both to the inclusion of surface and groundwater, and to the need to address water quality, water quantity and related environmental issues jointly. The need for integrated management stems from the interconnectedness in terms of territorial, hydrological and material scope of water management, both within and between these types of scope.[200] An overarching conceptual framework, which exceeds and captures the concept of integrated water resources management, can be found in the notion of water security, as this notion captures both the availability of good quality water (thus addressing the quantity and

Conference (International Law Association, London 2004) 334– 421. The Helsinki Rules represent the first instrument through which not only transboundary surface waters, but also transboundary groundwaters have been tackled. In the Berlin Rules, it was explicitly mentioned that confined groundwaters were also to be included in the scope. See, for a more elaborate discussion of the difference between the Helsinki and the Berlin Rules: Salman Ma Salman, 'The Helsinki Rules, the UN Watercourses Convention and the Berlin Rules: Perspectives on International Water Law' (2007) 23 International Journal of Water Resources Development 625.

193 Article 1 of the SADC Protocol on Shared Watercourse Systems (n 161).

194 Article 1(1) UNECE Water Convention.

195 This will be further discussed in the chapter on river basin management in the United States.

196 So-called "Compacts". E.g. Delaware River Basin Compact, Great Lakes St. Lawrence Compact, Colorado River Compact, Susquehanna River Basin Compact, and so forth. See Chapter IV.

197 Mees, Suykens and Crabbé (n 37).

198 Frank GW Jaspers, 'Institutional Arrangements for Integrated River Basin Management' (2003) 5 Water Policy 77, 79.

199 Global Water Partnership, 'Integrated Water Resources Management' (GWP 2000) TAC Background Papers 4/2000 < www.gwp.org/Global/GWP-CACENA_Files/en/pdf/tec04.pdf> accessed 20 February 2017.

200 The "scope" of water management is explained in detail in the following chapter on the EU legal framework for transboundary water management.

quality aspects) but also the risks for society associated with the presence of water in terms of natural disasters.[201]

51. Of course, there are dissenting opinions in literature, with authors criticising the river basin scale as the appropriate unit of governance. The difficulty in this regard is that pinpointing the relevant ecological scale can sometimes be ambiguous as it may not always cover every aspect of the issue at hand.[202] For example, the main disadvantage of the river basin approach is arguably its lack of full inclusion of land-based activities that have an impact on water management.[203] However, given the broad international support for the river basin approach for transboundary water management and the fact that it is also embedded in the UN Watercourses Convention and the Water Convention, which the relevant EU Directives aim to implement, it is the approach adopted in this book.

52. Implementing the "matching principle" and designating the river basin scale as the unit of governance for transboundary water management entail several legal and institutional and practical challenges and difficulties. For example, problems may occur with respect to institutional interplay: existing administrations may be reluctant to transfer sovereignty and competences, and the functional or "special purpose" entities may lack financial and human resources as a consequence; issues may arise related to democratic accountability; efficiency loss may occur due to institutional layering, and so forth.[204] An important conclusion is that such functional governance as embodied by river basin organisations is an incremental process and takes time to "mature".[205]

53. In line with nested management approaches and polycentric governance thinking, the appropriate scale can refer to the international hydrological scale but also to the lower-level scales such as the sub-basin and the sub-sub-basin.[206] As will

201 This concept will be explained in Chapter V.

202 Ruhl and Salzman discussing the difficulties surrounding the "matching principle": JB Ruhl and James Salzman, 'Climate Change, Dead Zones, and Massive Problems in the Administrative State: A Guide for Whittling Away' (2010) 98 California Law Review 59, 69.

203 Nigel Watson, Hugh Deeming and Raphael Treffny, 'Beyond Bureaucracy? Assessing Institutional Change in the Governance of Water in England' (2009) 2 Water Alternatives 448, 455.

204 Bruce Hooper, 'River Basin Organization Performance Indicators: Application to the Delaware River Basin Commission: Supplementary File' (2010) 12 1. Sander Meijerink and Dave Huitema, 'The Challenges and Pitfalls of Decentralisation in Water Resources Management' (2015) 5 Water Governance 16.

205 As also emphasised by Anne Schulte-Wülwer-Leidig, executive secretary of the International Commission for the Protection of the Rhine during the session related to transboundary water cooperation of the STAR-FLOOD end conference, which took place on 4 and 5 February 2016. See the Conference Plan that is attached in Annex to this book.

206 Dave Huitema and others, 'Adaptive Water Governance: Assessing the Institutional Prescriptions of Adaptive (Co-) Management from a Governance Perspective and Defining a Research Agenda' (2009) 14 Ecology and Society. Barbara Cosens, 'Transboundary River

be discussed in detail in the chapter on the Scheldt River, multi-level governance is omnipresent in transboundary water management. There is an element of subsidiarity inherent to (transboundary) water management, by acknowledging that not all water resources related matters should be governed at the higher levels.[207] It is important that the entity operating at the appropriate scale should have the means to act when necessary.[208] This includes the ability to take initiative when changing circumstances so require. Decision-making need not necessarily always be situated at the various hydrological scales for every issue, but there should be coordination and control at this scale in order to mitigate the (transboundary) externalities.[209] Coordination in this regard promotes inter-scalar links, which in turn are necessary for a proper flow of information, as will be discussed in the section below.[210]

54. The need for a fluent horizontal[211] and vertical[212] flow of information and coordination across scales is an inevitable consequence of the acknowledgment of the river basin scale as the appropriate unit of governance and the goal to achieve basin-wide coherence. Indeed, as much as water-related issues bring about important transboundary externalities, there is also an important local aspect to water management.[213] For example, large-scale floods can have impact in several countries in the course of one event, but the measures building up to the event can be rather local as well as the damage resulting from the floods. Reference can again be made to the plethora of competent authorities with respect to any given river basin, which will be explained in Chapters III and IV. This presents both opportunities for data production and challenges in ensuring that all the available data is properly used. Information gathered on the ground by, for example, local water managers should be fed to the authorities coordinating measures at the higher, not only national, levels and vice versa.[214] This also implies that at the international basin scale, there should be an awareness of what is taking place at the lower-level scales, e.g. in terms of cooperation initiatives and entities operating

Governance in the Face of Uncertainty: Resilience Theory and the Columbia River Treaty' (2010) 30 Journal of Land Resources and Environmental Law 229, 240.

207 This is also reflected in the EU Water Framework Directive and the Floods Directive, as will be discussed in the following chapter.

208 See e.g., Cosens (n 206) 257; Marleen van Rijswick and others, 'Ten Building Blocks for Sustainable Water Governance: An Integrated Method to Assess the Governance of Water' (2014) 39 Water International 725, 733.

209 Barbara Cosens, 'Legitimacy, Adaptation, and Resilience in Ecosystem Management' (2013) 18 Ecology and Society.

210 Tanya Heikkila, Edella Schlager and Mark W Davis, 'The Role of Cross-Scale Institutional Linkages in Common Pool Resource Management: Assessing Interstate River Compacts' (2011) 39 Policy Studies Journal 121, 122.

211 E.g. between agencies at the same level.

212 E.g. from the local level to the higher levels.

213 See, for a discussion of local entities, Mees, Suykens and Crabbé (n 37).

214 Joris Verwijmeren and Mark Wiering, *Many Rivers to Cross: Cross-Border Cooperation in River Management* (Eburon Delft 2007) 172; Barbara Cosens (n 206) 259.

at the cross-border sub-basin or sub-sub-basin scales. The sharing of information is important not only to remedy as much as possible the uncertainties discussed in the sections above, but also to ensure that decisions are based on the most accurate information and can be adopted in a coordinated manner.[215] There are specific techniques to help link several scales and remedy issues stemming from fragmentation, e.g. through the use of various types of bridging mechanisms.[216]

55. Considering the focus of this book on transboundary cooperation in shared waters, the assessment of this benchmark will mainly focus on the international scale governance mechanism. Especially in the context of the case studies on river basin management mechanisms of the Delaware River and the Scheldt River, the more local scales will come into play. These will especially be examined in light of their local cooperation processes, and the feedback mechanism (or lack thereof) with the main basin level. The case-study chapters of this book will analyse the functioning of two organisations operating at the river basin scale, i.e. river basin organisations or commissions.

4.5.2.3. Meaningful public participation

56. A well-designed and implemented public-participation mechanism is quintessential in the water management context, both from the perspective of resilience thinking and adaptive management principles, and from the perspective of integrated water resources management and integrated river basin management.[217] The term "meaningful" public participation has been used to indicate a two-way flow of information, i.e. where public authorities not only distribute information to the public, but also incorporate input acquired through participation processes early on.[218]

Especially in the context of flood risk management and when facing flood events, citizens' involvement is rather important and multi-faceted. Indeed, citizens can play an important role in mitigating flood risks by taking property-level risk-reduction measures, by following emergency procedures during the event, and so forth. This is also promoted through the discourse of "multi-layered water safety", which aims to share some of the burdens of flood risk management with citizens.[219] In countries that have a private or hybrid insurance

215 Green, Cosens and Garmestani (n 106); Van Rijswick and others (n 208) 734.

216 Gilissen and others (n 111). Different types of fragmentation require different bridging mechanisms. There are three broad categories of bridging mechanisms: transfer, coordination and cooperation.

217 Frank GW Jaspers, 'Institutional Arrangements for Integrated River Basin Management' (2003) 5 Water Policy 77, 82. Mees, Suykens and Crabbé (n 37).

218 Barbara Cosens (n 206) 261. Sarah Di Vittorio, Noelle Cole and Tamar Cooper, 'Accountability in Emerging Forms of Governance: A Comparison of the California Bay-Delta Process and the European Water Framework Directive' (2008).

219 See e.g. Dries Hegger and others, 'Flood Risk Management in Europe: Similarities and Differences between the STAR-FLOOD Consortium Countries' (STAR-FLOOD Consortium 2013) 26.

mechanism for compensating damage after a flood event has occurred, insurance companies also have an important role to play in encouraging such property-level measures, such as adaptive building and resilient reinstatement.[220] The public-private, government-population interplays are therefore inherent to flood risk management, and should be carefully considered in any flood risk management legal and policy framework.

57. A term that has been thrown around recently to describe a participation framework adapted to the trend of increasing involvement of private parties in water management is "coproduction".[221] Coproduction refers to a method of involving communities and private actors and citizens more closely in water management, in particular flood risk management, through empowerment, which presupposes that there is a high degree of awareness and perceived responsibility among these actors.[222] Mees and others. have identified a threefold meaning in the notion of coproduction: coplanning, which is the most well-known aspect of public participation and which relates to involving the public in the decision-making process; codelivery, which relates to the implementation process; and comprehensive coproduction, which relates to agenda setting, decision-making, implementation and evaluation of measures in the flood risk management realm.[223] A specific example of co-production is the "Flood Resilience Community Pathfinder Scheme", through which a substantial amount of money has been granted to certain communities to improve their response to floods.[224] Another specific technique to intensify public participation is the use of "citizens' juries", which will be explained in Chapter II of this book.

---

220 See Michael Faure, 'Insurability of Damage Caused by Climate Change: A Commentary' (2007) 155 University of Pennsylvania Law Review 1875; Paul Hudson and others, 'Incentivising Flood Risk Adaptation through Risk Based Insurance Premiums: Trade-Offs between Affordability and Risk Reduction' (2016) 125 Ecological Economics 1; Swenja Surminski and others, 'Reflections on the Current Debate on How to Link Flood Insurance and Disaster Risk Reduction in the European Union' (2015) 79 Natural Hazards 1451.Cathy Suykens and others, 'Dealing with Flood Damages: Will Prevention, Mitigation and Ex-Post Compensation Provide for a Resilient Triangle?' (2016) 21 Ecology and Society.

221 Hannelore Mees, Peter Driessen and Ann Crabbé (2017). Conditions for citizen co-production in a resilient, efficient and legitimate flood risk governance arrangement. A tentative framework. *Environmental Policy and Planning.* http://dx.doi.org/10.1080/1523908X.2017.1299623. Hannelore Mees and others, 'Coproducing Flood Risk Management through Citizen Involvement: Insights from Cross-Country Comparison in Europe' (2016) 21 Ecology and Society.

222 Mees, Driessen and Crabbé (n 221).

223 Hannelore Mees and others, 'Coproducing Flood Risk Management through Citizen Involvement: Insights from Cross-Country Comparison in Europe' (2016) 21 Ecology and Society.

224 £5m to 13 local authorities. See Kristina Ek and others, 'Strengthening and Redesigning European Flood Risk Practices: Towards Appropriate and Resilient Flood Risk Governance Arrangements Design Principles for Resilient, Efficient and Legitimate Flood Risk Governance; Lessons from Cross-Country Comparisons' (STAR-FLOOD Consortium 2016) 31.

58. Following the logic of the river basin approach, public participation should also be thought of – although not entirely of course – as a transboundary process. This becomes abundantly clear as public participation is an element of the (transboundary) Environmental Impact Assessment mechanism, which will be discussed in-depth in Chapter II of this book.[225] The evaluation of resilience in terms of the "meaningful public participation" benchmark will therefore be carried out from the transboundary perspective as well. For example, for the "Directives level" of evaluation, it will be analysed to what extent the applicable EU Directives promote transboundary public participation. For the case studies, it will be relevant to analyse whether participation processes are organised or coordinated at the main basin level, or, if not, whether information from such processes is exchanged between riparian States.

4.5.2.4. Monitoring and opportunities for learning

59. One of the key components of adaptive management relates to the ability of competent authorities to learn from updated insights and changed circumstances.[226] From a practical point of view, monitoring mechanisms and subsequent follow-up of the results of these exercises should be embedded into the legal and policy framework.[227] From a legal perspective, one of the terms used that resonates with this iterative process of rule-making is "reflexive law".[228] Monitoring in order to maintain a stream of up-to-date information is one of the elements supporting the ability to adapt.[229] Subsequently, the legal framework should enable the relevant authorities to effectively respond to the results of the monitoring exercise and the lessons learned.[230] From the resilience perspective, this ability to respond should ideally be tuned to real-time feedback, and not be tuned exclusively to a fixed-return period of planning.[231] Adaptive management therefore moves away from a linear approach of decision-making

---

225 Anne Glucker, Peter Driessen, Arend Kolhoff and Hens Runhaar, 'Public Participation in Environmental Impact Assessment: Why, Who and How?' (2013) 43 Environmental Impact Assessment Review 110.

226 Ahjond S Garmestani and others, 'The Integration of Social-Ecological Resilience and Law' (2014) Nebraska Cooperative Fish & Wildlife Research Unit – Staff Publications, 144 <http://digitalcomons.unl.edu/ncfwrustaff/144> accessed 20 February 2016.

227 Margot Hill Clarvis, Andrew Allan and David M Hannah, 'Water, Resilience and the Law: From General Concepts and Governance Design Principles to Actionable Mechanisms' (2014) 43 Environmental Science and Policy 98, 107.

228 Robin Kundis Craig and JB Ruhl, 'Governing for Sustainable Coasts: Complexity, Climate Change, and Coastal Ecosystem Protection' (2010) 2 Sustainability 1361; Catrien Termeer and others, 'The Regional Governance of Climate Adaptation: A Framework for Developing Legitimate, Effective, and Resilient Governance Arrangements' (2011) 2 Climate Law 159.

229 Keessen and van Rijswick (n 9) 8.

230 Van Rijswick and others (n 207) 732.

231 Olivia O Green and others, 'EU Water Governance: Striking the Right Balance between Regulatory Flexibility and Enforcement?' (2013a) 18 Ecology and Society.

toward a cyclical one, in which problem and objective-setting is followed by the selection of reference baselines, the implementation of actions and the evaluation of such implementation, before reaching full circle and returning to the problem and objective-setting point.[232] This brings us to the heart of the nexus of social-ecological resilience and law: the necessary balance between flexibility inherent to natural resources management and legal certainty, meaning that adaptive management should not come at the expense of legal certainty and effectiveness.[233]

60. As mentioned above, the equitable and reasonable utilisation standard is coherent with adaptive management principles such as monitoring, learning and flexibility. Indeed, both demand and supply in shared watercourses can vary greatly compared to the time of enactment of the joint agreement determining the translation of the equitable and reasonable utilisation standards. Reference can be made to an increase in the number of people relying on the water as a resource in correlation to a decrease of water, e.g. caused by salinisation. In that case, a modification of the allocation would be necessary to adapt to changing circumstances.[234] Allocation mechanisms can be made adaptable, for example, instead of allocating a specific quantity of water to be used by the riparian States; water use can also be allocated using percentages; and provisions can be made to allow reductions in periods when less water is available, to be compensated when there is an increase in availability, and so forth.[235]

4.5.2.5. Enforcement and effectiveness

61. This benchmark is closely related to the preceding criterion of flexibility. Effectiveness of a legal framework entails that it is provided with the necessary tools to achieve its aims, and this is an aspect of water management that should not be overlooked.[236] It presupposes that compliance can be monitored and that there

---

[232] McDonald and Megan C Styles, 'Legal Strategies for Adaptive Management under Climate Change' (2014) 26 Journal of Environmental Law 25, 28. Jb Ruhl, 'Regulation by Adaptive Management-Is It Possible?' (2005) 7 Minn. JL Sci. & Tech. 21 <http://heinonlinebackup.com/hol-cgi-bin/get_pdf.cgi?handle=hein.journals/mipr7&section=4> 34. Marleen van Rijswick and Andrea M Keessen, 'The EU Approach for Integrated Water Resources Management: Transposing the EU Water Framework Directive within a National Context – Key Insights from Experience' in Alistair S Rieu-clarke, Andrew Allan and Sarah Hendry (eds), *Routledge Handbook of Water Law and Policy* (Routledge 2017).

[233] Marleen van Rijswick and Willem Salet, 'Enabling the Contextualization of Legal Rules in Responsive Strategies to Climate Change' (2012) 17 Ecology and Society.

[234] Eyal Benvenisti, 'Collection Action in the Utilisation of Shared Freshwater: The Challenges of International Water Resources Law' (1996) 90 The American Journal of International Law 411.

[235] Mark Giordano and others, 'A Review of the Evolution and State of Transboundary Freshwater Treaties' (2013) 14 International Environmental Agreements: Politics, Law and Economics 245 <http://dx.doi.org/10.1007/s10784-013-9211-8> 4.

[236] Ibid 40. Michael G Faure, 'Effectiveness of Environmental Law: What Does the Evidence Tell Us?' (2012) 36 William & Mary Environmental Law and Policy Review 294. Jennifer K Poussin, WJ Wouter Botzen and Jeroen CJH Aerts, 'Effectiveness of Flood Damage Mitigation

are legal consequences in the event of failure to comply.[237] Enforcement is key in this regard, and empirical research has shown that this is often the 'Achilles heel' of flood risk management regimes, both at the national and the local levels.[238] For example, instruments which apply building conditions related to the preservation of the interests of the water system to new constructions, such as the Flemish water assessment, remain paper tigers if there are no follow-up mechanisms on the ground to ensure that these conditions have actually been fulfilled.[239] Standards, such as those related to water quality, need to be provided with enforcement mechanisms within the legislative framework.[240] Monitoring is a key enabler of enforcement, as it provides guarantees for parties, in this case States sharing a water body, that the freeriding issue for example, which is an inherent issue in the management of common-pool resources, will be addressed.[241]

62. A possible balance between the necessary flexibility linked to water management, taking into account the number of uncertainties which it is subject to, and effectiveness is exemplified by the adoption of flexible principles in combination with standards of protection mutually agreed on by the relevant parties.[242] For example, it has been argued that the Dutch legally enshrined safety

---

Measures: Empirical Evidence from French Flood Disasters' (2015) 31 Global Environmental Change 74. On the issue of ineffectiveness of CDMs in the context of gas flaring: Cathy Suykens, 'Gas Flaring in Developing Countries – Need for Kyoto Mechanisms or Sectoral Crediting Mechanisms?' (2010) 1 Carbon and Climate Law Review 42. On the lack of monitoring data, see Dave Huitema and others, 'Adaptive Water Governance: Assessing the Institutional Prescriptions of Adaptive (Co-) Management from a Governance Perspective and Defining a Research Agenda' (2009) 14 Ecology and Society. On the relationship between effectiveness and enforcement in the context of industrial risks, see Faure, 'The Complementary Roles of Liability, Regulation and Insurance in Safety Management: Theory and Practice' (2014) 17 Journal of Risk Research 689, 692.

237 Marleen van Rijswick and others (n 208) 736.

238 This may be caused e.g. by a lack of resources and a lack of expertise at the local levels. Kristina Ek and others, 'Strengthening and Redesigning European Flood Risk Practices: Towards Appropriate and Resilient Flood Risk Governance Arrangements Design Principles for Resilient, Efficient and Legitimate Flood Risk Governance; Lessons from Cross-Country Comparisons' (STAR-FLOOD Consortium 2016). See, for Belgium, Hannelore Mees, Cathy Suykens and others, 'Analysing and Evaluating Flood Risk Governance in Belgium Dealing with Flood Risks in an Urbanised and Institutionally Complex Country' (STAR-FLOOD Consortium 2016).

239 Hannelore Mees, Ann Crabbe and Cathy Suykens, "Flood risk governance in Belgium. Towards a resilient, efficient and legitimate arrangement?" in Corinne Larue and others, *Facing hydro-meteorological extreme events in Europe: a governance issue* (Wiley In Press). This will be discussed in Chapter III. See, for a general overview, Cathy Suykens, 'Water Management and Maritime Spatial Planning' (2013) 1 IUCNAEL 59.

240 Jonas Ebbesson, 'The Rule of Law in Governance of Complex Socio-Ecological Changes' (2010) 20 Global Environmental Change 414, 419.

241 Shlomi Dinar and others, 'Climate Change and State Grievances: The Resiliency of International River Treaties to Increased Water Variability' (2010) 3 Insights 1, 8.

242 Marleen van Rijswick and Willem Salet, 'Enabling the Contextualization of Legal Rules in Responsive Strategies to Climate Change' (2012) 17 Ecology and Society 2.

standards for flood risks are an example of such a balance, because they function in such a way that they are legally binding and enforceable but leave ample room for contextualisation taking local circumstances into account.[243] Moreover, in the context of environmental protection, legislation that encourages taking positive action by setting goals in this regard, instead of prohibiting deterioration of the environment, is likely to be more conducive to sustainable development.[244]

Effectiveness of transboundary water agreements can be measured, for example, by evaluating the cooperation and coordination following a flood event, by evaluating the development of water quality over time, where the baseline would be the time of entry into force of the institutional cooperation mechanism, and so forth.

243 Ibid. For a discussion of the Dutch safety standards, see M Kaufmann, HK Van Doorn-Hoekveld, W., Gilissen and M Van Rijswick, 'Analysing and Evaluating Flood Risk Governance in the Netherlands. Drowning in Safety?' (STAR-FLOOD Consortium 2016).

244 Anoeska Buijze, 'Promoting Sustainable Water Management in Area Development: A Regulatory Approach' (2012) 24 The Journal of Water Law 166, 173.

CHAPTER II
THE EU LEGAL FRAMEWORK FOR TRANSBOUNDARY WATERS WITH A FOCUS ON WATER QUANTITY MANAGEMENT

1. INTRODUCTION AND DELINEATION OF SCOPE

63. This chapter analyses the European legal framework for cooperation regarding transboundary waters primarily from the viewpoint of water quantity management. As explained in Chapter I, for the purposes of readability, water quantity management is used as an umbrella term encompassing scarcity, droughts and floods. The reasons from choosing water quantity as the main lens through which transboundary water management is viewed, have been explained in Chapter I as well.

It is noteworthy that although water quantity is teased out in this Chapter, it should not only be seen as a self-standing category but also in its relation to water quality.[245] In other words, this Chapter will set forth the role of water quantity management in the EU legal framework, as well as its link with water quality management. For example, in the discussion of the evolution of the legal basis for "quantitative management of water resources" pursuant to the EU Treaty, the comparison with the legal basis for water quality management will be highly relevant. In contrast, this Chapter will not delve into the EU-law requirements for water quality management, i.e. the requirements aimed at combating pollution of water bodies in the EU.

Analysing water quantity management in transboundary waters in the EU requires both an analysis of the EU primary and secondary law framework for the quantitative aspects of water resources, but also the mechanisms this framework provide to induce cooperation in these transboundary waters. Such a comprehensive, twofold, approach is necessary to identify the bottlenecks and in turn formulate recommendations at the end of this book.

---

[245] Namely, in the context of integrated water resources management.

64. The analysis will be guided by the five pillars of transboundary water governance as identified by the Global Water Partnership: (i) scope, (ii) substantive provisions, (iii) procedural provisions, (iv) institutional mechanisms and (v) dispute settlement.[246] These five pillars allow a comprehensive overview of all relevant aspects that form part of the framework. By using the pillars and the encompassing approach these enable to adopt, bottlenecks entrenched in said EU legal framework will be systematically identified. The chapter will both focus on the applicable EU regime for water quantity management, pursuant to primary and secondary law, and the manner in which this regime facilitates cooperation between EU Member States in shared waters. The pillars should not be equated with evaluation benchmarks, but rather be considered as a lens through which the issues at hand can be viewed and analysed.

Figure 2. GWP's Five Pillars of the Transboundary Water Legal Framework

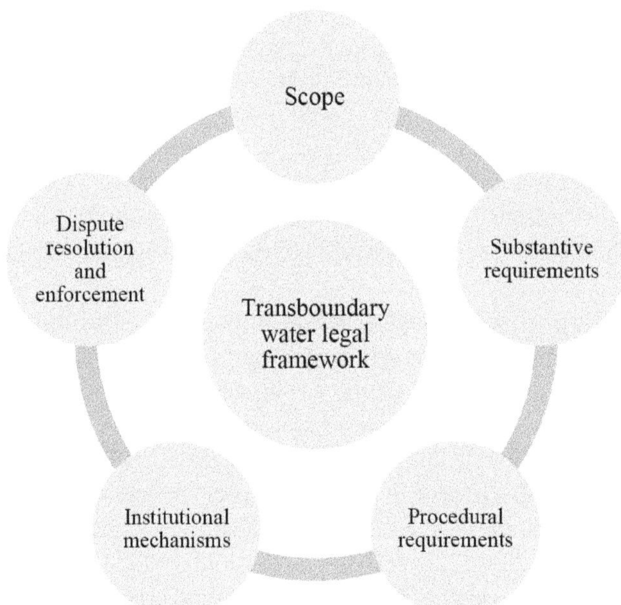

65. The chapter will firstly analyse the scope of the EU legal framework for water quantity management in transboundary waters, which includes the geographical, physical and material coverage of the regime. Then, the substantive and procedural requirements will be scrutinised, which include, e.g. the duty to exchange information with respect to flood risks in shared waters. The chapter will then dive into the institutional mechanisms for cooperation in shared waters, i.e. the manner in which Member States should formalise their cooperation

246 See Christophe Brachet and Daniel Valensuela, The Handbook for Integrated Water Resources Management in Transboundary Basins of Rivers, Lakes and Aquifers (GWP/INBO 2012).

requirements. Finally, the enforcement and dispute settlement mechanisms applicable to those shared waters will be looked into.

2. FIVE PILLARS OF THE TRANSBOUNDARY LEGAL WATER REGIME IN THE EU

2.1. PRELIMINARY OVERVIEW OF WATER LAW IN THE EU

66. The figure below illustrates the state of play of water law in the EU. The bulk of EU water law focuses on the qualitative aspects of water management, with the 2000 Water Framework Directive as the mother of EU Water Directives. The Water Framework Directive (WFD) also has a link with quantity, insofar as the quantitative aspects are necessary to support the qualitative requirements. Since 2007, quantitative aspects of water management have come more to the forefront, specifically since the entry into force of the Floods Directive (FD).

Figure 3. Overview of EU Water Law[247]

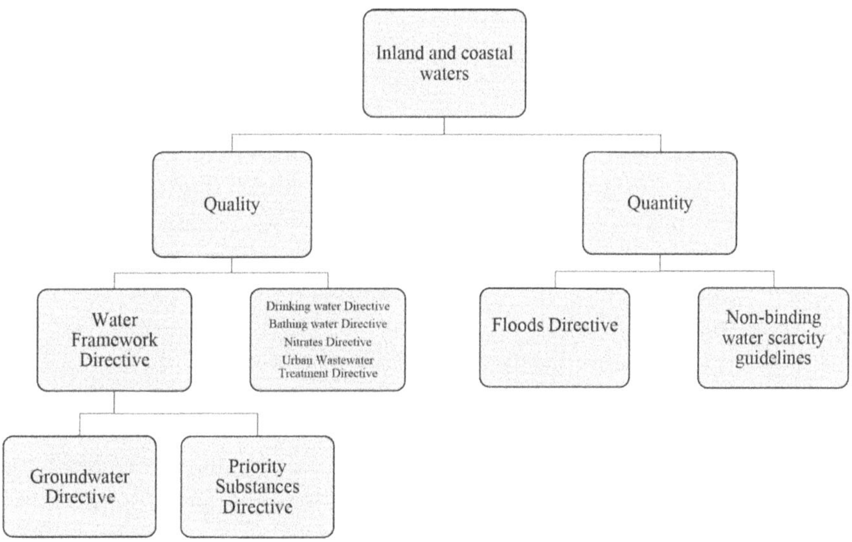

2.2. PILLAR I: SCOPE

67. The first pillar of the legal regime governing transboundary waters relates to the scope, which entails the geographical and functional coverage of the

[247] Similar figure can be found in Marleen van Rijswick and Herman Havekes, *European and Dutch Water Law* (Europa Law Publishing 2012).

transboundary water resource in question.[248] The question of scope relates to the identification of the water bodies covered by the relevant EU directives, and of which aspects of water management are included, i.e. mainly in terms of surface water and groundwater. It also relates to the extent to which these International River Basin Districts cover qualitative and quantitative aspects of water management. On the one hand, the concept of International River Basin Districts in the European Union will be studied, and on the other hand, the scope of the quantitative aspects of water resources management will be described. Table 2 illustrates three aspects of the pillar of "scope", which will be discussed in-depth in the sections below.

Table 2. Three Aspects of the Pillar Scope

Geographical coverage	Which geographical scales of governance have been adopted by the Directives and how have these been defined?
Physical coverage	What is the physical coverage of the legal regime in terms of surface water, groundwater, and aquifer inclusion?
Material scope	Which aspects of water management have been included into the scope with regard to water quality, quantity and risk management?

2.2.1. The geographical coverage

68. The cornerstone of the Water Framework Directive and the Floods Directive is the institution of River Basin Districts, which implies the focus on units based on hydrological boundaries instead of administrative and political ones. According to the EU Commission, the consultation process had revealed an "almost" universal support for using these River Basin Districts as the unit for governing water resources.[249]

A River Basin within the meaning of the Directives is "the area of land from which all surface run-off flows through a sequence of streams, rivers and, possibly lakes into the sea at a single river mouth, estuary or delta".[250] A District is considered as the area of land and sea constituted by one or more of these River Basins, including their associated groundwaters and coastal waters.[251] As mentioned above, the River Basin District is the main unit of governance, both in terms of the implementation of the Water Framework Directive and of the Floods

248 Patricia Wouters, 'International Law – Facilitating Transboundary Water Cooperation' (2012) TEC Background Papers n 17 <www.gwp.org/globalassets/global/toolbox/publications/background-papers/17-international-law---facilitating-transboundary-water-cooperation-2013-english.pdf> accessed 13 July 2017.

249 Commission, 'Proposal for a Council Directive establishing a Framework for Community Action in the Field of Water Policy, Explanatory Memorandum' COM(97) 49 final.

250 Article 2(13) WFD.

251 Article 2(15) WFD.

Directive. This focus is a reflection of the "principle of holistic water management at the catchment level"[252], namely the management of water from source to sea, and including both surface waters and associated groundwater. The shift from political boundaries to hydrological boundaries is the quintessential vantage point in European water law, and the cornerstone of this research. Indeed, whilst the EU aims to encourage hydrological scale governance, it should be examined whether its current legal framework has the right tools to facilitate such governance with respect to the relevant water bodies, and whether such governance is translated into practice in specific river basins.

69. Considering the subject of this research, the most relevant scope pertains to that related to transboundary waters. In this regard, the International River Basin District (IRBD) is the term used for the situation where a river basin spans the territory of several (Member) States. There are IRBDs that are located within the territory of the European Union, and Districts that extend beyond the territory of the EU. Needless to say, cooperation requirements in the latter case are more difficult to enforce than in the former case.

It is important to note that the concept "International River Basin District" does not explicitly feature in the Directives as a unit of governance. The term is not explicitly defined in the Directives. The WFD requires Member States to ensure that a river basin covering the territory of more than one Member State is assigned to an International River Basin District. Needless to say, the fact that the International River Basin District is not explicitly defined as a unit of governance in Article 2 WFD, contrary to the River Basin District, unavoidably has legal impacts on the enforceability of cooperation in these IRBDs. The substantive and procedural requirements applicable to Member States sharing IRBDs will be discussed in Sections 2.2 and following.

This rather loose character of the concept of IRBD is paired with the fact that it is often unclear which measures should be taken at the level of the IRBD, as opposed to the RBD – the RBD being the main unit of governance. For example, the WFD provides for the issuance of long-term forecasts of supply and demand for water in the River Basin District, as a basis of the economic analysis of water services.[253] The WFD does not stipulate whether this forecast should be made at the level of the IRBD, in case of transboundary waters.[254] This example

252 Commission, 'Report from the Commission to the European Parliament and the Council on the Implementation of the Water Framework Directive (2000/60/EC) River Basin Management Plans', COM(2012) 670 final, 9.

253 The concept of water services will be explained in the chapter on substantive provisions. See Recital 38 WFD.

254 The river basin approach was weakened in the run-up to the adoption of the WFD on other occasions as well, e.g. the Parliament had proposed an amendment to Article 1(a)(i) of the Directive, which sets out the goals of the Water Framework Directive. The Parliament had proposed as a goal of the WFD: "Promotes sustainable and efficient water use based on long-term protection of available water resources within a hydrological area or river basin". The

clearly illustrates (i) the link between the first pillar of the transboundary legal framework, i.e. scope, and the second and third pillars, i.e. substantive provisions and procedural provisions, and (ii) the importance of a clear scope of this framework.[255]

2.2.2. The physical coverage

70. The hydrological scope relates to the question which aspects of the hydrological cycle have been included in the legal regime, mainly in terms of surface and groundwater inclusion. The hydrological cycle is "the system in which water – solid, liquid, gas, or vapour – travels from the atmosphere to the Earth and back again in a constant cycle of renewal", as depicted in figure 4.[256] Surface waters and groundwater interact with one another in the hydrological cycle.

Figure 4. The Water Cycle

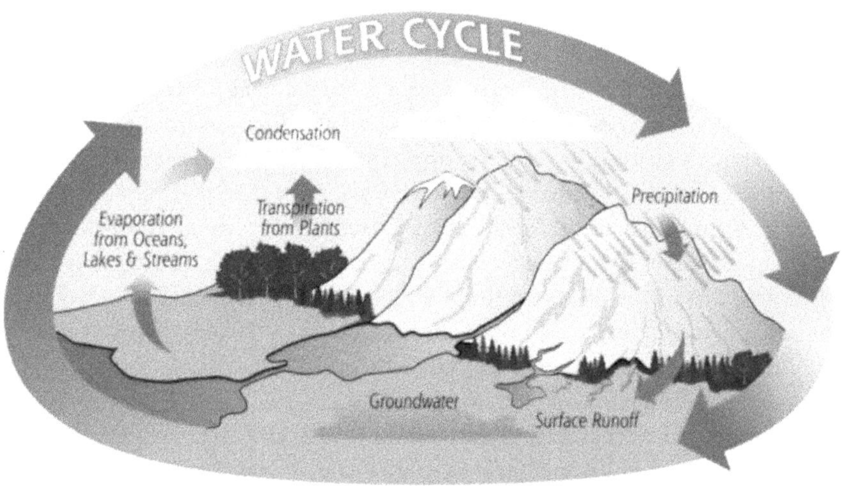

Source: NASA

The importance of legal consistency between the rules and regulations applicable to surface water on the one hand and groundwater on the other, is pivotal. A hypothetical example: If administrative requirements for the abstraction of surface water are stricter and more cumbersome than those applicable to the abstraction of groundwater, this may result in unwanted effects, i.e. over-abstraction of groundwater so as to avoid the administrative requirements related

words "within a hydrological area or river basin" do not appear in the final text of the WFD, namely, Article 1(b) WFD. In the final text of the WFD, this part of the purpose is formulated: "promotes sustainable water use based on a long-term protection of available water resources".

255 These pillars will be explained in the following sections.

256 Gabriel Eckstein, 'A Hydrogeological Approach to Transboundary Ground Water Resources and International Law' (2003) 19 American University International Law Review 201.

to the abstraction of surface water.[257] Moreover, considering the importance of groundwater in the hydrological cycle, its inclusion in any legal regime governing transboundary waters is paramount. This inclusion, however, is not self-evident. In international water law, transboundary groundwater has often been omitted from the relevant body of law.[258]

In the sub-section above, we established that the Directives apply both to surface water and groundwater, tackling these aspects in an integrated manner. Indeed, from the perspective of holistic water management at catchment level, the purpose of the WFD is to establish a framework for the protection of inland surface waters, transitional waters, coastal waters and groundwater in order to attain good status of these waters, mainly from a qualitative point of view supported by a quantitative approach.[259] In this regard, the WFD supports the concept of "Integrated Water Resources Management" (IWRM), which promotes "the co-ordinated development and management of water, land and related resources in order to maximise the resultant economic and social welfare in an equitable manner without compromising the sustainability of vital ecosystems".[260]

The inclusion of groundwater in the hydrological scope of the WFD is also illustrated by the existence of the Groundwater Directive as the daughter of the WFD. The Groundwater Directive includes specific measures to prevent and control groundwater pollution, with reference to Article 17(1) and (2) of the WFD. Groundwater is defined as "all water below the surface of the ground in the saturation zone and in direct contact with the ground or subsoil".[261] A groundwater body within the meaning of the WFD is defined as a distinct volume of groundwater within an aquifer or aquifers.[262] An aquifer constitutes "subsurface layer(s) of rock or other geological strata, with sufficient permeability to enable a significant flow of groundwater or the abstraction of significant quantities of groundwater".[263]

The question also arises whether confined aquifers are included in the scope of the Directives. Aquifers are important in the context of the quantitative aspects of

---

257 Kerstin Mechlem, 'Thematic Paper No. 6: Legal & Institutional Frameworks' (GWP 2012).

258 At the beginning of the 21st century, efforts were made at the global level to tackle transboundary groundwater resources. Scientific evidence has increasingly illustrated the link between groundwater and surface water, overriding the consideration that the transboundary character of surface water is more apparent in comparison to groundwater. Gabriel E Eckstein, 'Commentary on the UN International Law Commission's Draft Articles on the Law of Transboundary Aquifers' (2007) 18 Colorado Journal of International Environmental Law & Policy 537.

259 Article 1 WFD. See, for the importance of groundwater protection in the WFD, among other things Recitals 3, 10, 13.

260 Global Water Partnership Technical Advisory Committee, 'Integrated Water Resources Management' (GWP 2000) 22.

261 Article 2(2) WFD.

262 Article 2(12) WFD.

263 Article 2(11) WFD.

water management, as they contain significant amounts of groundwater used e.g. for abstraction.[264] It is therefore important that these aquifers should be subject of legal certainty.[265]

At the international level of transboundary groundwater law, a distinction is made between confined and unconfined aquifers. This difference is best illustrated by a figure[266]:

Figure 5. Confined and Unconfined Aquifers

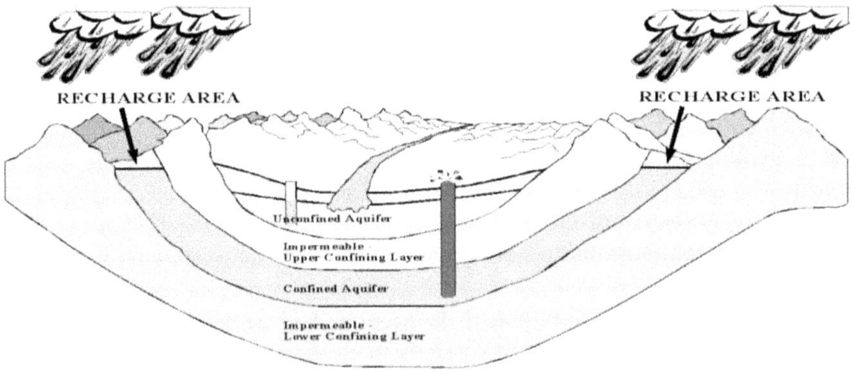

Source: International Water Law Project[267]

The unconfined aquifer is surrounded by an impermeable layer of rock or sediments and covered by permeable materials extending from the land surface.[268] Generally speaking, the unconfined aquifers are in direct contact with a body of surface water. The confined aquifer is considered as an aquifer that is contained between two impermeable layers.[269]

264 Philippe Quevauviller et al., The Water Framework Directive: Ecological and Chemical Status Monitoring (Wiley 2008) 496.

265 Indeed, for example, the UN Watercourses Convention does not include such contained aquifers in its scope, which has sparked criticism in the body of international doctrine. See Stephen McCaffrey, 'The International Law Commission Adopts Draft Articles on Transboundary Aquifers' (2009) 103 The American Journal of International Law 272, 273. The guidance document with regard to risk assessment for groundwater issued in the context of the CIS does not cover the issue of transboundary groundwater. The guidance document merely suggests that, for transboundary groundwater bodies, jointly agreed conceptual models are highly recommended. Commission, 'WFD CIS Guidance on Risk Assessment and the Use of Conceptual Models for Groundwater' (European Union 2010).

266 This figure originates from the following article: Gabriel Eckstein and Yoram Eckstein, 'A Hydrogeological Approach to Transboundary Ground Water Resources and International Law' (2003) American University International Law Review 201, 211.

267 Available on the Internet through: www.internationalwaterlaw.org/.

268 Gabriel Eckstein and Yoram Eckstein, 'A Hydrogeological Approach to Transboundary Ground Water Resources and International Law' (2003) American University International Law Review 201, 211.

269 Ibid, 212.

This distinction is made at the level of international law, as mentioned above, and it is considered to be a tricky one from the perspective of the governance of transboundary waters. Whereas an aquifer is a static geological formation, the groundwater in the aquifer is dynamic and it moves around. Indeed, the International Law Commission adopted draft articles on transboundary aquifers in 2008.[270] Without going into detail on the scope of these draft articles, it can already be stated that focusing on a static formation instead of on the moving water itself will consequently add more emphasis on the sovereignty of the states than in the latter case. This is in contrast to the 1986 Rules of International Groundwaters adopted by the International Law Association, which focus on the water within the international aquifers, as opposed to the rocks surrounding the waters.[271] At the level of the European Directives, this distinction between confined and unconfined aquifers is not made.

71. The UN Watercourses Convention defines a watercourse as including both surface water and the groundwater that interacts with it, meaning that surface waters and groundwaters constitute a unitary whole by virtue of their physical relationship. Groundwater that is not hydrologically connected to the surface water is therefore not covered by this definition of watercourses.[272] The unit of management adopted by the WFD and FD is the River Basin District, which is defined as the area of land and sea, made up of one or more river basins with their associated groundwaters and coastal waters. The WFD does not refer to the concept of confined transboundary aquifers, and it is therefore not clear whether or not these are covered by the scope of the legal regime governing transboundary waters.

2.2.3. *The material scope*

2.2.3.1. Quantity in its relation to quality water management

72. The third element of the first pillar "Scope" relates to the balance between the focus of the EU legal framework on water quality management, i.e. point source and diffuse pollution, and water quantity management, i.e. droughts, scarcity and floods. This relates back to the one of the reasons this book focusses on quantity in the first place: the consideration that this aspect has been underdeveloped in the EU legal framework. The following section will provide the data to support this statement.

---

270 International Law Commission, 'Report of the International Law Commission on the Work of its Sixtieth Session' (5 May – 6 June and 7 July-8 August 2008) UN Doc A/63/10).

271 International Law Association, Report of the Sixty-Second Conference (International Law Association, Seoul 1986).

272 Article 2 UN Watercourses Convention. See Stephen McCaffrey, 'The International Law Commission's flawed Draft Articles on the Law of Transboundary Aquifers: the way forward' (2011) 36 Water International 566, 567.

Indeed, the WFD primarily focuses on the qualitative aspects of water management.[273] The quantitative aspects are, taking into account the issue with regard to the legal basis for quantitative water management, generally an accessory to the qualitative aspects, as the centre of gravity in the WFD is "water quality". This has also been criticised by parliamentary committees[274], who stated that quantity-related concerns being subordinate to quality measures in the Directive is not in line with the realities of the 21st century[275], and is superseded by events that have taken place in the recent past.[276] These committees argue that it seems inappropriate to focus on the protection of the environment, without taking into account or including an adequate quantitative analysis of water resources.[277]

However, in its recent "Blueprint to Safeguard Europe's Water Resources", the EU Commission emphasises the intimate connection between water quality and quantity in the context of achieving "good status".[278]

The link between water quality and water quantity can be illustrated by the following example: On the basis of Article 4 (3) of the Water Framework Directive, Member States may designate certain bodies of surface water as "artificial or heavily modified", which results in certain leniency in obtaining the target of "good status".[279] Such designation and subsequent leniency may occur when the changes to the hydromorphological characteristics of the body of water, which would be necessary in order to obtain said "good status" would have significant adverse effects on measures such as flood protection.[280]

Another example where the intrinsic bond between quantity and quality is apparent is related to artificial water recharge, where water that originates from a surface water source is stored in the ground, thus increasing groundwater resources. The quality of the original surface water benefits from this process.[281] One of the manners in which this artificial water recharge can take place, is surface water infiltration, which in turn is one of the important measures in the context of flood risk management.

---

273 Monika Ambrus, Herman Kasper Gilissen and Jasper JH Van Kempen, 'Public Values in Water Law: A Case of Substantive Fragmentation?' (2014) 10 Utrecht Law Review 8, 18.

274 Committee on Research, Technological Development and Energy, Report on the proposal and the amended proposals for a Council Directive on establishing a framework for Community Action in the field of water policy (1998) A4-0261/98, 73.

275 Which UNESCO refers to as "the century of water".

276 In terms of droughts and floods.

277 Committee on Research, Technological Development and Energy, Report on the proposal and the amended proposals for a Council Directive on establishing a framework for Community Action in the field of water policy (1998) A4-0261/98, 73.

278 Commission, 'A Blueprint to Safeguard Europe's Water Resources' (Communication) COM (2012) 673 final, 6.

279 On the basis of biological, chemical and morphological criteria. See Ecologic, WFD and Agriculture Linkages at the EU Level – Analysis of the Policy and Legal Linkages between CAP and WFD, FP6, 2006, 3.

280 Article 3 (a) (iv) of the Water Framework Directive.

281 EEA, 'Water Resources across Europe – confronting Water Scarcity and Drought' (EEA, 2009), 16.

For the purposes of environmental protection there is a need for stronger integration of qualitative and quantitative aspects of both surface waters and groundwaters, taking into account the natural flow conditions of water in the hydrological cycle.[282]

As the link between quantitative and qualitative water management is becoming increasingly clear, it should be examined what the limits are of management of water quantity as such, i.e. without it being subordinate to the quality aspects of water management. In order to find these limits, it is important to look at the legal basis for water quantity management and water quality management in the EU, as these are subject to different decision-making processes.

2.2.3.2. The legal basis for water quantity at EU level

(a) Subsidiarity and proportionality

73. In accordance with the principle of conferral, the EU may act within the parameters of the competences granted in the Treaties.[283] These competences can either be exclusive or shared or they can be geared toward supporting, coordinating or supplementing Member States' actions.[284] For areas where the Union does not have exclusive competence, all actions need to pass the subsidiarity and proportionality test.[285] This is true for environmental policy, which is a shared competence and which constitutes the legal basis on which the WFD and FD have been construed.[286] Competences with respect to the environment policy, on which EU action with regard to the WFD and FD is based, are shared competences and any such action should therefore pass the subsidiarity test.[287] When the subsidiarity principle was first introduced, in the 1987 Single European Act, it was only relevant to measures related to environmental protection.[288] The Treaty of Maastricht broadened the scope of the principle, and the Treaty of Amsterdam provided a protocol to provide guidelines on its application.[289]

---

282 Recital 34 WFD.

283 Article 4(1) TEU and 5(1) TEU.

284 Articles 3, 4 and 6 TFEU.

285 Article 5.3 TEU.

286 In areas with shared competences, Member States may take action where the EU has not exercised its competence. See Article 2 TFEU, Article 4(2)(e) TFEU.

287 Geert Van Calster and Leonie Reins, *EU Environmental Law* (Edward Elgar Publishing 2017) 7.

288 Article 130r(4) of the SEA stated: "The Community shall take action relating to the environment to the extent to which the objectives referred to in paragraph 1 can be attained better at Community level than at the level of the individual Member States. Without prejudice to certain measures of a Community nature, the Member States shall finance and implement the other measures".

289 David Langlet and Said Mahmoudi, *EU Environmental Law and Policy* (Oxford University Press 2016) 47. Protocol on the Application of Subsidiarity and Proportionality Principles [1997] OJ C 340/105.

A new protocol addressing subsidiarity[290] was added to the Treaty of Lisbon, which has also embedded the principle in Article 5(3) TEU. On the basis of this provision, the EU may only act insofar as the objectives of the action cannot be sufficiently achieved by the Member States, either at central, regional or local level, meaning that the EU level would be better equipped to tackle the issue because of the scale or the effects of the action. The transboundary nature of the issue at hand is a major element in determining whether the subsidiarity hurdle has been passed.[291] The 1997 protocol is still relevant in determining whether a certain act passes the subsidiarity test, because the currently applicable protocol mainly addresses the procedural requirements associated with the principle. Relevant indicators are that there are transnational aspects to the issue at hand, that actions initiated by Member States alone might run contrary to Treaty requirements or harm Member States' interests and that there are clear benefits to EU-level action for reasons related to scale and effects.[292]

In addition to the subsidiarity test, EU action must also pass the proportionality test, which entails that the content and the form of said action must not go beyond what is necessary to achieve the objectives of the Treaties.[293] The action should therefore be appropriate to reach the goal and should be necessary to do so, meaning that the goals would not be reached by initiating another type of action. In this context, the type of instrument used is relevant, where a well-known instrument in EU environmental law is the framework directive, e.g. the WFD, which leaves substantial room for the Member States to design and operationalise their proper policies.[294] Framework Directives have been promoted as part of the EU's Better Regulation Strategy, which embraces flexibility and co-regulation.[295]

74. These principles and their indicators are paramount in the water management landscape. The importance of careful scrutiny in this regard can be illustrated by the subsidiarity principle. There is an abundantly clear necessity for EU action with respect to water (and river) management, due to its inherent transboundary nature reflected in the clear upstream-downstream inter-jurisdictional interactions in combination with the fact that 60% of EU territory is

290 And also proportionality.
291 Nicolas De Sadeleer, 'Principle of Subsidiarity and the EU Environmental Policy' (2012) 9 Journal for European Environmental & Planning Law 63, 64.
292 Paragraph 5 of the 1997 Protocol.
293 Article 5(4) TEU.
294 David Langlet and Said Mahmoudi, *EU Environmental Law and Policy* (Oxford University Press 2016) 48.
295 Co-regulation implies involving more stakeholders in the legislative process. This is linked to the discussion on public participation in this chapter. Mireille Bogaart, 'The Emergence of the Framework Directive in EU Environmental Policy: An Exploration of Its Function and Characteristics', *EU Environmental Legislation Legal Perspectives on Regulatory Strategies* (Edward Elgar Publishing 2014) 48, 56.

situated in International River Basin Districts.[296] At the same time, however, water management has a vitally important local aspect. For example, "keeping people away from the water" should be done through smart spatial planning decisions and local zoning plans.[297] This delicate balance is kept in mind throughout this book, and in the formulation of normative recommendations in Chapter V.

75. Now that the general background on EU decision-making has been explained, the following sections will address the issues related to the different legal bases relevant to water management.

(b) Different decision-making procedures

76. Water quantity management and water quality management are subject to different regimes on the basis of primary EU law. The policy-making on aspects related to water quality has traditionally been part of one of the core topics of EU environmental law.[298] On the other side of the spectrum is water quantity management, with regard to which Member States have ceded their sovereignty to a lesser extent. Moreover, a relevant restriction with respect to flood risk management relates to the position of land use planning in EU primary law.[299] As will be explained below, the threshold for issuing legislation at the EU level with regard to water quantity is higher than on water quality.

However, there is legal uncertainty with regard to the legal basis and scope for the issuance of legislation on the quantitative aspects of water management. Indeed, measures with regard to "the quantitative management of water resources" are subject to the unanimity requirement pursuant to the TFEU. It is not entirely clear what constitutes "quantitative management of water resources", as will be discussed in the following sections. The material scope in the sense of this pillar therefore mainly pertains to the legal basis for the issuance of the WFD and FD. In the context of the following pillar, which pertains to substantive provisions, the obligations set forth in the relevant Directives with regard to quantitative water management will be explained.

77. As mentioned above, the management of water resources is subject to two different decision-making processes in the European Union. On the one hand, there is the generally applicable decision-making procedure with respect to environmental measures, and on the other the special decision-making procedure that entails unanimity voting.

---

296 *Infra*, Chapter I.

297 See, for an explanation of this narrative, translated in the "risk prevention strategy": DLT Hegger and others, 'A View on More Resilient Flood Risk Governance: Key Conclusion of the STAR-FLOOD Project' (STAR-FLOOD Consortium 2016) 4.

298 Jan Jans, *European Environmental Law* (Europa Law Publishing, 2000) 48.

299 I.e. requiring unanimity voting.

The relevant legal basis for environmental protection measures is Article 192(1) TFEU, which states that "the European Parliament and the Council, acting in accordance with the ordinary legislative procedure and after consulting the Economic and Social Committee and the Committee of the Regions, shall decide what action is to be taken by the Union in order to achieve the objectives referred to in Article 191." Article 191 TFEU relates to the Union policy on the environment, which aims at preserving, protecting and improving the quality of the environment, protecting human health, prudent and rational utilisation of natural resources and the promotion of measures at the international level to deal with regional or worldwide environmental protection. On the basis of the ordinary legislative procedure, the Parliament acts as co-legislator with the Council, whereby the two institutions, in accordance with Article 249 TFEU, adopt the relevant acts at first or second reading, if need be with intervention of a Conciliation Committee. Under the ordinary legislative procedure, acts are adopted on the basis of qualified majorities.

On the other hand, the special legislative procedure applies to specific cases covered by the Treaties and forms an exception to the ordinary legislative procedure. It entails that the Council is the only legislator, where the Parliament may either be consulted or may approve of the relevant Act, and unanimity voting is applicable. Needless to say, the latter decision-making procedure entails a significantly higher threshold for measures to be adopted. The TFEU provides for a special legislative procedure applicable to measures affecting the quantitative management of water resources or affecting, directly or indirectly, the availability of these resources, referred to in Article 192(2) TFEU:

> "By way of derogation from the decision-making procedure provided for in paragraph 1 and without prejudice to Article 114, the Council acting unanimously in accordance with a special legislative procedure and after consulting the European Parliament, the Economic and Social Committee and the Committee of the Regions, shall adopt:
> (a) provisions primarily of a fiscal nature;
> (b) measures affecting:
> - town and country planning,
> - quantitative management of water resources or affecting, directly or indirectly, the availability of those resources,
> - land use, with the exception of waste management;"

The Council should adopt these Acts unanimously after consulting the Parliament, the Economic and Social Committee and the Committee of the Regions. These measures are considered to be the most sensitive in the context of environmental protection. With regard to these types of measures, in analogy with town and country planning measures, which have traditionally always been considered as part of the sovereignties of the Member States, it is considered that the decision centres should be in the Member States, rather than at the level of the European Union.[300]

[300] Jean-Claude Piris, *The Future of Europe: Towards a Two-Speed EU?* (Cambridge University Press 2012) 23.

The TEU provides for the possibility, where legislative acts need to be adopted by the Council in accordance with the special legislative procedure, for the European Council to adopt a decision allowing for the adoption of such acts in accordance with the ordinary legislative procedure.[301]

78. Both the WFD and the FD have been based on Article 192(1) TFEU and adopted through the ordinary legislative decision-making procedure. This fact deserves further examination, as both Directives, to varying degrees, cover aspects of water quantity management. This means that the unanimity requirement has been circumvented.

2.2.3.3. The legal basis for water quantity management through time

79. The entry into force of the Maastricht Treaty resulted in a shift from consequently adhering to unanimity voting to majority decision-making in the context of Article 192 TFEU.[302] Indeed, the applicable Article 130s(1) (now Article 192(1) TFEU) of the Maastricht Treaty provided for majority decision-making, whereas Article 130s(2) Maastricht Treaty provided for unanimity with regard to measures related to "management of water resources". At the time, there was confusion as to the scope of this derogation in the field of water management. Indeed, some language versions of the Treaty referred to "quantitative" management of water resources in the context of the unanimity requirement, such as the Dutch and the French, whereas other versions (the majority) only generally referred to "management" of water resources.[303]

For example, the French version of the Treaty referred to the "gestion des ressources hydrauliques". This expression is said to pertain to the control of the flow of the watercourses, the use of the quantities of water, and the exploitation of the water resources for purposes such as energy generation.[304]

80. In *Spain v Council*, Spain argued that the Convention on Cooperation for the Protection and Sustainable Use of the River Danube (The Danube Convention), to which the EU had adhered through its Decision 97/825, should have been based on Article 130s (2) (unanimity) (now Article 192(2) TFEU) as opposed to Article 130s(1) (majority) (now Article 192(1) TFEU).[305] Spain argued that water is a natural resource absolutely essential to human life, and

---

301 Article 48(7) TEU.

302 Geert Van Calster and Kurt Deketelaere, 'Amsterdam, the IGC and greening the EU Treaty' (1998) 7 European Environmental Law Review 1, 12.

303 For example, the German ('die Bewirtschaftung der Wasserressourcen"), the Italian ('la gestione delle risorse idriche'), the Swedish ('Förvaltning av vattenresurser'). Also see Ludwig Krämer, *EU Environmental Law* (7th edn, Sweet & Maxwell 2012) 252.

304 Case C-36/98, *Spain v Council of the European Union* [2001] ECLI:EU:C:2001:64, para. 25.

305 Convention on Cooperation for the Protection and Sustainable Use of the River Danube, (adopted 29 June 1994, entered into force 22 October 1998) < https://www.ecolex.org/details/

therefore merits special attention. Even in cases where the general objectives of the EU's environmental policy would be at stake, the unanimity procedure set out in Article 192(2) TFEU should apply.[306] Measures with regard to the management of water inevitably take into account the protection and quality of water resources. The Council, however, made a distinction between "water management" on the one hand and "management of water resources" on the other hand.[307] Measures to promote the quality of water were considered to fall into the first category, and measures relating to the quantitative management of water resources into the second category. The Commission also emphasised that "management of water resources" is limited to the quantitative aspects of the management of water, as, like town and country planning and land use, these measures have a direct impact on the manner in which States use their land for infrastructure projects.[308]

The Court firstly emphasised that different language versions should be compared, in order to interpret specific provisions. Where these language versions did not correspond, the purpose and general scheme of the rules of which they formed part should be considered.[309] Looking at the other measures covered by Article 192(2) TFEU[310], the Court concluded that these measures, i.e. town and country planning and land use with the exception of waste management, regulate the use of the territory of the Member States, considering that the territory and land of the Member States, together with their water resources, are limited resources. The second paragraph of the provision therefore relates to the management of limited resources in its quantitative aspects, and not to the resources aimed at improving and protecting the quality of the resources. The Court of Justice then emphasised the fact that the Danube Convention pertained to the qualitative aspects of water management, which, in accordance with the purpose and general scheme of Article 192(2) TFEU was not covered by this provision.[311]

Following this judgment in *Spain v Council* the Treaty of Nice explicitly reserved this unanimity requirement to the quantitative management of water resources now included in Article 192 (2) TFEU.[312]

81. The main element to decide whether an Act should be based on Article 192(1) or 192(2) TFEU, is whether the Act puts stronger focus on qualitative or

---

treaty/convention-on-cooperation-for-the-protection-and-sustainable-use-of-the-danube-river-tre-001207/ > accessed 17 June 2017.

306 Then Article 130s(2) TEC. Case C-36/98, *Spain v Council of the European Union* [2001] ECR I-779, para. 11.

307 Ibid para. 17.

308 *Spain v Council of the European Union* (n 304) para. 37.

309 Case C-420/98 *W.N. and Staatssecretaris van Financiën* [2000] ECLI:EU:C:2000:209, para. 21.

310 Then article 130s(1) TEC.

311 *Spain v Council of the European Union* (n 304).

312 Ibid.

quantitative management.[313] Taking it one step further, it should be noted that, at this point of time, it is not crystal clear what exactly constitutes *quantitative* management of water resources.

82. Generally speaking, the protection of the water relates to the qualified majority procedure, whereas the use of water or the economic exploitation of water is often indicated with the term "water resources" and is said to fall under the more restrictive treatment of the unanimity umbrella applicable to quantitative water management.[314]

83. The Water Framework Directive, the Groundwater Directive and the Floods Directive are based on Article 175(1) TEC (ex Article 130s(1) TEC, now Article 192(1) TFEU). As for the latter, this seems counter-intuitive, as floods are naturally a quantitative water-related issue.[315] Both directives will be discussed below.

2.2.3.4. The struggle for balance of the Water Framework Directive

84. In the run-up to the adoption of the WFD, there constantly was a fine balance between water quality and water quantity, where it was important to keep the focus on water quality, in order to be able to resort to the environmental protection article, 130s(1) TEC, and avoid the stringent unanimity procedure of the water quantity management article, 130s(2) TEC.

85. The *ratio legis* for basing the WFD on Article 130s(1) TEC relates to the goal of the Directive: maintaining and improving the aquatic environment in the EU, establishing the objective of achieving good status of surface waters and groundwater.[316] The quantitative aspects are included to support the objectives related to the quality of water – hence the choice for Article 130s(1) TEC rather than Article 130s Treaty. Indeed, the Commission emphasised that the factor of water quantity is an important parameter in safeguarding water quality. In fact, the WFD represents one of the first instruments at EU level that really touches on quantitative water aspects. As will be explained below in the chapter on the second pillar on substantive provisions, the WFD contains, among other things, provisions on prior authorisation of water abstraction.

86. In drafting the Water Framework Directive, the Commission treaded carefully to ensure that Article 130s(1) TEC (now Article 192(1) TFEU) could

---

313 Nicolas de Sadeleer, *EU Environmental Law and the Internal Market* (Oxford University Press 2014) 156.

314 *Spain v Council of the European Union* (n 304), para. 28.

315 See *supra*, in the section on flood risk management.

316 Commission, 'Proposal for a Council Directive establishing a framework for Community action in the field of water policy' COM (1997) 49 final.

be justified as the correct legal basis for the Directive, and not Article 175(2) of the Treaty, which would require unanimity voting. To this end, the quantitative aspects of water management are ancillary to the quality-related measures. In the course of the drafting of the WFD, amendments were proposed that pertained specifically to the management of water resources.[317] The Commission rejected these amendments, as they would require the legal basis of Article 175(2) TEC (now Article 192(2) TFEU), and would therefore shift the balance of legal basis of the whole Directive from Article 175(1) TEC to Article 175(2) TEC.

With regard to the management of water resources, the Commission has rejected several amendments proposed by the Parliament, on the basis of the concern that these would shift the balance too much to quantitative water management, and trigger the unanimity requirement.[318] For example, Amendment 18 of the Parliament pertained to the goal of the WFD, and implied adding the promotion of efficient water use within a river basin as one of the goals of the Directive.[319] This addition of "efficient water use" was not included as one of the goals of the Directive, but was replaced by "sustainable water use". Similarly, the suggestion of the Parliament to add as a requirement into the scope of the programme of measures an obligation to implement measures for "more efficient use of water in all water-use sectors in those cases where water demand exceeds the quantity sustainably available within a river basin" was not upheld in the final version of the Directive.[320] Another example, in Amendment 16, the Parliament had proposed that human water demands in a hydrological area or river basin should be satisfied with the available water resources in the basin.[321]

Another amendment touching on quantitative water management and rejected by the Commission pertained to inter-basin water transfers. The amendment implied that the programme of measures should include as a requirement the prior authorisation of such transfers. Moreover, on the basis of the amendment, these transfers should be limited to those particular cases where the receiving river basin had reduced demand.[322]

These amendments all seem to fall into the category of "quantitative management of water resources or affecting, directly or indirectly, the availability of those resources", as set out in Article 192(2) TFEU, and subject to the unanimity requirement.

---

317 Commission, 'Amended proposal for a European Parliament and Council Directive establishing a framework for Community action in the field of water policy' COM (1999) 271 final, 5.

318 For example, Amendments 16, 18, 21, 36, 73, 74, 75, 76.

319 European Parliament, 'Legislative resolution embodying the Parliament's opinion on the proposal and amended proposals for a Council Directive establishing a framework for Community action in the field of water policy' (1999) OJ C 150, Amendment 18.

320 Amendment 73.

321 Committee on Research, Technological Development and Energy, 'Report on the proposal and the amended proposals for a Council Directive on establishing a framework for Community Action in the field of water policy' (1998) A4-0261/98.

322 Amendment 75.

2.2.3.5. The Floods Directive: based on the Environment Title

87. Contrary to what one would first believe, the Floods Directive is not based on the primary Treaty article for water quantity management. In some sense, this is counterintuitive, as floods relate to water quantity. In contrast, the Floods Directive is based on the environmental Article 175(1) TEC, as one of the core goals of the Directive is environmental protection.[323] In addition to the environment, the FD also aims to protect human health, cultural heritage and economic activity. Certain authors wonder whether the fact that the protection of the environment is just one of the goals of the FD, as opposed to being the primary goal, may justify the reliance on the environmental Article 192(1) TFEU.[324] Indeed, in the context of flood risk management, the aspect of civil protection forms a crucial part. However, the difference between the area of environmental protection and that of civil protection, among other things, lies in the competence of the EU. Whereas the former constitutes shared competence between the EU and the Member States, the latter constitutes an area where the EU can only take action to support, coordinate or supplement the actions of the Member States.[325]

There are several arguments that are in favour of the FD relying on the environmental article of the Treaty. Firstly, as the Floods Directive is the sister Directive of the WFD, and considering the strong link between the two, especially from a procedural point of view – the time schedule for the implementation of the two Directives has been synchronised – the respective plans issued on the basis of both Directives need to be coordinated, if not integrated, and so on. In this regard, it seems logical to base both Directives on the same primary Treaty article, namely Article 192(1) TFEU. Secondly, even starting from the premise that the environmental article may not be fully compatible with the scope of the FD, other articles included in the Treaty are even less compatible. For example, the suggested civil protection area, in my opinion, certainly does not do the trick in terms of meeting the demands of Flood Risk Management. Indeed, flood awareness and warning systems only form part of Flood Risk Management, together with flood risk prevention, flood defence, mitigation and flood recovery. These strategies taken together make up Flood Risk Management.[326] Basing the Directive, tackling floods and managing the related risks, on the area of just one of these strategies, would fail to do Flood Risk Management justice. Another primary Treaty article that comes to mind is Article 192(2) TFEU, which deals

---

[323] Recitals 12 and 13 FD, which refer to environmental pollution as a result of floods. See Benjamin Richardson, *Local Climate Change Law: Environmental Regulation in Cities and Other Localities* (Elgar 2012) 295.

[324] Ibid 295.

[325] Article 6 TFEU.

[326] Peter PJ Driessen and others, 'Toward More Resilient Flood Risk Governance' (2016) 21 Ecology and Society; Corinne Larrue, Dries Hegger and Jean-Baptiste Trémorin, *Researching Flood Risk Governance in Europe: Background Theories* (STAR-FLOOD Consortium 2013).

with water quantity management. However, as will be explained in the chapters on the second and third pillars below, the Floods Directive does not include any substantive provisions or objectives for Member States to comply with; it only sets out certain procedural paths the Member States should follow. In this sense therefore, the FD far from touches on the organisation of the territories or the resources of Member States.

2.2.3.6. The scope of quantitative water management

88. We can deduce, albeit merely approximately, from the scope and legal bases of the relevant EU water Directives, what actually constitutes "water quantity management" in the sense of Article 192(2) TFEU.

Firstly, the measures need to truly affect the physical territories of the Member States, or the availability of water resources. The former aspect includes in its scope infrastructural measures that alter the physical outlook of the national territories. The latter aspect includes in its scope the use of resources.[327] Specific provisions on how the water resources should be managed, how the water should be allocated, whether the water can be diverted from basin to basin etc. would have triggered the unanimity requirement. It is clear that these limits pertain to the shortage of water rather than to an excess of water.

This leads us to the second aspect of the question of scope: whether only substantive requirements can trigger the unanimity requirement. Flood risk management may potentially fall under several of the categories subject to the unanimity requirement pursuant to Article 192(2) TFEU: (i) measures affecting town and country planning and (ii) quantitative management of water resources. Indeed, an important part of flood risk management relates to spatial planning, for example the delineation of flooding zones. However, the Floods Directive is based on the first paragraph of the environmental article of the TFEU. As will be discussed in the section below, the FD does not contain any substantive obligations. It merely requires Member States to produce cartographical information and Flood Risk Management Plans, thereby strongly focusing on the procedural aspects. Which strategies Member States choose to adopt in order to comply with these procedural requirements, is completely at their own discretion. Therefore, the Directive does not *directly* affect the physical territory of the Member States, although one cannot deny that there is an indirect impact on the territories. It would probably be a different story if the Floods Directive were to determine various safety standards or zoning requirements with respect to flood risks, which countries would have to comply with.[328] It can therefore be tentatively concluded that a body of procedural requirements with large discretionary powers for the

[327] Ludwig Krämer, *EC Treaty and Environmental Law* (Sweet & Maxwell 1998) 99.

[328] Although in the case of safety standards, this could still be justified on the basis of transboundary necessity.

Member States, not directly influencing the territories or the quantities of water in the Member States, does not necessarily trigger the unanimity requirement.

Finally, it should be noted that the Commission has traditionally favoured a more restrictive interpretation of the exception regime for water management.[329] Since the derogation provision related to the management of water resources was inserted in the Treaty of Maastricht in 1992, it has yet to provide the legal basis for a legislative instrument issued at EU level.[330] Moreover, the CJEU has issued a number of landmark judgments, clearly favouring a broad approach to environmental protection.[331] Furthermore, the insertion of the reference to the fight against climate change into the TFEU might provide the Commission with further legitimacy to promulgate legislative initiatives related to water shortages.[332]

89. Taking into account the looming predicaments in the light of climate change, the continuing harmonisation efforts in the environmental field in the EU, and the realisation that integrated water resources management will become an increasingly important approach to water management in general, the question arises whether Article 192(2) TFEU should not be interpreted more and more restrictively in the future. This question is quintessential in developing an EU-wide regime for transboundary water management. It will be addressed in Chapter V of this research.

2.3. PILLAR II: SUBSTANTIVE PROVISIONS

90. For the sake of clarity, a distinction will be made between the substantive provisions and the requirements pertaining to surface water and to groundwater.

2.3.1. Surface water

91. Substantive obligations regarding floods, droughts and scarcity of surface water pursuant to EU water law are mainly based on two instruments: the EU

---

329 Jan Jans, *European Environmental Law* (Europa Law Publishing, 2000) 48.

330 Teresa Calderon, 'La Politique de l'Eau de l'Union Européenne: Vers une Gestion Quantitative des Ressources Hydriques?' (2010) 51 Les Cahiers de Droit 862, 873.

331 For example, Titanium Dioxide Case, see Case 300/89 *Commission v Council* [1991] ECLI:EU:C:1991:244, where the Court ruled that measures to reduce and eliminate titanium dioxide pollution could be based on the internal market provision of the Treaty (Article 100 A of the Treaty of Rome, requiring QMV), and not necessarily on the environmental article (then requiring unanimity – Article 130R of the Treaty) as harmonisation measures could also pursue environmental protection.

332 Article 191(1) § 4 TFEU. See also Teresa Calderon, 'La Politique de l'Eau de l'Union Européenne: Vers une Gestion Quantitative des Ressources Hydriques?' (2010) 51 Les Cahiers de Droit 862, 873.

WFD and the FD. As mentioned before, the former mainly tackles water quality related issues, although droughts and scarcity concerns are also indirectly addressed. The latter focuses solely on floods. Specifically, for droughts and scarcity, there is a Strategy for Water Scarcity and Droughts, but currently no legislative framework at EU level exists in this regard.

Even though the WFD mainly tackles pollution-related concerns, quantitative aspects are to some extent subject to binding obligations. The WFD also provides for explicit substantive requirements with regard to the water management measures of Member States.[333]

The fact that quantitative aspects are to some extent subject to binding provisions is also apparent from the goals of the directive. Article 1 WFD states that the purpose of the Directive is to establish a framework for the protection of inland surface waters, transitional waters, coastal waters and groundwater, as well as for terrestrial ecosystems and wetlands dependent upon these aquatic systems. This framework then, among other things, is meant to further promote sustainable water use based on long-term protection of available water resources[334] and to contribute to the mitigation of the effects of floods and droughts.[335]

The ultimate objective of the WFD is the provision of a *sufficient* supply of *good quality* surface water and groundwater as needed for sustainable, balanced and equitable water use. This is logical, as a healthy water status implies an inextricable bond between quantity and quality. Below, these limited requirements for quantitative water management will be discussed.

2.3.1.1. Good status for surface water

92. The majority of obligations in the context of the WFD is included in Article 4(1) of the Directive, which sets out the specific environmental objectives.[336] From a substantive point of view, good surface water status, which relates to its ecological[337] and chemical status[338], must be attained. The CJEU has ruled that the environmental objectives included in the WFD are legally binding on the Member States. For example, the requirements included in Article 4(1)(a)(i) to (iii) imply that Member States must refuse the authorisation of projects which may cause the deterioration of surface water bodies' status or may compromise achieving good surface-water status.[339]

---

333 Lasse Baaner, 'The Programme of Measures of the Water Framework Directive – More than just a Formal Compliance Tool' (2011) JEEPL 82, 94.

334 Article 1(b) WFD.

335 Article 1(e) WFD.

336 In conjunction with Article 7 WFD.

337 This relates to the quality of the functioning and structure of aquatic ecosystems associated with surface waters. The classification in this regard is included in Annex V of the Directive.

338 This relates to the determination that the concentration of pollutants does not exceed the environmental quality standards set out in the Annex.

339 Case C-461/13 *Bund v Germany* [2015] ECLI:EU:C:2015:433, para 51.

93. The Directive provides for numerous exemptions and exceptions to this general obligation to achieve good status[340]:

- The first example clearly illustrates the link between Good Ecological Status and Flood Risk Management, and thus between the WFD and the FD. The Member States may designate a body of surface water as artificially or heavily modified. Once this designation has been given, these bodies may be exempted from obtaining good ecological status if the changes to the hydromorphological characteristics of that body to achieve good ecological status would have significant adverse effects on, among other things, flood protection.[341] Heavily Modified Water Bodies are not required to achieve a Good Ecological Status, but instead need to attain Good Ecological Potential.[342] It should be noted that good chemical status must also be attained for artificial and heavily modified surface water bodies.
- Another important exception is when new sustainable human development activities lie at the basis of the failure to fulfil the obligations.
- They may also extend the deadline for achieving the WFD objectives a number of times until 2027 at the latest, on the condition that no further deterioration occurs, or they may plead for a temporary deterioration of the water bodies.[343]
- They may aim to achieve less stringent environmental objectives, if attaining these objectives is infeasible or disproportionately expensive.
- Finally, the Directive provides for Force Majeure: A Member State may determine that the additional measures are not feasible if the causes of the pollution are the result of circumstances that were not foreseeable or exceptional due to natural causes or force majeure, particularly substantial floods or prolonged droughts.[344]

94. It should also be noted that for protected areas, an obligation exists for Member States to comply with all the applicable norms and goals within fifteen years following the entry into force of the Directive.[345] Protected areas within the meaning of the WFD for example include the bodies of waters that are used for the abstraction of drinking water, or for the protection of economically significant aquatic species.[346] For these protected areas, a register must be created indicating these areas in each River Basin District for which special protection would be required.[347] It is not unequivocally clear whether or not any of the exemptions

340 Andrea Keessen, Annelies Freriks and Marleen van Rijswick, 'The Clash of the Titans: The Relation between the European Water and Medicines Legislation' (2010) 47 Common Market Law Review 1435.

341 Article 4(3) WFD.

342 Good Ecological Potential in accordance with Annex V.

343 Article 4(4) WFD.

344 Article 4(6) WFD.

345 Article 4(1)(c) and Article 4(2) WFD.

346 Annex VI WFD.

347 Article 6 WFD.

mentioned above can apply to protected areas.[348] Van Rijswick argues that it would be possible to invoke such exemptions, as the legislation creating the protected areas provides for exemptions to the applicable requirements.[349]

95. As is clear from the above, quantitative aspects, such as flood risk management, are not only relevant insofar as they support the provisions with regard to water quality, but also as they provide for exceptions and exemptions to the attainment of the substantive requirements set forth in the Directive. The WFD therefore embodies a double link between water quantity and quality management.

96. It should be examined to what extent the availability and flow of water plays a role in the achievement of Good Ecological Status (GES). These factors are taken into account to some extent. For example, for rivers, hydromorphological elements that determine the biological elements to analyse the ecological status are the hydrological regime, e.g. the quantity and dynamics of the water flow[350], the river continuity[351], and the morphology, e.g. the structure of the riverbeds.[352] The values of the hydromorphological elements are not generally used to determine the ecological status of a water body, but these are used to determine High Ecological Status.[353] In terms of good, moderate, poor and bad water status, the hydromorphological quality variable only relates to "conditions consistent with the achievement of the values specified for the biological quality elements".[354]

In this context, reference can again be made to the concept of "environmental flow", which as explained in Chapter, relates to the "amount of water required for the aquatic ecosystem to continue to thrive and provide the services we rely on".[355] This definition implies the acknowledgment of the intrinsic bond between quality and quantity in the objective of achieving good status.

The concept of environmental flow is not generally considered as an important impact factor in terms of achieving Good Ecological Status in the context of the WFD. For example, the fact that the flow regime of a water body is altered because of the abstraction of water, provided the biology is not influenced, does not exclude obtaining Good Ecological Status, although in reality significant

348 Marleen van Rijswick and Herman Havekes, *European and Dutch Water Law* (Europa Law Publishing 2012) 287.
349 Ibid 308.
350 This concerns the question whether and to what extent the flow quantity and dynamics and the relations to the groundwater bodies are disturbed.
351 This concerns the question whether the continuity of the river is disturbed by human activities.
352 Annex V WFD.
353 Commission, 'WFD CIS Guidance Document No. 3: Analysis of Pressures and Impacts' (European Union 2003) 26.
354 Annex V of the WFD. See Rafael Sanchez and Guido Schmidt, 'Environmental Flows as A Tool to Achieve the WFD Objectives' (INTECSA-INARSA 2012).
355 See also *supra* and Chapter I. COM (2012) 673 final, 6.

alteration of the flow regime will compromise the GES.[356] Also, alterations to environmental flow do not necessarily justify the designation of this water body as Heavily Modified Water Body (HMWB). The environmental flow regime has been subject of a Common Implementation Strategy guidance document.[357]

This means that, currently, quantitative aspects form a small part of the analysis of good ecological status. In the future, the concept of environmental flow may potentially constitute a powerful bridging mechanism between water quantity and water quality management, and it will therefore be a recurring feature in this book.

97. The CJEU has scrutinised the question whether diversion of water from one River Basin District to a neighbouring district in Greece is to be seen as compatible with the WFD and on the basis of which conditions.[358] With respect to the case in question, the River Basin Management Plans had not yet been adopted by the competent national authorities. Indeed, the Directive does not contain provisions related to the diversion of water, so there is no absolute prohibition in this regard, and the fact that diversion of water is an action included in Annex I of the EIA Directive does not detract from this conclusion.[359] The Court concluded that, indeed, there was no a priori prohibition of such diversion. Such a transfer may not have the effect of seriously jeopardising the realisation of the objectives of the Directive and even when the transfer has negative effects, it can still be allowed under the exemption regime, provided the related conditions have been met.[360] Another interesting observation is that such a transfer does not necessarily need to be based on the impossibility of the receiving river district to meet its needs related to the supply of drinking water, electricity production or irrigation.[361] AG Kokott stressed that this is the case if the conditions of Article 4(1), 4(7), (8) and (9) of the Directive have been met.

- *Programmes of measures*

98. Substantive obligations also exist through the programme of measures, which is an instrument used in the WFD, but not in the FD.[362] Each Member State must draw up a programme of measures for each River Basin District,

---

356 M.C. Acreman, 'Environmental Flows and the European Water Framework Directive' (2010) 55 Freshwater Biology 32, 36.

357 Commission, 'WFD CIS Guidance Document No. 31: Ecological Flows in the Implementation of the Water Framework Directive' (European Union 2015).

358 Case C-43/10 *Nomarchiaki Aftodioikisi Aitoloakarnanias and Others v Ypourgos Perivallontos, Chorotaxias kai Dimosion ergon and Others* [2012] ECLI:EU:C:2012:560.

359 The fact that it is mentioned in Annex I implies that its effect on the environment should be scrutinised, not that it is permitted or prohibited as such. See the section below on the third pillar, more specifically, on notification requirements.

360 *Nomarchiaki Aftodioikisi Aitoloakarnanias and Others v Ypourgos Perivallontos, Chorotaxias kai Dimosion ergon and Others* (n 358) para 69.

361 Ibid.

362 Also in the Marine Strategy Framework Directive.

using which the objectives included in Article 4 WFD are to be achieved cost-effectively.[363] The WFD does not include the requirement that these programmes of measures should be coordinated across the International River Basin District. Indeed, the WFD provides that each Member State, individually, must establish programmes of measures for each RBD, or for the part of an International River Basin District in its territory. These programmes of measures may ensue from national legislation, and may therefore be applicable to all RBDs or parts of IRBDs falling within its territory.[364] Such coordination requirement does exist for River Basin Management Plans.[365]

99. The measures that should be included in the programmes of measures in accordance with the WFD consist of basic measures, all of which are listed in Article 11 and constitute obligatory elements in the programmes, and supplementary measures, for which a non-limitative list has been included in Annex VI of the Directive.[366]

Several requirements with regard to the quantitative aspects of water management are included in the scope of the basic measures of the programmes of measures, e.g. measures to promote efficient and sustainable water use in order to avoid compromising the achievement of the objectives specified in Article 4.[367] Also, measures to control the abstraction of fresh surface water and groundwater, and impoundment of fresh surface water, are included in the scope.[368] To this end, Member States should set up a register of water abstraction and should comply with a requirement of prior authorisation for abstraction and impoundment. The Member States may exempt abstractions and impoundments that do not have a significant effect on the water status, from these control measures.

The Directive thus provides a framework to manage water use in an indirect manner, namely through the programmes of measures, where all the discretionary power with regard to the determination of the limits of this use is left to the Member States.

2.3.1.2. Water services and the cost-recovery principle

100. An important aspect of the material scope of the EU legal framework for transboundary quantitative water management relates to the requirements with regard to water services.[369] Indeed, depending on the interpretation of the

---

363 Article 11 WFD.
364 Article 11(1) WFD.
365 See *infra*.
366 More specifically, Annex VI part B.
367 Article 11(3)(c) WFD.
368 Article 11(3)(e) WFD.
369 Article 2(38) in conjunction with 9 WFD.

concept of "water services", measures related to inter alia flood protection and hydroelectric power generation are covered by the scope of the Directive or not.[370] The distinction between water use and water services is important, as the WFD attaches concrete obligations to the concept of water services.[371]

(A) The cost-recovery principle

101. The issue of water pricing constituted one of the most controversial topics in the run-up to the adoption of the WFD.[372] The cost-recovery principle applied to water implies the efficient use of water resources, and therefore touches on the quantitative aspects of water management. In this regard, a horizontal complaint was filed to the Commission by eleven Member States with regard to the interpretation of the term "water services" as included in Article 2(38) WFD.[373] The term has also been the subject of proceedings before the Court of Justice, which will be discussed below.

Article 9 WFD stipulates that Member States must take into account the cost-recovery principle with regard to water resources, including environmental and resource costs.[374] Through the application of full cost recovery, the consumer would be informed of the complete picture of the costs of their consumption, i.e. reflecting all financial and other costs borne by the producer.[375]

102. The delicate character of the issue of cost recovery is apparent from the legislative history of the WFD. The Parliament amended Article 9(1) of the WFD proposal with regard to the cost-recovery principle, so that instead of the requirement for Member States to "take account" of the recovery of costs of water services, they should "ensure", by 2010, that water-pricing policies provide

---

370 Petra Lindhout, 'A Wider Notion of the Concept of Water Services in EU Water Law' (2012) 8 Utrecht Law Review 86.

371 On a side note, when considering the topic of water use and the management of water use, the role of the agricultural sector becomes prominent. Indeed, about 24% of the total water abstraction rate in Europe stems from agriculture (in some parts of Europe: 80%). The interplay with the Common Agricultural Policy ("CAP"), and the fact that the CAP does not follow the river basin approach and, for the transboundary aspects, the International River Basin Districts, will be kept in mind in the analysis and evaluation of the institutional mechanisms for cooperation further along. See EEA report, 2009.

372 Maria Kaika, 'Water for Europe: The Creation of the European Water Framework Directive' in Julie Trottier and Paul Slack (eds), *Managing Water Resources Past and Present* (OUP 2004) 102.

373 Commission, 'Report from the Commission to the European Parliament and the Council on the Implementation of the Water Framework Directive (2000/60/EC) River Basin Management Plans', COM(2012) 670 final, 9.

374 Whilst having regard for the economic analysis included in Article III WFD.

375 Herwig Unnerstall, 'The Principle of Full Cost Recovery in the EU-Water Framework Directive – Genesis and Content' (2007) 19 Journal of Environmental Law 1, 31.

adequate incentives for an efficient use of water resources.[376] The Parliament thereby added a binding character to the cost-recovery principle. However, in later amendments, the Council weakened the wording of the provision by replacing the words "shall ensure" by "take account of".[377]

The Parliament, in its first reading of the Commission Proposal of 1997, also added a provision, Article 12(1)(a), on the inclusion of charges if it is not possible to calculate the full environmental costs of water use.[378] Moreover, a provisional Article 12(1)(b) was added to introduce a charging mechanism on the basis of which users that need to treat their water due to polluting activities of another party, are able to fully recover the additional costs incurred for the treatment of the water, from the polluter.[379] The charging mechanism covered pollution activities not included in the scope of the definition of "water use", among other things diffuse pollution.[380] Eventually, the amendments of Article 12(1)(a) and 12(1)(b) were thrown overboard.

Concretely, by the year 2010, Member States needed to make sure that water-pricing policies stimulate users to make efficient use of water resources, and to contribute to the environmental objectives of the Directive. Moreover, Member States needed to provide an adequate contribution of the different water uses, i.e. industry, households and agriculture, to the recovery of costs of water services.[381] In order to follow up on these requirements, the Directive stipulates that the planned measures in order to realise these objectives included in Article 9 WFD, must be integrated into the River Basin Management Plans.

These cost-recovery requirements apply to water services. As mentioned above, there is a differentiation in the WFD between water services and water use: water use encompasses water services, but not vice versa. This will be explained in the following sections.

---

376 European Parliament, 'Legislative resolution on the common position adopted by the Council with a view of the adoption of a European Parliament and Council Directive establishing a framework for Community action in the field of water policy' (2000) OJ C 339.

377 Article 9 WFD states: "Member States shall take account of the principle of recovery of the costs of water services, ...".

378 European Parliament, 'Legislative resolution embodying Parliament's opinion on the proposal and the amended proposals for a Council Directive on establishing a framework for Community action in the field of water policy (1999) OJ C 150, Amendment 67. See also Unnerstall (n 375) 30. The exact wording of the provision: "Where it is not possible, or impractical, to calculate the full environmental costs of water use, charges shall be set at a level which encourages the attainment of the environmental objectives of this Directive".

379 The exact wording of the provision: "In accordance with the polluter pays principle, Member States shall ensure that the charging system provides that users faced with a need to treat their water as a result of another's polluting activities, can fully recover their additional costs from the polluter."

380 Unnerstall (n 375) 31.

381 Article 9(1) WFD.

(b) Water services

103. In the initial draft of the WFD presented by the European Commission in 1997, the term water use explicitly encompassed the following activities[382]:
(a) Abstraction, distribution and consumption of surface water and groundwater;
(b) Emission of pollutants into surface water and waste-water collection and treatment facilities which subsequently discharge into surface water;
(c) Any other application of surface water or groundwater having the potential of a significant impact on the status of water.

104. The draft of the Commission did not define the term "water services". The differentiation between "water use" and "water services" was introduced by the European Council. The term "water services" is defined in Article 2(38) WFD and now encompasses bullets (a) and (b), namely the abstraction, impoundment, storage, treatment and distribution of surface water or groundwater, and the waste-water collection and treatment facilities, which subsequently discharge into surface water.[383] The term "water use" now resembles the content of the initial Article 2(38), but now has a broader scope, i.e. it refers to any activity, including water services, having significant impact on the status of the water.[384] Water services can relate to actions resulting in hydromorphological changes, such as reservoirs in the context of flood risk management.[385]

105. The interpretation of the concept of water services has been scrutinised by the CJEU, as the Commission referred Germany to the CJEU on 31 May 2012.[386] The main question in *Commission v Germany* relates to whether any service related to each and any of the activities listed in Article 38(a) WFD[387] should be subject to the cost-recovery requirements, or only the service of water supply.

The Commission argued that the German interpretation of the term water services was too narrow, i.e. excludes important services from the scope. Indeed, Germany interprets the cost-recovery principle as only applicable to the supply of water and the disposal of wastewater.[388] With respect to the supply of water, the term water services would refer to the supply of water as a whole, and not to the

---

382 Commission, 'Proposal for a Council Directive establishing a framework for Community action in the field of water policy' COM (1997) 49 final. Article 2(32).

383 Article 2(38) WFD.

384 Article 2(39) WFD states: "water services together with any other activity identified under Article 5 and Annex II having a significant impact on the status of water".

385 Unnerstall (n 375) 36.

386 Whilst the whole discussion already started in 2006, when a complaint was submitted to the Commission that Germany misinterpreted the concept of water services.

387 In addition to waste-water treatment as set forth in Article 38(b), which was not a subject of the discussion.

388 Denmark, Hungary, Austria, Finland, Sweden and the UK intervened in the proceedings and supported the arguments set forth by the German government.

activities listed in Article 2(38)(a) separately. However, the Commission argued that water services should be interpreted much more widely, to include water abstraction, flood protection, hydropower production and so forth. In essence, the Commission argued that all activities listed in Article 2(38) should be subject to the cost-recovery principle as set forth in Article 9 WFD.

For example, referring to subsection (a) of Article 2(38) WFD, abstracting, distributing and consuming surface water and groundwater, Germany had interpreted this in a cumulative way, as opposed to the Commission, which held that it is sufficient that any one of the activities listed is present in the water service, not all of them.[389] Germany on the other hand, links the concept of water services to the definition included in Article 57 TFEU.[390] Given that Article 57 TFEU implies a bilateral relationship in order for an activity to be categorised as a water service, activities such as flood-protection measures are excluded from the scope, whereas activities such as water supply and waste-water treatment are included.[391] This interpretation stands in contrast with the interpretation where the services included in Article 38 WFD are viewed as environmental services within the meaning of the Environmental Liability Directive, where it should be noted that these services do not require a bilateral or contractual relationship between actors.[392]

The Court followed the opinion of AG Jääskinen, adopting a restrictive interpretation of Article 2(38) in conjunction with Article 9 WFD, namely that the EU legislator aimed to provide for two types of pricing, i.e. for the supply of water on the one hand and for waste-water treatment on the other hand.[393] As for the supply of water, the different activities of "abstraction, impoundment, storage, treatment and distribution" are listed to ensure that each of them is taken into account in calculating the costs of the supply of water. The Court established that the absence of pricing for activities with regard to certain uses of water does not necessarily imply a failure to achieve the objectives of the Directive.[394] This seems coherent with the "no size fits all" approach adopted by the WFD. Pricing mechanisms may be vital to activities such as water supply in order to reflect the cost of water, but may be awkward to apply to measures in the context of flood

389 Case 525/12 *European Commission v Germany* [2014] ECLI:EU:C:2014:2202.

390 Services shall be considered to be 'services' within the meaning of the Treaties where they are normally provided for remuneration, in so far as they are not governed by the provisions relating to freedom of movement for goods, capital and persons. 'Services' shall in particular include: (a) activities of an industrial character; (b) activities of a commercial character; (c) activities of craftsmen; (d) activities of the professions.

391 *European Commission v Germany* (n 389), para 39.

392 As argued by the Commission. See Directive (EC) 2004/35 on environmental liability with regard to the prevention and remedying of environmental damage [2004] OJ L143/56. Article 2(13) of this Directive defines services and natural resources services as 'functions performed by a natural resource for the benefit of another natural resource or the public'.

393 Case 525/12 *European Commission v Germany* [2014] ECLI:EU:C:2014:449, Opinion of AG Jääskinen, para 5.

394 *European Commission v Germany* (n 389), para 56.

risk management. On a side note, a consequence of this interpretation adopted by the Court seems to be that activities that do not imply a bilateral relationship within the meaning of Article 57 TFEU, but that should be made subject to the cost-recovery principle, such as the abstraction of water for self-supply, are not covered by the scope of Article 2(38) WFD.

In *Commission v Germany*, the AG criticised the Commission for indirectly trying to broaden the scope of the WFD towards more quantitative water management, by interpreting the concept of water services as including more than just the supply of drinking water and waste-water treatment.[395] By broadly interpreting this concept, activities that would affect "directly or indirectly, the availability of water resources" would be subject to pricing obligations, which would circumvent the unanimity requirement included in Article 192(2) TFEU.

The judgment of the Court has now brought some degree of clarity with respect to the interpretation of the notion "water services" in light of the cost-recovery principle, in the sense that pricing should not be applied to all of the activities listed in the definition of water services. It is especially with respect to this issue of pricing that the Court has ruled that the notion of water services should be interpreted narrowly. Certain questions with regard to this notion, however, still remain as the Court has not sufficiently clarified its meaning.[396] These questions are not necessarily related to the pricing issues ensuing from the concept of water services. For example, this judgment has not tackled the question to what extent ecosystem services can constitute WFD water services. However, it can be deduced from the judgment that quantity ecosystem services such as water regulation and flood protection will not *à priori* constitute WFD water services that should be subject to generalised pricing obligations.[397]

106. The Court again ruled on the water services regime in 2016, in the context of a preliminary ruling procedure and with respect to the methodology of calculating the price of water services.[398] The Court stipulated that it is up to the Member States to ensure that water-pricing instruments provide the right incentives for efficient water resources use by water users.[399] Member States are allowed to include both variable components connected with the actually

395 Case 525/12 European Commission v Germany [2014] ECLI:EU:C:2014:2202, Opinion of AG Jääskinen, para 70.

396 Petra Lindhout and Marleen Van Rijswick, 'The Effectiveness of the Principle of Recovery of the Costs of Water Services Jeopardized by the European Court of Justice – Annotations on the Judgment in C-525 / 12' (2015) 12 Journal for Environmental and Planning Law 80.

397 Petra Lindhout, 'A Wider Notion of the Scope of Water Services in EU Water Law: Boosting Payment for Water-Related Ecosystem Services to Ensure Sustainable Water Management?' (2012) 8 Utrecht Law Review 86, 99.

398 Case C-686/15, *Vodoopskeba i odvodnja d.o.o. v Željka Klafurić* [2016] ECLI:EU:C:2016:927.

399 Ibid para. 21.

consumed volume of water and a fixed component that is not connected to the volume of water.[400]

2.3.1.3. Substantive obligations for floods

107. As will be explained in more detail in the section on the "procedural pillar", the Floods Directive is very much focused on procedural obligations. The FD aims to provide a framework for the assessment and management of flood risks, with the goal of reducing the adverse consequences for human health, the environment, cultural heritage and economic activity resulting from floods.[401]

Central to the premise of the Floods Directive is the idea that floods cannot be totally prevented. Therefore, the Directive revolves around the risks of floods, which are seen as the combination of the probability of the flood event and the potential adverse consequences for the stakes mentioned above.[402]

108. In the spirit of flexibility, the determination of objectives and the formulation of measures with a view to flood risk management should take place at the level of the Member States.

This focus on the procedural aspects is mainly due to the consideration of the European Commission that there is a variety of causes and impact of floods throughout Europe, and because of these diverse conditions, no "one size fits all" approach can be adopted, also taking into account the proportionality and subsidiarity principles. Different solutions should be applied to different situations, in terms of e.g. flood vulnerability, leading to the necessity to provide a flexible legal instrument.[403] This is the reason why the option of promulgating a prescriptive legislative instrument, in addition to the option of "doing nothing", was rejected in an early stage of development of the EU action on floods.

When weighing the further options for the development of an EU policy related to the management of flood risks, there were two possibilities that were seriously considered. The first possibility consisted of a strictly voluntary approach, with voluntary commitments by Member States and the existing international river basin commissions. In the context of this option, Member States would be encouraged to exchange information and to exchange best practices.[404] However, this option, for good reasons, was considered to be too much of a lightweight, considering the urgent need to develop a more harmonised approach in light of the ever-increasing flood risks in Europe. The option that was perceived to be the most appropriate and cost-effective was the approach in which a flexible

---

400 Ibid para. 22.

401 Article 1 FD.

402 Article 2(2) FD.

403 Commission, 'Commission Staff Working Document: Annex to the Proposal on the Assessment and Management of Floods' SEC(2006)66.

404 Ibid.

legislative instrument would be adopted, supporting voluntary and cooperation measures carried out by the Member States.

109. Based on this point of view, the FD includes various procedural requirements, the three phases of which will be explained below, and in the context and on the basis of these procedural requirements, Member States, pursuant to Article 7(2) FD, should set the appropriate objectives for the management of the existing and future flood risks in their River Basin Districts.[405] These objectives should be determined on the basis of local and regional circumstances.[406] These objectives should serve the goals set in Article 1 of the Floods Directive: the reduction of adverse consequences for human health and the environment, cultural heritage and economic activity caused by floods.

The areas for which appropriate objectives should be set, are those areas for which Member States, individually, have concluded that potential significant flood risks exist or might be considered likely to occur. These areas are to be defined for each River Basin District, or unit of management pursuant to Article 3(2)(b) FD or the part of the IRBD located in their territory.[407] Even though there is a general obligation to cooperate, and an obligation of best efforts to coordinate measures throughout the IRBDs, from the joint reading of Article 7(2) and Article 5(2) FD, it seems that Member States are subject to the procedural requirement of setting objectives individually and for their own territories.

One concrete example of this lack of objective-setting at EU level is the fact that Member States may choose to apply their own safety standards, or not to adopt safety standards at all.[408] In this scenario, it is entirely possible that within the same International River Basin District, one country applies a very high and legally anchored safety standard, another one applies low standards, and still another does not apply such standards at all. The Netherlands has included specific legal safety standards in its Water Act, as will be explained in Chapter III, whereas the Flemish and Walloon Regions in Belgium do not have legally embedded safety standards for flood risks. The question then arises whether we can still speak of local and regional varieties within a same IRBD, such as the Scheldt or the Meuse, considering the focus of the European water directives on the ecological unit of governance as opposed to the geo-political one.

110. In addition to this concern on the discrepancy between the various objectives in the same International River Basin District, there is also the question

405 Andrea Keessen, 'In Search of a European Legislative Approach to Adaptation to Climate Change' in Marjan Peeters and Rosa Uylenburg (eds), *EU Environmental Legislation: Legal Perspectives on Regulatory Strategies* (Elgar 2014).

406 Recital 10 and Article 7(2) FD.

407 See Article 5(1) FD.

408 Andrea Keessen, 'In Search of a European Legislative Approach to Adaptation to Climate Change' in Marjan Peeters and Rosa Uylenburg (eds), *EU Environmental Legislation: Legal Perspectives on Regulatory Strategies* (Elgar 2014) 203.

whether the fulfilment of the procedural requirements will actually lead to a satisfactory substantive outcome.[409]

2.3.2. *Groundwater*

- WFD: Explicit quantitative inclusion: Good quantitative status

111. Groundwater forms a vital part of the hydrological cycle. Indeed, groundwater has a volume that exceeds the volume of surface water by 60%.[410] Despite its importance, groundwater has not been systematically integrated into the legal and policy analyses of water governance, neither at the international nor at the European level. Generally speaking, there is often uncertainty as to whether or not groundwater is included in the scope of the legal agreements. For example, under the Water Resources Development Act, which was enacted in the United States Congress in 1986[411], it was impossible to transfer water from the Great Lakes Basin without unanimous consent of each of the governors of the Great Lakes States.[412] However, the Water Resources Development Act did not include provisions regarding groundwater, which lead to a plethora of discussions on the scope of this diversion requirement.[413]

However, recent years have shown a spiked interest into legal research on the issue of inclusive groundwater management, among other things due to the development toward integrated water management. At the international level, in 2008 the International Law Commission adopted the "Draft articles on the Law of Transboundary Aquifers".[414] These draft articles are limited to unconfined aquifers, rather than also including confined aquifers.[415]

112. The WFD, consistent with its holistic approach, includes numerous requirements related to groundwater bodies, attributing importance to the quantitative aspects. Generally speaking, the WFD provides for the obligation for Member States to protect, enhance and restore all bodies of groundwater, while balancing abstraction and recharge rates. The ultimate goal for the Member

409 Marleen Van Rijswick, Herman Kasper Gilissen and Jasper van Kempen, 'The Need for International and Regional Transboundary Cooperation in European River Basin Management as a Result of New Approaches in EC Water Law' (2010) 11 ERA Forum 129, 135.

410 The larger part of the fresh water on earth is stored below the surface. Preamble 2 of the Groundwater Directive. See Maude Barlow and Tony Clark, Blue Gold: The Fight to Stop the Corporate Theft of the World's Water (The New Press 2001).

411 Water Resources Development Act, 42 U.S.C. § 1962d-20(b)(3) (2003).

412 Namely, Illinois, Indiana, Michigan, Minnesota, New York, Ohio, Pennsylvania and Wisconsin.

413 MK Scanlan, JH Sinykin and J Krohelski, 'Realizing the Promise of the Great Lakes Compact: A Policy Analysis for State Implementation' (2006) 8 Vt. J. Envtl. L. 39.

414 The law of transboundary aquifers (UNGA Resolution A/RES/63/124, 11 December 2008).

415 Stephen McCaffrey, 'The International Law Commission Adopts Draft Articles on Transboundary Aquifers' (2009) 103 The American Journal of International Law 272, 273.

States is to attain a good groundwater status.[416] This status pertains both to the achievement of a good chemical status and a good quantitative status.[417]

In Annex V, the meaning of "good quantitative status" is further specified. In accordance with Annex V to the WFD, more specifically Table 2.1.2, the average yearly rate of abstraction of water cannot exceed the available groundwater resource in the long term. Indicators for achieving the quantitative status have also been set, and relate to the consideration that the groundwater body in question is not influenced by anthropogenic alterations to such an extent that the ecological quality objectives for associated surface waters would be affected, or that the status of these waters would deteriorate significantly, or that significant damage is caused to dependent terrestrial ecosystems.[418] Furthermore, another indicator related to quantitative groundwater levels relates to flow direction, where an alteration may occur on a temporary basis but such alteration must not cause salt water or any other form of intrusion. The WFD connects the obligation to maintain a certain quantitative level of groundwater to the ecological objectives for surface waters. Indeed, the term "available groundwater resource" is explained as "the long-term annual average rate of overall recharge of the body of groundwater less the long-term annual rate of flow required to achieve the ecological quality objectives for associated surface waters".[419] This means that the obligation with regard to the quantitative status of groundwater relates to the ecological status of the water and to avoiding damage to associated terrestrial ecosystems.

Finally, controls are included, including a requirement for prior authorisation of artificial recharge or augmentation of groundwater bodies.[420] The WFD regulates beyond the balance of abstraction and recharging, by linking this quantitative balance with ecological requirements with regard to surface waters.[421] Monitoring forms an integral part of the groundwater good-status requirements, with Member States, i.e. the designated competent authorities, being required to set up groundwater-monitoring networks. In case of transboundary groundwater flows, the density of monitoring points should be adequate to estimate the rate and direction of the cross-border flows.[422] For International River Basin Districts, monitoring programmes can be coordinated, which has been done for example under the auspices of the Danube Convention, which has introduced the Transnational Monitoring Network, a joint monitoring programme for the Danube basin States.[423]

---

416 Article 4(ii) WFD.

417 Article 2(2) WFD.

418 Philippe Sands and Paolo Galizzi (eds), *Documents in European Community Environmental Law* (Cambridge University Press) 957.

419 Article 2(27) WFD.

420 Article 11(3)(f) WFD.

421 A. Allan, F. Loures and M. Tignino, 'The Role and Relevance of the Draft Articles on the Law of Transboundary Aquifers in the European Context' (2011) JEEPL 231, 238.

422 Annex V, Article 2(2)(1) and 2(2)(2) WFD.

423 Allan, Loures and Tignino (n 421) 245.

113. The Groundwater Directive further stipulates that threshold values can either be established at the national level, or at the level of the River Basin District, or the part of the International River Basin District falling within the territory of a Member State, or finally at the level of a body or a group of bodies of groundwater.[424] The Directive therefore does not explicitly provide for the establishment of threshold values at the level of the International River Basin District, even though this would technically be possible, from a hydrological point of view. The same goes for the summary of the assessment of groundwater chemical status in the RBMPs, which should, for IRBDs, only be established for the part of the IRBD falling within the territory of a Member State, or be established at the level of the River Basin District.[425]

It is clear that the fulfilment of the requirements with regard to the quantitative groundwater status by Member States in IRBDs will be influenced by the actions by upstream States in these districts, for example when the latter abstract water.[426]

2.4. PILLAR III: PROCEDURAL RULES

2.4.1. Proceduralisation in the EU Water Landscape

114. As mentioned above, procedural rules are put front and centre by the Water Framework Directive and, especially, the Floods Directive. Indeed, the totality of procedural requirements constitutes the backbone of these Directives. It has been stated in the literature that the WFD and FD represent a shift from the "government" approach to the "governance" approach, which entails a shift from detailed and specific norms and requirements to more broadly formulated and vague norms and goals, with more emphasis on procedural requirements.[427] The government approach generally relates to command-and-control measures widely used in the first waves of European environmental legal instruments, for example emission limit values. The governance approach represents including a wide array of stakeholders[428], more discretionary powers and flexibility for the Member States implementing the relevant Directives, and wider use of non-binding tools

---

424 Article 3(2) Groundwater Directive.

425 Article 4(4) Groundwater Directive.

426 Jasper Van Kempen, *Europees Waterbeheer: Eerlijk Zullen We Alles Delen?* (Boom Juridische Uitgevers 2012) 165.

427 Ibid 93. Described by Van Holten and Van Rijswick as "Governance Mode Directives". Marleen van Rijswick and Herman Havekes, European and Dutch Water Law (Europa Law Publishing, 2012) 107. Saskia van Holten and Marleen van Rijswick, 'The Governance Approach in European Union Environmental Directives and Its Consequences for Flexibility, Effectiveness and Legitimacy', *EU Environmental Legislation: Legal Perspectives on Regulatory Strategies* (Cheltenham 2014) 16; William Howarth, 'Aspirations and Realities under the Water Framework Directive: Proceduralisation, Participation and Practicalities' (2009) 21 Journal of Environmental Law 391, 396.

428 I.e. increased public participation.

and guidelines.[429] This approach represents a stronger focus on procedural requirements, which entails that the EU level prescribes certain actions that need to be taken in the course of the decision-making process, without defining the final outcome.[430]

This "proceduralisation" of the water-related directives and the consequences thereof in terms of the five pillars of the legal regime governing transboundary waters will be discussed below.

115. The four main procedural requirements with regard to quantitative water management that will be discussed in these sections are the following: (i) the duty to draw up plans and programmes, (ii) the obligation to notify other riparian States on planned measures, (iii) the exchange of data and information, and (iv) public participation.

The latter three procedural obligations do not all stem directly from the WFD and FD, as they ensue from other directives in the area of European environmental law, such as the EIA and SEA Directives and the Public Participation Directive.[431] Some of these procedural requirements also ensue from international customary water law, such as the obligation to notify States of planned measures.[432] As an example, reference can be made to Article 11 of the 1997 UN Watercourses Convention, which stipulates that Watercourse States should exchange information and consult with one another, and even negotiate on the effects of planned measures on the transboundary watercourse. Moreover, the UNECE Water Convention includes several procedural obligations related to notification and consultation, which are relevant in this context.

This section therefore aims to provide an overview of the specific procedural obligations applicable to EU Member States with regard to quantitative aspects of water management, including flood risk management, in International River Basin Districts.

2.4.2. Procedural requirements with regard to water quantity management, including flood risk management

2.4.2.1. The duty to cooperate – general

116. The precise institutional mechanisms that exist to warrant cooperation between Member States sharing IRBDs will be explained in the fourth pillar.[433]

429 For example, the Common Implementation Strategy, which will be further discussed in the section on "institutional mechanisms".
430 Howarth (n 427) 397.
431 This will be explained in the sections below.
432 Alistair S. Rieu-Clarke, A Survey of International Law relating to Flood Management: Existing Practices and Future Prospects, (2008) International Law and Flood Management 649, 660.
433 The fourth pillar "Institutional Mechanisms".

Member States in IRBDs are subject to several "duties to cooperate" throughout the WFD and FD.[434]

The WFD stipulates a general duty to achieve a coordinated implementation of the environmental objectives included in Article 4, and in particular in all programmes of measures established to comply with the Directive. This obligation to coordinate can be seen as an "obligation of result".[435] The FD does not provide for a similar duty to coordinate as stipulated by Article 3(4) WFD. One of the explanatory factors is that the FD does not provide for specific objectives similar to Article 4 WFD. Indeed, the FD mainly contains procedural obligations, for which specific cooperation provisions exist. These will be discussed in the following chapters.

In addition to this general coordination duty, the cooperation paradigm is mainly reflected in the coordination requirements for specific procedural requirements, such as adopting River Basin Management Plans and Flood Risk Management Plans. As will be discussed below, cooperation requirements with regard to these plans are merely obligations of effort.[436]

117. The stringency of cooperation requirements was diminished in the run-up to the adoption of the WFD. The Parliament had proposed to include a provision ensuring that the competent management authorities for IRBDs would have the standing and capacity to effectively manage the transboundary waters in question, so as to simplify the transboundary aspects.[437] In the final version of the WFD, there is a requirement to appoint the competent authorities for the IRBDs in question[438], however, no further specifications are set as to the powers that should be granted to these authorities.

118. As regards the Floods Directive, the European Parliament had, in its first reading, proposed a provision requiring Member States in IRBDs to ensure overall coordination of the requirements of the Directive, possibly making use of existing structures stemming from international agreements.[439] However, this general coordination requirement was not upheld. As mentioned before, the cooperation

---

434 Cooperation within the meaning of this research also includes "coordination".

435 Ellen Hey and Marleen van Rijswick, 'Transnational Water Management' in Oswald Jansen and Bettina Schöndorf-Haubold (eds), *The European Composite Administration* (Intersentia 2010) 231, 243.

436 Ibid 13.

437 European Parliament, 'Legislative resolution embodying Parliament's opinion on the proposal and the amended proposals for a Council Directive on establishing a framework for Community action in the field of water policy (1999) OJ C 150, Amendment 37.

438 Article 3(3) WFD.

439 European Parliament, 'Legislative resolution on the proposal for a directive of the European Parliament and of the Council on the assessment and management of floods' (2006) <www.europarl.europa.eu/sides/getDoc.do?pubRef=-//EP//NONSGML+TA+P6-TA-2006-0253+0+DOC+PDF+V0//EN> accessed 10 July 2017.

requirement is mainly at the level of the Flood Risk Management Plans, although the legal value of this requirement is questionable.[440]

2.4.2.2. Plans, programmes and objectives setting

119. An important part of the implementation of the WFD and the Floods Directive consists of the drafting and submission of respectively River Basin Management Plans (RBMPs), in the context of the WFD, and Flood Risk Management Plans (FRMPs), in the context of the FD. The deadline for submission of these second-cycle RBMPs and first-cycle FRMPs was 22 December 2015.[441]

120. These RBMPs should include the elements that are included in Annex VII WFD, e.g. mapping of the location and boundaries of groundwater bodies and surface water bodies, protected areas, a map of the monitoring networks, a list of the environmental objectives established on the basis of Article 4 WFD and a summary of the economic analysis of water use as required by Article 5.[442]

121. With regard to flood risks specifically, the FD requires Member States to draw up FRMPs. To this end, Member States are required to set objectives for the management of flood risks for these areas that have been identified on the basis of either their preliminary flood risk assessments or their maps.[443] These objectives should focus on the goal of the Directive, namely the reduction of adverse consequences of flooding for human health, the environment, cultural heritage and economic activity.[444] If considered appropriate by the Member States, the FRMPs should entail non-structural initiatives and/or the reduction of the likelihood of flooding. The fact that the latter element is included as an ancillary option at the discretion of the Member States is in line with the *ratio legis* of the Floods Directive, namely that floods cannot be completely avoided, but their consequences should be mitigated as far as possible.

122. The Directive sets out a compulsory set of elements to be included in the FRMPs.[445] The first-cycle FRMPs should include five components: (i) the conclusions of the preliminary flood risk assessment, indicating the areas that

---

440 See *infra*.

441 Failing to comply with this may result in Commission action. The Commission has referred Denmark to the Court for its failure to submit its RBMPs on time. This was partly due to the fact that the Danish Courts considered that the consultation period set by the authorities with regard to the RBMPs were too short under Danish law, resulting in an annulment of the existing RBMPs in 2012. The RBMPs should have been submitted in 2009, and the Danish only intended to submit the plans mid-2014.

442 Annex VII WFD.

443 Article 7(2) FD.

444 Article 7(2) FD.

445 Part A of the Annex FD.

are the subject of the FRMPs in question, (ii) the flood-hazard maps and flood risk maps that have been prepared, (iii) the description of the objectives set by the Member States, (iv) a summary of the measures aimed to achieve these objectives and also flood-related measures taken on the basis of other EU Acts, and (v) *when available*[446], for shared river basins or sub-basins, a description of the methodology, defined by the Member States in question, of the cost-benefit analysis used to assess measures with transnational effects.

It is problematic that there are coordination requirements for the FRMP as a whole, but that the Directive does not stipulate that Member States should cooperate on all the specific elements that constitute the FRMP, e.g. setting the objective, preparation of the maps, and so forth.

123. For IRBDs located entirely within the boundaries of the European Union, Member States must ensure coordination with the aim of submitting one single international RBMP.[447] Where Member States cannot reach agreement and are therefore not able to produce such a single RBMP, they must produce an RBMP for the parts of the IRBD falling within their territory, in order to achieve the objectives of the WFD. The FD stipulates the same requirements for coordination in order to produce a single FRMP for international districts.[448] For IRBDs that extend beyond the territory of the European Union, Member States must endeavour to produce one single RBMP. Where this is not possible, Member States should adopt plans that cover the portion of the IRBD lying within the territory of the Member State in question.[449]

124. Referring to the principle of solidarity, the FD stipulates that the Flood Risk Management Plans may not include measures that would result in a significant increase of flood risks upstream or downstream of other countries sharing the river basin or sub-basin.[450] This prohibition can be linked to the well-known principle of international water law, the "no harm" rule, which has been explained in Chapter I. No further explanation is given as to the precise meaning of the word "significant", although it is a word that has often been used in international water law, which can be helpful in interpreting its reach. The term "significant" is generally seen as less than "substantial" but more than representing "a minor inconvenience".[451] An exception to this prohibition exists when such measures have been coordinated and the Member States have reached an agreed solution.

446 Emphasis added.
447 Article 13(2) WFD.
448 Article 8(1) FD.
449 Article 13(3) WFD and Article 8(2) FD.
450 Article 7(4) FD.
451 Commentary to the Draft Articles on the Law of the Non-Navigational Uses of International Watercourses, in ILC, 'Report of the International Law Commission on the work of its 46th session' (2 May-22 July 1994) UN Doc A/49/10, 94, para 15.

The FD does not attach legal consequences to the inclusion or lack of inclusion of such measures in the FRMPs, nor does it determine the platform or provide for the instruments through which this coordination may occur. Finally, States sharing an IRBD are not obligated to conclude agreements on the meaning and application of "significant increase in flood risks" with respect to the district in question.[452]

2.4.2.3. Prior exchange of information in the context of cartography

125. One of the cross-border procedural requirements included in the Floods Directive is the obligation of Member States to exchange information on (the preparation of) flood-hazard maps and flood risk maps[453] for areas identified as areas where potential significant flood risks exist or might be considered likely to occur.[454] The Member States that share the areas are required to exchange information.

The flood hazard maps and flood risk maps should be prepared at the level of the River Basin District or the unit of management within the meaning of Article 3(2)(b) FD. The FD does not require Member States to coordinate the drawing up of these maps, but only states that these should be prepared at the "most appropriate scale".[455]

2.4.2.4. Prior notification of planned measures

126. In international water law, the prior notification of potentially to be affected states of planned measures is an explicit obligation. Neither the WFD nor the FD stipulate explicit requirements related to prior notification mechanisms regarding Member States in the same IRBD. However, the obligation to notify (potentially) affected parties of a planned measure prior to the execution or implementation of this measure is covered by other directives.

The Environmental Impact Assessment (EIA) Directive 2011/92/EU takes into account cases where certain projects in one Member States may affect the environment of another Member State, and the Strategic Environmental Assessment (SEA) Directive 2001/42/EC applying to a range of plans and programmes.[456] In order to grasp the scope of the prior notification requirement in terms of EIA and SEA mechanisms, it should be examined which measures

---

452 In Chapter III, the question whether states have actually done so in a specific IRBD will be tackled.

453 Article 6(2) FD.

454 I.e. the areas identified under Article 5 FD.

455 Article 6(1) FD.

456 Directive (EU) 2011/92 on the assessment of the effects of certain public and private projects on the environment [2012] OJ L 26/1, Article 7.

related to flood risk management, droughts and scarcity are included in the scope of the EIA and SEA Directive.

(A) Quantitative water management measures included in the EIA scope

127. Below is an overview of the measures and works in the context of the quantitative management of water resources subject to EIA requirements. Measures that are included in Annex I are subject to Article 4(1), which implies that the Member States are required to provide EIA provisions. For measures included in Article 4(2), the Member States have the discretion to determine whether they require assessments for these projects. For example, in the Flemish Region, Annex-II measures still require an EIA, but motivated applications for exemptions can be submitted to the competent authorities.[457]

Table 3. Water Quantity Measures Included in the EIA Directive

Annex I (Article 4(1))	Annex II (Article 4(2))
– The abstraction of groundwater, or the schemes for recharging artificial groundwater where the annual volume of water abstracted o recharged equals or exceeds 10 million cubic metres.[458] – Inter-basin water transfers: • Preventing water shortages and the amount of water transferred exceeds 100 million cubic metres per year. • All works for the transfer of water resources between basins when the multi-annual average flow of the basin of abstraction exceeds 2000 million cubic metres/year and the amount of water transferred exceeds 5% of that flow. – Dams (or other installations) aimed at holding back or permanently storing water	– Water management projects for agriculture (including irrigation & land drainage projects) – Inland-waterway construction not included in Annex I, canalisation and flood-relief works – Dams and other installations designed to hold water or store it on a long-term basis, if these projects are not included in Annex I – Groundwater abstraction and artificial groundwater recharge schemes not included in Annex I – Works for the transfer of water resources between river basins not included in Annex I

128. Table 3 indicates that the EIA Directive covers a substantial part of the measures affecting the quantity of water, whether in the context of droughts, floods or scarcity.

In the context of Annex-II interpretation, the measures "canalisation and flood-relief works" have already been scrutinised by the Court of Justice.[459] The

457 Order of the Flemish Government of 10 December 2004 determining the categories of projects subject to environmental impact assessment, Belgian Official Journal 17 February 2005.
458 Annex I, point 8 of the EIA Directive.
459 Case C-72/95 *Aannemersbedrijf P.K. Kraaijeveld BV and others* [1996] ECLI:EU:C:1996:404.

Dutch Council of State had asked the Court for a preliminary ruling with regard to the interpretation of these measures in terms of the scope of the EIA Directive. More specifically, it was unsure whether this category of measures pertains to the construction of a new dyke, the replacement of a dyke, or the widening of an existing dyke.

A private company had taken legal steps to annul a zoning plan for the reinforcement of a dyke. In this context, the Dutch government had argued that there was a difference between flood-relief and canalisation works on the one hand and works on a dyke on the other hand. Whereas flood-relief and canalisation works have an impact on the quantity and quality of water, and therefore alter the character of the watercourse itself, dyke reinforcement work consists of heightening the embankment with sand or clay, and, in accordance with the Dutch Law on Rivers, does not affect the regulated level of the watercourse.[460]

The Court referred to the definition of "project", which means the "execution of construction works or of other installations or schemes", where the Directive is aimed at projects likely to have significant effects on the environment by virtue inter alia of their nature, size or location.[461] It therefore appears that the Directive is broad in scope, resulting in the conclusion that point 10(e) of the Directive covers all works for retaining water and the prevention of floods, including dyke works. The second issue pertained to the question whether modifications to development projects included in Annex II were to be covered by the EIA Directive. Indeed, whereas the Directive stipulates that modification to development projects included in Annex I should be subjected to the same system as the projects included in Annex II, the Directive is silent as to whether the Directive covers such modifications to projects included in Annex II. Once again, the Court referred to the broad scope of the Directive, and stated that the exclusion of such modifications to projects included in Annex II from the assessment mechanisms would undermine the purpose of the Directive.[462] Following this line of reasoning, the scope of the Directive covers not only the construction of a new dyke, but also modifications to an existing dyke, including its relocation, reinforcement or widening, replacement of a dyke by constructing a new dyke in situ[463], or a combination of the above.[464]

The fact that Article 4(2) in correlation with Annex II of the Directive provides a degree of discretion with regard to the inclusion of projects in assessment requirement schemes in the Member States was also tackled by the Court. If a Member State were to establish thresholds and criteria in such a way that all projects would a priori be exempted from these schemes, the Member State in

460 Ibid para 24.

461 *Aannemersbedrijf P.K. Kraaijeveld BV and others* (n 459) para 30. Article 1(2) and 2(1) EIA Directive.

462 *Aannemersbedrijf P.K. Kraaijeveld BV and others* (n 459) para 39.

463 Regardless of whether or not the new dyke is stronger or wider than the old one.

464 *Aannemersbedrijf P.K. Kraaijeveld BV and others* (n 459) para 42.

question would have exceeded the limits of its discretionary powers in the sense of Article 4(2). Member States can, for example, specify certain capacity thresholds for floodplains triggering the application of the respective annexes, e.g. in the Flemish Region, floodplains with a volume capacity equal to or exceeding 250,000 m^3 or dikes with a length equal to or exceeding 500 metres are subject to Annex II, for which in principle an EIA is required with the possibility of applying for an exemption to this obligation.[465]

129. The measures in the context of flood risk management that must definitely be subjected to EIA requirements in the Member States pertain to dams and other installations aimed at holding back or permanently storing water. The measures in the context of flood risk management that may be subjected to EIA requirements in the Member States pertain to inland-waterway construction, canalisation and other flood-relief works, or dams and other installations to the extent that they are not covered by Annex I. The measures with regard to flood risk management, and more specifically in the context of the strategies of flood risk prevention, flood defence and flood risk mitigation include, among many other things, the following: man-controlled flood zones, weirs, retention basins, etc.[466] It seems that these actions, where they affect the flood risk management policies of other Member States, can be excluded from obligatory prior notification rights and obligations.

(b) Effect on the environment

130. It is also unclear whether the impact on flood risk management in one Member State classifies as having "significant effects on the environment". Article 3 of the EIA Directive states that the effect of the project on the following factors should be evaluated: (a) human beings, fauna and flora; (b) soil, water, air, climate and the landscape, (c) material assets and cultural heritage, and (d) the interaction between the above factors.[467] Now, both an excess of water and a shortage of water are likely to trigger a significant, possibly simultaneous interaction between factors (a), human beings and fauna and flora, (b) soil, water and the landscape, and (c) material assets and cultural heritage.

The Commission has emphasised that the effects on the environment should be assessed regardless of the social objective of the measure in question. The fact that the construction of river dykes is aimed at protecting the population against

[465] Annex II, section h) of the Flemish EIA Order.

[466] For an explanation of flood risk management strategies and associated measures, see Dries Hegger and others, 'Flood Risk Management in Europe: Similarities and Differences between the STAR-FLOOD Consortium Countries' (STAR-FLOOD Consortium 2013). Dries Hegger and others, 'A View on More Resilient Flood Risk Governance: Key Conclusions of the STAR-FLOOD Project' (STAR-FLOOD Consortium 2016).

[467] Article 3 EIA Directive.

floods, is irrelevant to the determination that the dykes have an impact on the environment.[468]

(c) Transboundary aspect

131. In terms of the transboundary aspect of EIA requirements with regard to water quantity management, things are more complicated.[469] Article 7 of the EIA Directive pertains to the assessment of transboundary environmental effects, and determines that Member States are required to notify the Member State whose territory might be significantly affected by the measure in question (whether or not the latter has requested to be informed).[470] The Member State in whose territory the project will be carried out should submit to the affected Member State a description of the project, including all available information on the possible transboundary impact of the project, and information on the nature of the decision which may be taken.

An Affected Member State in the sense of the EIA Directive and in the quantitative water management context, can be both a Member State sharing an IRBD with the Member State in whose territory the project is being carried out, and a neighbouring Member State that does not necessarily share its waters.

132. In the spirit of the WFD and FD, Member States sharing IRBDs may well use the existing international structures to notify the affected Member States. However, the WFD and FD do not refer to the content of Article 7 of the EIA Directive. Annex VI stipulates that the programme of measures should include measures taken pursuant to the Environmental Impact Assessment Directive.

133. To sum up, the WFD and the FD do not stipulate specific obligations of prior notification from one Member State to another. Consequently, there is no platform for this procedural right and obligation. However, certain measures with regard to quantitative water management, be it floods, droughts or scarcity, are to some extent covered by the EIA Directive.

134. The Strategic Environmental Assessment (SEA) allows for an analysis of environmental effects of strategic plans.[471] Plans that are likely to have significant environmental effects should be subjected to an assessment exercise. The SEA

---

468 *Aannemersbedrijf P.K. Kraaijeveld BV and others* (n 459) para 27.

469 On transboundary EIA's: Cathy Suykens, 'Globalisation of the Nuclear Fuel Cycle and Maritime Carriage of Radioactive Materials: Review of the Legal Regime' (2011) 26 The International Jounral of Marine and Coastal Law 385, 401.

470 Article 7 EIA Directive.

471 Jeremy Carter and Joe Howe, 'The Water Framework Directive and the Strategic Environmental Assessment Directive: Exploring the Linkages' (2006) 26 Environmental Impact Assessment Review 287, 290.

Directive explicitly applies to water management plans, and therefore to River Basin Management Plans as well.[472] Furthermore, the WFD determines that RBMPs should include an analysis of the impact of susceptibility of the status of surface water to human impact.[473] The susceptibility of the WFD plans to the SEA Directive has also been confirmed by AG Kokott, stating that management plans pursuant to the WFD should be subjected to an SEA assessment.[474] Of course, if a plan falls within the scope of the WFD, the SEA and the EIA, there is no need to conduct an independent SEA if the substantive and procedural requirements of the SEA Directive have been met.[475] Flood Risk Management Plans should equally include the summary of measures resulting from the SEA. In the context of a SEA of a relevant plan or programme, a transboundary consultation requirement exists vis-à-vis potentially affected Member States.[476] The recipient Member State may subsequently decide whether it wishes to enter into consultations prior to the adoption of the plan or programme on the transboundary environmental effect of its implementation.[477] The respective States should involve the relevant authorities with environmental responsibilities as well as the public in this consultation process.[478]

2.4.2.5. (Transboundary) public participation

135. One of the principles of the WFD and FD is the involvement of interested parties in their implementation.[479] According to certain authors, the explicit inclusion of public participation in the context of the drafting of respectively the River Basin Management Plans and the Flood Risk Management Plans constitutes an important part of the legal anchoring of the shift from "government to governance".[480]

Public participation on the basis of the WFD and the FD consists of three levels, which are hierarchically ordered: (i) the supply of information, (ii) consultation, and (iii) active involvement. Whereas the first two levels must be ensured by the Member States, the active involvement of stakeholders should be

472 Article 3(2) of Directive (EC) 2001/42 on the assessment of the effects of certain plans and programmes on the environment (SEA Directive) [2001] OJ L 197/30.

473 Annex II, 1.5 of the WFD.

474 "It must therefore be stated that management plans within the meaning of Article 13 of the Water Framework Directive must, in principle, be subjected to an environmental assessment under the SEA Directive." Case C-43/10 Nomarchiaki Aftodioikisi Aitoloakarnanias and Others [2011] ECLI:EU:C:2012:560, para 157.

475 Case C-43/10 *Nomarchiaki Aftodioikisi Aitoloakarnanias and Others* [2011] ECLI:EU: C:2011:651, Opinion of AG Kokott, para 178.

476 Article 7 SEA Directive.

477 Article 7(2) SEA Directive.

478 In accordance with Article 7 in conjunction with Article 6(3) and 6(4) SEA Directive.

479 Article 9(3) and Article 10 FD, and Article 14(1) WFD.

480 Ben Page and Maria Kaika, 'The EU Water Framework Directive: Part 2. Policy Innovation and the Shifting Choreography of Governance' (2007) Eur. Env. 13, 2.

encouraged, meaning that it is not a strict requirement.[481] The WFD states that it is necessary to provide information of measures in the pipeline, and the report on all progress with regard to the implementation of these measures, so as to involve the general public prior to the adoption of final decisions on the measures.[482]

136. Article 14 of the WFD[483] further defines the specific requirements in terms of public participation in the context of the drafting of the River Basin Management Plans. Indeed, the WFD stipulates that Member States, for each River Basin District, must submit inter alia a timetable and work programme for the production of the plan, an overview of water management issues within the river basin, draft copies of the river basin management plan, no later than one year before the beginning of the period to which the plan refers. Moreover, if citizens request background documents used for the development of the RBMPs, these should be provided as well.[484] AG Kokott has emphasised that there are no exemptions or derogations from the public participation requirements in Article 14 WFD, e.g. in relation to the production of RBMPs.[485]

137. The FD equally provides that the Member States must encourage the active involvement of interested parties in the process of production, review and updating of the flood risk management plans as referred to in Chapter IV.[486] Furthermore, the FD provides, in quite general terms, that the Member States must make available to the public (i) the preliminary flood risk assessment, (ii) the flood hazard maps, (iii) the flood risk maps, and (iv) the Flood Risk Management Plans.[487] The FD further provides that Member States must coordinate, as appropriate, the active involvement of all interested parties under Article 10 of the Directive, with the active involvement of interested parties under Article 14 of the WFD. Therefore, the FD does not set out the specific data that must be made available in the context of flood risk management, but refers to the documents and information that must be made available in the context of drawing up the River Basin Management Plans.

The Court of Justice ruled in its landmark case C-32/05 that Article 14 of the WFD does subject Member States to specific obligations "intended to confer on individuals and interested parties a right to be actively involved in the implementation of the directive and, in particular, in the production, review and

481 Commission, 'WFD CIS Guidance Document on Public Participation in relation to the Water Framework Directive' (European Union, 2014).

482 Recital 46 of the WFD. The FD does not include such recital in its scope.

483 Also relevant: Annex VII A of the WFD.

484 Article 14 WFD.

485 Case C-43/10 *Nomarchiaki Aftodioikisi Aitoloakarnanias and Others* [2011] ECLI:EU:C:2011:651, Opinion of AG Kokott, para 126.

486 Article 10(2) FD.

487 Article 10(1) FD.

updating of the river basin management plans".[488] The Court further emphasises that individuals should be able to ascertain, well in advance, the full extent of their rights in the context of the procedures set out in Article 14(1) and (2) WFD. The fact that the legislative framework in Luxembourg did not comply with the time schedules and deadlines included in Article 14, compromised this right. Therefore, the Court considered that, by failing to adopt the laws necessary to comply with Article 14[489] of the WFD within the prescribed period, Luxembourg had violated Article 24 of the WFD.[490] Considering the similar wording of the Floods Directive related to the active involvement of the public, this judgment can be applied mutatis mutandis to the relevant provision.[491]

138. The active involvement of the public, as required by the FD, is important in the context of, among other things, the discourse on "multi-layered water safety", which represents a general shift toward sharing responsibility between the government and its citizens.[492] This shift is also framed in the recognised importance of diversifying flood risk management strategies, where, for example, individual household measures such as adaptive building and resilient reinstatement are important factors.[493] The underlying prerequisite for this greater emphasis on individual responsibility, however, relates to citizens' awareness of the risks faced and proper involvement in decision-making procedures. As explained in Chapter I, one way of reaching this involvement is through comprehensive co-production.[494] It could be considered for the directives to issue a mandate to

---

488 Case 32/05 *Commission of the European Communities v Grand Duchy of Luxemburg* [2006] ECLI:EU:C:2006:749, para 80.

489 And also Article 2, 7(2) WFD.

490 *Commission of the European Communities v Grand Duchy of Luxemburg* (n 488).

491 Article 10 FD. With the three aspects of public participation included in the Directives, these are in line with the Aarhus Convention and its implementing Directives at EU level. Convention on Access to Information, Public Participation in Decision-Making and Access to Justice in Environmental Matters (Aarhus, 25 June 1998) 2161 UNTS 447. Elisa Morgera, 'Water Management and Protection in the EU' in M Alberton and F Palermo (eds), *Environmental Protection in Multi-Layered Systems: Comparative Lessons from the Water Sector* (Martinus Nijhoff 2012) 273.

492 As opposed to the government being solely responsible for managing flood risks.

493 Diversification of flood risk management strategies is related to public participation in the sense that traditional hard infrastructure such as dykes are clearly a government responsibility, whereas there is a tendency in Europe to deepen the promotion of adaptive measures at individual household levels, thereby diversifying the instruments available to policy makers to protect society and deal with flood risks. Peter Driessen and others, 'Toward More Resilient Flood Risk Governance' (2016) 21 Ecology and Society; Hannelore Mees, Cathy Suykens and Ann Crabbé, 'Evaluating Conditions for Integrated Water Resource Management at Sub-Basin Scale. A Comparison of the Flemish Sub-Basin Boards and Walloon River Contracts' (2017) 27 Environmental Policy and Governance 2. P Matczak and others, 'Comparing Flood Risk Governance in Six European Countries: Strategies, Arrangements and Institutional Dynamics' (STAR-FLOOD Consortium 2016).

494 Hannelore Mees and others, 'Coproducing Flood Risk Management through Citizen Involvement: Insights from Cross-Country Comparison in Europe' (2016) 21 Ecology and Society.

the Member States to create a form of co-production mechanism, as reference is made to the *active* involvement of the public in the *production* of the RBMPs and FRMPs.[495] Also the Guidance Document on the implementation of the Water Framework Directive in relation to public participation refers to shared decision-making and self-determination as good practices in water management.[496] However, research into Member State practice has shown that many States mainly focus on disseminating the plans at the end of the decision-making process, when meaningful and substantial input is more difficult.[497]

(a) Transboundary public participation in the water directives?

139. It is clear that public participation is enshrined in the WFD and FD. However, neither Directive includes explicit provisions on cross-border public participation for measures and planning processes in International River Basin Districts. However, the phenomenon of cross-border participation has been introduced in other European legal instruments. The IPPC Directive for example, emphasises the importance of transboundary consultation in situations where applications relate to the licensing of new installations or changes to existing installations that are likely to have significant environmental effects. These applications should then be available to the public of the Member State likely to be affected.[498] More specifically, Article 18 of the IPPC Directive is dedicated to such situations where transboundary effects occur. Article 18 stipulates that, in the framework of bilateral relations between the Member States in question, these should ensure that the applications are made available for an appropriate period of time to the citizens of the Member State likely to be affected, to allow the public to submit comments on the application before the relevant competent authority makes a decision in this regard.[499] This transboundary public consultation requirement is in fact a reflection of what is required by the ESPOO Convention in terms of public participation.[500]

The EIA Directive and the SEA Directive provide for similar transboundary involvement of the public. Article 7 of the EIA Directive stipulates that, with regard to projects likely to have significant effects on the environment in another

---

495 Article 14 WFD and Article 10 FD.

496 Commission, 'WFD CIS Guidance Document on Public Participation in relation to the Water Framework Directive' (European Union, 2014), 13.

497 See Kristina Ek and others, 'Strengthening and Redesigning European Flood Risk Practices Towards Appropriate and Resilient Flood Risk Governance Arrangements Design Principles for Resilient, Efficient and Legitimate Flood Risk Governance; Lessons from Cross-Country Comparisons' (STAR-FLOOD Consortium 2016). Maria Pettersson and others, 'How Legitimate Is Flood Risk Governance in Europe? Insights from Intra-Country Assessments' (in review) Water International.

498 Recital 29 of Directive (EC) 2008/1 of 15 January 2008 concerning integrated pollution prevention and control [2008] OJ L 24/8 (the IPPC Directive).

499 Article 18(2) IPPC Directive.

500 James Anderson, Liam O'Dowd, Thomas M. Wilson, *New Borders for a Changing Europe: Cross-Border Cooperation and Governance* (Psychology Press 2003).

Member State, the Member States involved must ensure that the public of the Member State whose territory will likely be influenced by the project, is given the opportunity to express its opinion on the project before the competent authority in the Member State in whose territory the project would be carried out, reaches its decision.[501] The SEA Directive includes the obligation to allow the public to provide input in the course of a transboundary consultation process for plans or programmes with transboundary environmental effects.[502] This is highly relevant because the scrutiny of the effect of human activity on the water system in the context of the RBMP falls under the SEA Directive, which triggers the public participation procedure included there.[503]

140. The WFD and FD do not explicitly provide for this kind of transboundary public participation in IRBDs.[504] It is, however, unclear whether this is implicitly implied by the Directives, and to what extent such participation could be enforceable. With regard to consultation and access to information, reference is made to the term "public", which has not been defined in the Directives, but it has the same meaning as the notion used in the SEA Directive and Aarhus Convention, namely "one or more natural or legal persons, and, in accordance with national legislation or practice, their associations, organisations or groups".[505] With regard to active involvement, Article 14 WFD refers to *all interested parties.* Assuming that "interested parties" include all parties in the same River Basin District, or International River Basin District, this provision may well be purposively interpreted.[506] This purposive approach to interpretation has been supported by the CJEU on many occasions, stating: "The function of the words in question must be examined in the light of the intention and purpose of the regulations in questions".[507] Moreover, in CILFIT, the Court stated that

---

501 Article 7(3)(b) of the EIA Directive.

502 Article 7(2) of the SEA Directive.

503 Eg early availability of the environmental report, involvement of the Member States affected by the plan, and so forth. Elisa Morgera, 'Water Management and Protection in the EU' in M Alberton and F Palermo (eds), *Environmental Protection in Multi-Layered Systems: Comparative Lessons from the Water Sector* (Martinus Nijhoff 2012) 265, 273.

504 Neither does the Aarhus Convention on public access to environmental information and its implementing EU Directives. Directive (EC) 2003/4 on public access to environmental information and repealing Council Directive 90/313/EEC, Directive (EC) 2003/35 of the European Parliament and of the Council providing for public participation in respect of the drawing up of certain plans and programmes relating to the environment and amending with regard to public participation and access to justice Council Directives 85/337/EEC and 96/61/EC [2003] OJ L 41/26.

505 See Article 2(4) of the Aarhus Convention and Article 2(d) of the SEA Directive. Commission, 'WFD CIS Guidance Document on Public Participation in relation to the Water Framework Directive' (European Union, 2014) 15.

506 James Anderson, Liam O'Dowd, Thomas M. Wilson, *New Borders for a Changing Europe: Cross-Border Cooperation and Governance* (Psychology Press 2003), 73.

507 Case C-803/79 *Criminal Proceedings Against Gérard Roudolff* [1980] ECLI:EU:C:1980:166. See Gerard Konway, *The Limits of Legal Reasoning and the European Court of Justice* (Cambridge Studies in European Law and Policy 2014).

"every provision of Community law must be placed in its context and interpreted in the light of the provisions of Community law as a whole, regard being had to the objectives thereof and to its state of evolution at the date on which the provision in question is to be applied".[508]

The Guidance Document developed by the WFD CIS Drafting Group on Public Participation provides some insights into the meaning of "interested parties", which is described as any person, group or organisation with a stake in the issues concerned, either because they will be directly affected or because they may have some influence on its outcome. Moreover, the definition also refers to members of the public who are not yet aware that they will be affected. The Guidance Document is therefore silent on whether the concept of "interested parties" includes persons or a group of persons in Member States across the border.

Again helpful in this regard is a reference to the Aarhus Convention. In the context of the interpretation of this Convention, it has been made clear that persons, natural or legal, should not be excluded from the participation process based on nationality, citizenship or seat.[509] In a case involving the decision-making process related to the construction of a deep-water navigation canal in Ukraine, with a Romanian NGO seeking access to relevant documentation, the Compliance Committee expressly stipulated that foreign or international NGOs that have expressed interest in the decision-making process, fall under the definition of the public.[510]

The above-mentioned purposive approach to interpretation would therefore lead to the conclusion that the Directives implicitly confer transboundary participation rights to the public to the extent that Member States enter into transboundary management of the IRBD, for example, in drafting International RBMPs.[511] This conclusion, however, is limited by the consideration that failure to cooperate in drafting management plans for the entirety of the IRBD, results in a situation where Member States draft the plans for the parts of the IRBD within their own respective territories. Considering the fact that a lack of transboundary cooperation does not have explicit legal repercussions on the Member States

508 Case C-283/81 *CILFIT v. Ministry of Health* [1982] ECLI:EU:C:1982:335, para 20.

509 Serhiy Vykhryst, 'Public Information and Participation under the Water Convention' in Attila Tanzi and others (eds), *The UNECE Convention on the Protection and Use of Transboundary Watercourses and International Lakes* (Brill Nijhoff 2015) 271.

510 The Compliance Committee stated: "The communicant is a non-governmental organization working in the field of environmental protection and falls under the definitions of the public and the public concerned as set out in article 2, paragraphs 4 and 5, of the Convention. Foreign or international nongovernmental environmental organizations that have similarly expressed an interest in or concern about the procedure would generally fall under these definitions as well." see UNECE, 'Report on the Seventh Meeting of the Parties to the Convention on Access to Information, Public Participation in Decision-Making and Access to Justice in Environment Matters' (14 March 2005) ECE/MP.PP/C.1/2005/2/Add.

511 James Anderson, Liam O'Dowd, Thomas M. Wilson, *New Borders for a Changing Europe: Cross-Border Cooperation and Governance* (Psychology Press 2003) 73.

involved, it seems questionable that the cross-border public participation rights would indeed be upheld in such situations.

141. Practices that encourage transboundary public participation have been initiated at the level of the IRBD, on the basis of the requirements of Article 14 WFD. The Danube River Basin (DRB) was the first of all IRBDs in Europe to include public participation for river basin management planning at the level of the international district. The International Commission for the Protection of the Danube River coordinates these public participation efforts and, to this end, promulgated a Strategy for Public Participation in River Basin Management Planning for the period of 2003-2009.[512] The Strategy aims to safeguard public participation in the implementation of the WFD in the Danube River Basin, and to put in place structures and mechanisms for public participation in this basin, beyond the first-cycle RBMPs, which had to be submitted to the European Commission by 2009. Other International River Commissions have also included public participation at the level of the international district, e.g. the International Commission for the Protection of the Rhine ("ICPR").[513]

Given that plans do not have to be issued at the level of the international district, and in practice the unit of planning activities is still very much the national level, the national legislative frameworks would have to include provisions on the inclusion of citizens of other countries sharing the IRBDs in question. The Flemish Decree Integrated Water Policy (DIWP) does not require that the draft RBMPs and FRMPs should be made available to citizens of e.g. the Netherlands or France. It merely states that the draft documents should be made available to "the public".[514] However, during the six-month public consultation period, the first-cycle River Basin Management Plans were submitted to the competent authorities of the neighbouring countries in the IRBDs Meuse and Scheldt. Moreover, the RBMP referred to the website of the International Meuse Commission and the International Scheldt Commission respectively. The Water Code applicable in the Walloon Region does stipulate that, in the context of the six-month public consultation period of the RBMP and FRMP and in light of producing one single management plan for the IRBD, the results of the public inquiry should be sent to the other States or regions of the international hydrological district in question.[515] The Water Code does not specify that these plans should then be made available

512 ICPDR, Danube River Basin Strategy for Public Participation in River Basin Management Planning 2003-2009 (ICPDR 2003) < https://www.icpdr.org/main/sites/default/files/PP%20 Strategy%20-%20FINAL.pdf> accessed 10 July 2017.

513 Information on the website of the ICPR addressed to "all inhabitants of the Rhine river basin". Convention on the Protection of the Rhine (adopted 22 January 1998, entered into force 1 January 2003) <http://faolex.fao.org> accessed 14 June 2017.

514 Article 37 Decree Integrated Water Policy of 18 July 2003, Belgian Official Journal 14 November 2003.

515 Article D.26 § 3 Water Code of 27 May 2004, Belgian Official Journal 23 September 2004.

to the citizens of the other countries in question. The Walloon Region equally submitted its part of the RBMP to its neighbouring countries.

In conclusion, despite the lack of explicit reference to transboundary public participation in the WFD and FD, Member States make efforts to accommodate transboundary public participation in the context of the finalisation of the RBMPs and FRMPs for IRBDs, although to a limited extent and not on a structural and systematic basis.

(B) Citizens' Juries: a useful tool?

142. Several projects have been launched to enhance public participation in the implementation of the Water Framework Directive. One of these instruments is promising in terms of transboundary public participation, and will be briefly discussed below. The instrument of Citizens' Juries in the context of the WFD has been referred to by the CIS Working Group on Public Participation, which, as mentioned above, drafted a Guidance Document "Public Participation" in relation to the WFD. In this Guidance Document, the instrument of "Citizens' Juries" has been proposed as a possible facilitator of public participation in the implementation of the Directive.

The Citizens' Jury is an instrument in the context of public participation that was originally developed in the United States in the 1970s.[516] It has been described as a situation where a group of citizens is brought together to discuss a specific issue before making a public recommendation.[517] The difference between the instrument of Citizens' Jury and the general participation method outlined by the WFD, is that for the former, the stakeholders are specifically targeted and recruited, whereas the latter entails an open invitation to the general public.[518] They are asked to look at specific policy issues, and present a number of recommendations. In this sense, they operate as a sort of representation of their respective communities, constituting a bridge between stakeholders and policy makers, and between groups of stakeholders.[519]

The tool of the Citizens' Jury has already been utilised as an experiment in the context of the implementation of the Water Framework Directive. There might be opportunities for Citizens' Juries to constitute innovative tools to enhance transboundary public participation. However, the instrument should be adapted to the specific needs of transboundary public participation in water management.

---

516 Leontien Bos-Gorter, Dave Huitema, Marleen van de Kerkhof, 'Public Participation on its Own Barricades: Citizens' Jury on Water Management from Experiment to Instrument', available on the Internet at: <www.macaulay.ac.uk/pathconference/outputs/PATH_abstract_5.1.3.pdf> accessed 10 July 2017.

517 Ibid.

518 Wendy Kenyon, 'A Critical Review of Citizens' Juries: How Useful are they in facilitating Public Participation in the EU Water Framework Directive?' (2007) 48 Journal of Environmental Planning and Management 431, 432.

519 Ibid, 431.

For example, in the case of IRBDs, the traditional model of Citizens' Juries, i.e. a group of 12 to 24 people that should represent a certain population, would be less suitable considering the geographical coverage of the districts. Indeed, in the case of the River Scheldt, there are six national and regional governments involved in the coordination of the management of the Scheldt[520] at the highest levels of governance[521]: France, Belgium, the Walloon Region, the Flemish Region, the Brussels Capital Region[522] and the Netherlands. Moreover, the issue of the governance of a freshwater body involves a myriad of aspects, e.g. nature conservation, water supply, water pollution, tourism, and so forth. Because of the fact that the river runs through different countries and different sectors, legitimacy concerns may arise if only an average of two or three citizens per country or region in the International River Basin District were to take part in the jury mechanism. The fact that only such a small number of representative citizens from every region would discuss the issues at stake in international districts, would add to the already existing concerns related to the representative credibility of Citizens' Juries for questions on purely national issues.

To remedy these inherent legitimacy concerns, the geographical and material scope of the Citizens' Juries vis-à-vis the implementation of EU water legislation in transboundary waters, could be expanded. For example, instead of launching one Citizens' Jury, a series of Juries could be introduced with a broad geographical coverage but a narrow material scope. This idea of "network juries" entails that different, regional juries are composed, and the results of these respective juries are then brought together. This solution to some extent alleviates the concerns with respect to the doubtful representative character of one set of juries covering a large geographical scale.[523] In the case of the River Scheldt, firstly Citizens' Juries in the Netherlands, Flanders, Wallonia, Brussels and France respectively could be organised. Secondly, another Jury, the so-called "Syndicate Jury", might be organised to bring these findings of the different regional Juries together for the entire IRBD.

143. This concept of Citizens' Juries resembles solutions offered by other authorities, such as the European Economic and Social Committee (ESC). The ESC found the participatory process included in the proposed directive insufficient. In order to improve the participation, the ESC suggested that this Directive should require setting up consultative bodies within the competent authorities designated. These consultative bodies should then represent consumers, and organisations from the socio-economic, agricultural and environmental

[520] In the context of the International Scheldt Commission.

[521] Without taking into account sub-national, sub-regional or local levels.

[522] In Belgium, the Regions (the Walloon Region, Flemish Region, Brussels Capital Region), as opposed to the federal level, are the competent levels of government for water management and environmental protection.

[523] Wendy Kenyon (n 518) 439.

perspective. These consultative bodies would then monitor the preparation and implementation of the RBMPs, supporting the Commission, and would also play a role in information provision and sensibilisation.[524] It boils down to measures affecting the territory of the Member States.

144. The passive and active involvement of the citizens in flood risk management measures has proven to be very difficult in the national context. Actors at the level of the cities preparing plans to increase water infiltration capabilities of the urban environment are reluctant to involve or consult their citizens, and indicate that there is very little interest from citizens to participate, e.g. in the context of the first-cycle River Basin Management Plans.[525] In this regard, a well-functioning mechanism in which one riparian EU Member State were to consult the citizens of the other riparian in the same IRBD would seem rather unrealistic at this point.

2.4.2.6. Monitoring and reporting

145. Adequate monitoring mechanisms are paramount in environmental law.[526] The Directives provide for monitoring and reporting requirements to follow up on the efforts of the Member States to achieve good status of their water bodies.

For surface waters, the monitoring programmes that Member States must set up not only refer to the quality of the water bodies, but also to the volume and the flow level or rate to the extent relevant to the ecological[527] and chemical status.[528] For groundwater bodies, the quantitative status must also be monitored.[529] Member States are required to set up a surface-water monitoring network, in order to provide an overview of the ecological and chemical status within each river basin. Furthermore, the Directive provides for surveillance monitoring programmes and operational monitoring programmes.[530] The former aim to provide information inter alia for the assessment of long-term changes in natural conditions and changes caused by anthropogenic activity. The latter should be set up when, on the basis of impact assessments, bodies of water are identified that are at risk of failure to comply with the environmental objectives. In certain cases,

---

524 Economic and Social Committee, 'Opinion on the Proposal for a Council Directive establishing a framework for Community action in the field of water policy' (1997) OJ C 355/83.

525 See the section on the Antwerp case study in the following report: Hannelore Mees, Cathy Suykens and others, 'Analysing and Evaluating Flood Risk Governance in Belgium Dealing with Flood Risks in an Urbanised and Institutionally Complex Country' (STAR-FLOOD Consortium 2016).

526 Barbara Beijen, Marleen van Rijswick and Helle Tegner Anker, 'The Importance of Monitoring for the Effectiveness of Environmental Directives: A Comparison of Monitoring Obligations in European Environmental Directives' (2014) 10 Utrecht Law Review 126.

527 Or the ecological potential.

528 Article 8(1) WFD.

529 This is because groundwater bodies are measured in terms of their quantitative status.

530 Annex V WFD.

Member States should carry out investigative monitoring, e.g. when it is likely that good status will not be achieved, and to evaluate the impact of accidental pollution. The results from the monitoring exercise must be incorporated in the subsequent planning cycle for the RBMPs and intermediate amendment of the programmes of measures. The adaptability lies in the obligation to adjust the programme of measures on the basis of the monitoring data.[531]

Another procedural provision anchored in the WFD and FD relates to the duty to report on the progress of certain measures. The duty to report is much more clearly present in the WFD than in the FD, e.g. Member States should publish an interim report on the progress in the implementation of the programme of measures included in the RBMPs, three years after publication.[532] No equivalent reporting obligation exists for the implementation of the FD, likely due to the lack of programmes of measures included there.

2.5. PILLAR IV: INSTITUTIONAL MECHANISMS

2.5.1. Scope of application

146. Institutional mechanisms can also be considered as implementation mechanisms.[533] These mechanisms can be described as instruments through which cooperation between States, in this case Member States of the European Union with regard to quantitative water management, can be safeguarded.

Institutional mechanisms in the context of transboundary water management generally refer to the establishment of joint bodies. These may constitute important tools for facilitating coherence in the implementation of the relevant agreements between the parties sharing watercourses, and in light of effective dispute avoidance.[534]

147. The recourse to institutional mechanisms for the governance of transboundary waters has been promoted at the international level for a long time. For example, the Declaration of Madrid of 1911 regarding the Use of International Watercourses for Purposes other than Navigation recommends the appointment of joint commissions for parties sharing transboundary watercourses, where these commissions were to render decisions, or provide their opinion, in those

---

531 Beijen, Van Rijswick and Tegner Anker (n 526) 131.

532 Article 15 WFD.

533 Alistair S. Rieu-Clarke, A Survey of International Law relating to Flood Management: Existing Practices and Future Prospects (2008) International Law and Flood Management 649, 660.

534 Patricia Wouters and others, 'Sharing Transboundary Waters – An Integrated Assessment of Equitable Entitlement: the Legal Assessment Model' (2005) IHP Technical Documents in Hydrology, 25 <http://unesdoc.unesco.org/images/0013/001397/139794e.pdf> accessed 17 July 2017.

cases when the building of new establishments or the making of alterations to existing establishments have serious consequences for the stream in the territory of the other States.[535] Also, the Helsinki Rules of 1966 on the Uses of Waters of International Rivers recommend basin States to refer a question or dispute to a joint agency, which should survey the international drainage basin and formulate plans in order to facilitate the fullest and most efficient use of the waters to the benefit of all the basin States.[536]

This means that it is widely accepted that proper institutional mechanisms are essential in the success of any legal regime governing transboundary waters. The sections below will address the institutional mechanisms embedded in the European legal framework for transboundary water management, which will then be tested in the Scheldt case study in Chapter III. First, the topic and functioning of International River Basin Districts determined on the basis of the WFD and FD will be studied as well as the mediating role of the Commission on the basis of these Directives. Second, Member States' coordination efforts will be analysed as well as the existing international structures on which the Member States rely. Third, the Common Implementation Strategy introduced by the WFD will be analysed and evaluated. Finally, on the basis of the above, conclusions will be drawn on the adequacy of these institutional mechanisms for EU water quantity management.

2.5.2. International River Basin Districts

148. The core of the requirement to cooperate for the implementation of the WFD and FD relates to the very existence of the river basin approach, i.e. the shift of the level of governance from administrative boundaries to ecological boundaries, in itself necessitating cooperative behaviour. The question then arises to what extent the WFD and FD carry this approach through, by establishing the necessary mechanisms and frameworks within which this cooperation can take place effectively.

In the following sections, the constellations of International River Basin Districts will be examined, and what they entail exactly in terms of cooperation requirements and mechanisms ensuing from the Directives.

- *Appropriate administrative arrangements in IRBDs*

149. The EU WFD and FD stipulate that Member States should make the appropriate administrative arrangements for the implementation of the Directives.

---

535 See the 1911 resolution on the international regulations regarding the use of international watercourses. International Regulation regarding the Use of International Watercourses for Purposes other than Navigation (Annuaire de l'Institut de Droit International 1911) Vol. 24, 365-367.

536 Article 31 of the Helsinki Rules on the Uses of the Waters of International Rivers. International Law Association Report of the Fifty-Second Conference (International Law Association, Helsinki 1967).

More specifically, on the basis of Article 3.1 WFD, Member States are to assign the individual river basins within their territory to individual River Basin Districts. Subsequently, competent authorities are to be assigned for the application of the provisions of the WFD within each River Basin District lying in their territory.[537]

150. In terms of appropriate administrative arrangements for the watercourses in their national territory, the Floods Directive largely refers to the provisions of the WFD. However, for the implementation of the Floods Directive, Member States may designate different competent authorities than those designated in the context of Article 3(2) WFD.[538] Moreover, coastal areas or individual river basins may be assigned to a different unit of management than those assigned in the context of Article 3(1) WFD.[539]

151. In terms of management of transboundary watercourses, the EU WFD and FD have not created an overarching body or institution for the management of International River Basin Districts in the territories of Member States. Instead, the Directives have opted to leave the sculpting of administrative arrangements and the designation of competent authorities to the Member States.[540]

– *First step: designation of an International River Basin District*

152. The first step in the process entails that a river basin, which covers the territory of more than one Member State, is to be assigned to an IRBD. The concept of International River Basin District has been explained above, in the context of the first pillar "Scope".

– *Second step: appointment of competent authority*

153. Then, the appropriate competent authority should be assigned for the application of the rules of the Directive within the part of the IRBD in its territory. Member States may either opt to install a new institution or to make an existing national or international body the competent authority.[541] Both in the first and the second step, Member States may request the Commission to act as a facilitator. For IRBDs, Member States are not obligated to designate a body ensuing from existing international agreements as competent authority for the international district in question.[542] Member States are not obligated to designate

537 Article 3(2) WFD.
538 Article 3(2) (a) Floods Directive.
539 Article 3(2) (b) Floods Directive.
540 Article 3(3) WFD.
541 Article 3(6) WFD.
542 In contrast to what was proposed by the European Parliament, see European Parliament, 'Legislative resolution embodying Parliament's opinion on the proposal and the amended

the entity acting at the IRBD level as the competent authority vis-à-vis the EU level. The formal competent authorities stem from the RBD levels.

The provisions with regard to the powers that should be granted to these competent authorities in the context of IRBDs were weakened in the process of drafting the WFD. Whereas the original proposal for the WFD stipulated that Member States should ensure that these competent authorities are granted with sufficient clout to carry out their tasks[543], the final version of the Directive does not provide for such explicit empowerment of the bodies in question. Also, the EP had suggested inserting a provision requiring the Member States to encourage cooperation between the competent authorities to simplify the management of cross-border aspects in IRBDs.[544] This amendment was not maintained in the final version either.

This means that Article 3 remains rather vague on the tasks and competences of these competent authorities. The attributes of these authorities are partly included in Annex I to the Directive.[545]

154. As will be explained below, there are great differences in how Member States have gone about the designation of competent authorities for RBDs and IRBDs, for example, several authorities for one River Basin District, or one authority for several River Basin Districts. Moreover, where the existing authorities for transboundary waters ensuing from international agreements were designated as the competent authority for the IRBD, significant discrepancies exist in terms of powers granted to these authorities in the respective IRBDs. For example, whereas the Rhine Commission to some extent has binding powers regarding the IRBD Rhine, the Scheldt and Meuse Commission hardly have any powers and are restricted to activities such as providing non-binding advice.

– *Third and continuous step: administrative arrangements ensuring coordination in implementation of the Directive*

155. As mentioned above, the WFD stipulates that the requirements of the Directive with regard to the achievement of the environmental objectives included in Article 4 and the programmes of measures are coordinated for the whole of the River Basin District. Moreover, for IRBDs, Member States are required to ensure

proposals for a Council Directive on establishing a framework for Community action in the field of water policy (1999) OJ C 150, Amendment 5.

543 Commission, 'Proposal for a Council Directive establishing a Framework for Community Action in the Field of Water Policy' COM (97) 49 final.

544 See Amendment 123 of the Parliament's opinion on the proposal and legislative proposal of the WFD [1999] OJ C150.

545 The fact that Article 3 does not specify these attributes, but that they are only limitedly mentioned in Annex, has been criticized by the European Parliament. Parliament, Report on the Proposal and the Amended Proposals for a Council Directive on Establishing a Framework for Community Action in the Field of Water Policy (1998) A4-0261/98.

this coordination together. The general obligation for Member States to cooperate is included in Article 3(4) of the Water Framework Directive.[546] The requirement for a coordinated implementation of these relevant provisions included in the WFD pursuant to Article 3(4) WFD can be viewed as an obligation of result.[547] It should be noted that the Floods Directive does not refer to Article 3(4) WFD, and does not establish such coordination requirement equivalent to the one included in this article. This can be explained by the fact that the Floods Directive does not establish specific objectives equivalent to those included in Article 4 WFD.[548]

To this end, Member States may make use of existing structures pursuant to international law, i.e. structures ensuing from international agreements.[549] Whilst the European Parliament favoured explicitly prioritising this recourse to existing structures from international agreements to coordinate efforts in IRBDs, the choice as to whether the Member States make use of existing international agreements is left entirely up to them, for concerns that such prioritisation would breach the subsidiarity principle.[550]

156. Some authors have reflected on the question whether it was the intention of the European legislator to exclude the possibility for Member States to make use of international structures for the realisation of the provisions of the Floods Directive, taking into account that the Floods Directive has not adopted the coordination mechanism included in Article 3(4) WFD.[551] This consideration will not be further examined in this book, taking into account (i) the explicit wish of the European legislator to integrate and synchronise the implementation efforts of the Member States of the WFD and FD respectively, in conjunction with (ii) the fact that coordination is necessary in light of the implementation of the procedural requirements.

– *A Trans-European Water Management Plan?*

157. In the run-up to the adoption of the WFD, the possibility of providing a European Water Plan was considered. This option was especially promoted by the Committee on Research, Technological Development and Energy, which argued

546 Herman-Kasper Gilissen, Jasper van Kempen and Marleen van Rijswick, 'The Need for International and Regional Transboundary Cooperation in European River Basin Management as a Result of New Approaches in EC Water Law' (2010) 11 ERA Forum 129, 131.

547 Ellen Hey and Marleen van Rijswick, 'Transnational Water Management' in Oswald Jansen and Bettina Schöndorf-Haubold (eds), *The European Composite Administration* (Intersentia 2010) 231, 243.

548 Herman-Kasper Gilissen, *Internationale en Regionaal Grensoverschrijdende Samenwerking in het Waterbeheer* (Sdu Uitgevers 2009) 21.

549 Article 3(4) WFD.

550 Commission, 'Opinion on the European Parliament's Amendments to the Council's Common Position regarding the Proposal for a European Parliament and Council Directive establishing a Framework for Community Action in the Field of Water Policy' COM (2000) 219 final.

551 Gilissen (n 548) 21.

for the importance of studying the viability of infrastructure projects with the goal of durable joint management through a European Water Plan.[552]

In essence, the Committee criticised the parochial character of the Directive, where it provides for the coordination of measures through River Basin Districts, and does not refer to the integrated management of water between these Districts.

To this end, the Parliament had proposed the following addition to Article 3 WFD, which pertains to the coordination of administrative arrangements within River Basin Districts: "Various International River Basin Districts may be grouped together in order to carry out integrated action and joint physical planning at European level designed to draw up a trans- European water management plan".[553] This amendment, however, was not maintained in the final version of the Directive.

- *IRBDs extending beyond the EU*

158. The WFD further provides that for the RBDs extending beyond the territory of the European Union, the Member State(s) in question "shall endeavour to establish appropriate coordination with the relevant non-Member States, with the aim of achieving the objectives of this Directive throughout the River Basin District".[554] Once again, the Floods Directive does not include this provision of the WFD nor does it include a similar one for the application of the requirements of the Directive in IRBDs extending beyond the territory of the EU.

Needless to say, the power of the EU to legislate with regard to IRBDs insofar as third, non-EU, countries are concerned, is limited.

- *The institutional constellations for IRBD: Peculiar dynamics*

159. The situation brought about by the Directives, namely the fact that the European framework does not provide for specific cooperation mechanisms for the joint implementation of the Directives, and the default recourse to international agreements and mechanisms for this implementation, does have consequences that might not necessarily be constructive for the coordinated efforts of Member States in this regard. States in general have a broad margin of freedom to act and decide in the context of the implementation of international agreements, in comparison with the leeway bestowed on them in the context of the implementation of EU law.[555] However, this is in contrast with the exclusive

---

552 Committee on the Environment, Public Health and Consumer Protection, 'Report on the proposal and the amended proposals for a Council Directive on establishing a framework for Community action in the field of water policy' (1998, A4-0261/98).

553 Ibid, Amendment 36.

554 Article 3(5) WFD.

555 Simone Borg, 'The Influence of International Case Law on Aspects of International Law Relating to the Conservation of Living Marine Resources beyond National Jurisdiction' (2012) 23 Yearbook of International Environmental Law 44; Hey and Van Rijswick (n 435) 243.

jurisdiction of the European Court of Justice over all disputes between Member States on the basis of Article 344 TFEU. Moreover, national legislation that is in violation with European law can be declared inapplicable by the European or national judiciaries. This forces Member States into a situation where they apply international agreements, but, in terms of the implementation of the relevant Directives, are not granted the amount of discretion that usually goes with the implementation of international agreements.

2.5.2.1. Member States' coordination efforts

160. Chapter III will zoom in on Member States' coordination efforts with respect to the River Scheldt. It is relevant to describe, more generally, the coordination arrangements Member States have made with respect to the adoption of the relevant plans and programmes. In the context of the first generation of RBMPs, 128 River Basin Districts (RBDs) were designated in the European Union. Of these 128 RBDs, 49 RBDs are international.[556] For these IRBDs, most Member States have produced national RMBPs for the part of the international RBD in their territories, amounting to 63 plans. Certain Member States have produced joint international RBMPs for the entire international RBD.[557]

The Commission also reported that certain countries did not adopt RBMPs per River Basin District, but did adopt them at the level of the units according to the national institutional situation. In this regard, the Commission stated, "there is no one-to-one relationship between the number of RBDs and the number of RBMPs reported".[558] For example, Germany adopted 16 RBMPs, at the level of the Federal State, but the Commission assessed these plans as one RBMP per RBD. In Belgium, the drafting of these plans falls under the competence of the three Regions, i.e. the Walloon Region, the Brussels Capital Region and the Flemish Region, and is therefore not based on the RBD at national level. The Commission also mentioned the fact that the three Regions adopted different timetables for the implementation of the Directive.[559]

With regard to the designation of competent authorities, on the basis of the WFD, the Member States were required to report to the Commission which authorities they had designated in 2004. At the time of evaluation by the Commission, most Member States had not adapted their existing structures in terms of competent authorities to facilitate the implementation of the Directive.

556 Commission, 'Report from the Commission to the European Parliament and the Council on the Implementation of the Water Framework Directive (2000/60/EC) River Basin Management Plans', COM(2012) 670 final, 9.

557 Ibid.

558 Ibid 31. The Commission further states that, counting each national part of an international RBD separately, the total number of RBDs is 170.

559 Ibid. N.B. the submission of the Walloon RBMPs and those of the Brussels Capital Region were substantially delayed.

For example, in several Member States, there are different authorities for surface waters and groundwater, several competent authorities for one IRBD or one competent authority for several IRBDs.

Only 23 of 112 RBMPs stipulate that one single authority is responsible for implementing all provisions and requirements included in the WFD. However, the Commission noted that in most Member States the responsibilities for the different categories of watercourses were not split up, which is considered a positive thing in light of the integration of the management of these categories.[560]

2.5.2.2. The Common Implementation Strategy

161. The most apparent mechanism that the European Commission developed to facilitate a coordinated implementation is the Common Implementation Strategy (CIS). The CIS has been referred to in the context of the discussion of the pillar on procedural provisions, and its mechanism is further discussed below.

The CIS is an informal cooperation platform set up for the implementation of the WFD. It is primarily led by the Water Directors, who are the representatives of the Water States with responsibility on water policy, and the Commission. The CIS represents a possibility for Member States to share best practices and information in a structured and systematic, although very informal, way. Indeed, the relevant Directives do not refer to the CIS as an instrument for cooperation or implementation.

The CIS revolves around three levels of organisation:

(a) The Technical Groups, which are divided into topic-based categories.
(b) The Strategic Coordination Group, which coordinates the different working groups.
(c) The Water Directors, as mentioned above the representatives of the Member States with competences in the water sector, who take final responsibility for the deliverables published.[561]

The CIS can be considered as forming part of an already existing instrument existing in the European order, namely the Open Method of Coordination (The OMC), which was introduced in 2000 at the Lisbon Summit.[562] The OMC has been sculpted as a tool for fixing guidelines for the achievement of goals in the short term and the long term, establishing quantitative and qualitative indicators

---

560 Commission, 'Report from the Commission to the European Parliament and the Council on the Implementation of the Water Framework Directive (2000/60/EC) River Basin Management Plans', COM(2012) 670 final, 34.

561 Elisa Morgera, 'Water Management and Protection in the EU' in M Alberton and F Palermo (eds), *Environmental Protection in Multi-Layered Systems: Comparative Lessons from the Water Sector* (Martinus Nijhoff 2012) 265, 279.

562 Sabrina Regent, 'The Open Method of Coordination: A New Supranational Form of Governance?' (2003) 9 European Law Journal 190.

and benchmarks to compare best practices, to set out measures and targets to implement European guidelines, and to monitor the progress on setting up mutual learning processes between Member States.[563]

162. Although generally seen as a helpful instrument in coordinating compliance with the Directives, the CIS has been criticised in doctrine. Various legal scholars have expressed concerns pertaining to the role and the effectiveness of the CIS.[564] It has been argued that the CIS may undermine the implementation of the relevant Directives due to, amongst others, its consensual nature, resulting in the lowest common denominator phenomenon.[565] Indeed, the very consensual nature of the CIS process has in some cases led to compromises as to what should be included in the final version of one of the Guidance Documents, leaving little of the intention of the guidance papers: providing and explaining best practices.[566] This is especially the case when it comes to quantitative water management, where some Member States assume a high degree of sovereignty. Moreover, the accountability process in the context of the CIS has been questioned, as it hardly seems guaranteed that the Water Directors[567] can keep a complete overview of all the issues reflected in the CIS.[568] Finally, it has been stated that the CIS might detract from public participation, as documentation ensuing from CIS discussions, which reflect "expert participation", might not be very accessible to the general public.[569]

2.6. PILLAR V: DISPUTE SETTLEMENT AND ENFORCEMENT

163. The fifth and final pillar of the legal regime governing transboundary waters relates to enforcement and dispute settlement mechanisms, in case the

---

563 Parallel to the organisation of the Common Implementation Strategy, other fora have been developed, namely Expert Advisory Forums. These fora are shaped as multi-stakeholder platforms involving the European Commission, Member States, NGOs, industrial representatives, etc. Joanne Scott and Jane Holder, 'Law and New Environmental Governance in the European Union' in G De Burca and Joanne Scott (eds), *Law and New Governance in the EU and the US* (Hart Publishing 2006). Joanne Scott, *Environmental Protection* (Oxford University Press 2009).

564 Joanne Scott, *Environmental Protection* (Oxford University Press 2009) 48.

565 Ibid 36.

566 European Environmental Bureau, 'Tips and Tricks for Water Framework Directive Implementation' (2004) <www.rivernet.org/general/docs/200403_EEB_WWF_Tips&Tricks.pdf> accessed 20 October 2014.

567 The representatives of the Member States.

568 Scott and Holder (n 563) 47.

569 Elisa Morgera, 'Water Management and Protection in the EU' in M Alberton and F Palermo (eds), *Environmental Protection in Multi-Layered Systems: Comparative Lessons from the Water Sector* (Martinus Nijhoff 2012) 265, 283.

dispute avoidance instruments embodied by the second, third and fourth pillars fail. This pillar relates to the aspect of enforcement of the provisions of the relevant Directives in the context of IRBDs.

164. As mentioned above, Member States should cooperate to ensure that the obligations to achieve the environmental objectives and the programmes of measures are coordinated throughout the District. However, the International River Basin District has not been included as an autonomous unit of governance. The classical EU responsibility mechanism applies: Member States should individually comply with the Directives for their part of the territory, i.e. the part of the IRDB situated in their territories.

Generally speaking, enforcement of environmental directives is difficult, because of the discretion granted to Member States. The objective parameter is whether the Directive has been transposed into national law. This is especially so with regard to the Floods Directive, which does not stipulate any substantive obligations. When it comes to transboundary cooperation in IRBDs, enforcement will be even more difficult, as the cooperation requirements promulgated by the WFD and FD are easily circumvented.[570]

The general enforceability of provisions strongly depends on the question whether the provisions can be considered as obligations of result or of best effort. Specifically in terms of cooperation requirements, it was already described above that there is a limited number of solidly enforceable provisions. One of the provisions that can be considered as an obligation of result is Article 3(4) WFD, which includes the duty for Member States sharing an IRBD to achieve a coordinated implementation of the environmental objectives included in Article 4, and in particular all programmes of measures established to comply with the Directive.[571] However, the FD does not provide for a similar duty to coordinate as included by Article 3(4) WFD, mainly because the FD does set specific objectives similar to those included in Article 4 WFD. The compliance with the procedural requirements in terms of cooperation, however, is very weak. As mentioned above, if Member States "fail to produce" one single River Basin Management Plan or Flood Risk Management Plan, it is sufficient for them to adopt plans for the part of the IRBD located in their territories. Needless to say, enforcement of cooperation in this regard is inevitably set to be equally weak.

2.6.1. The Commission as mediator

165. In addition to the traditional role of the European Commission as guardian of the Treaty, the WFD and Floods Directive also gives the Commission the

570 E.g. with regard to drafting international RBMPs and FRMPs.

571 Hey and Van Rijswick (n 435) 243.

possibility to act as mediator between the Member States in the implementation of the Directives.[572] However, it should be noted that this legal basis does not grant the Commission a binding decision-making power with regard to the issue at hand.[573]

166. As for the procedure to be followed in the context of cooperative efforts, the WFD and FD grant an explicit role to the European Commission in facilitating cooperation between Member States. Indeed, the Commission may assign water bodies to river basins, when Member States make such a request.[574] Furthermore, the Commission, again at the request of the Member States involved, may facilitate the establishment of the programmes of measures in the context of International River Basin Districts.[575]

167. Finally, the Directives have also introduced a mediation mechanism for transboundary issues that might arise in the course of their implementation. Indeed, whenever issues arise in one Member State that may influence the situation in another Member State, the former has the option, not the obligation, to report the issue to the EU Commission. The Commission may then make recommendations for the resolution of the issue.[576] This aspect of intervention by the Commission is the most relevant to this fifth pillar.

168. This Commission intervention included in the WFD is not restricted to the qualitative aspects of water management; any "issue which has an impact on the management of the water by a Member State" can be submitted to the Commission for mediation. The Floods Directive has equally included this mediation procedure 'starring' the Commission.[577]

This mediation mechanism can be relevant in those cases where a Member State in an IRBD is aware of a planned project in another Member State in that District, which might influence the attainment by the former of its obligations under the Directives. Indeed, formally, a Member State may not refer to a certain action or lack thereof by another Member State on the one hand or the failure of cooperation on the other hand to justify non-compliance with the Directives. The Directives do not take into account the consequences of the failure of cooperation in terms of attainment of their objectives.[578]

---

572 On the basis of Article 12 WFD and Article 8(5) FD.

573 Olivia O Green and others, 'EU Water Governance: Striking the Right Balance between Regulatory Flexibility and Enforcement?' (2013) 18 Ecology and Society 10. Andrea Keessen and Marleen van Rijswick, 'Adaptation to Climate Change in European Water Law and Policy' (2012) 8 Utrecht Law Review 38, 48.

574 Article 3 WFD.

575 Article 3(4) WFD.

576 Article 12(1) WFD.

577 Article 8(5) FD.

578 Gilissen, Van Kempen and Van Rijswick (n 546) 137.

The Directives do not provide for a mechanism or guidance on the procedure to be followed other than stipulating that the Commission should respond to this application within a period of six months.[579] Considering the fact that this mechanism is completely voluntary for the Member States, the Commission lacks the legal basis to settle the disagreement in a binding manner.[580] Of course, the lack of explicit legal status of the reply of the Commission does not equal a lack of authority, as the Commission has the power to initiate infringement proceedings vis-à-vis the Member States in question at a later stage.[581]

169. A solid and clear dispute settlement mechanism for potential conflicts in transboundary waters is of paramount importance. As for International River Basin Districts, several ways forward are imaginable. For example, the way forward for mediation in IRBDs could be part of a broader development toward a comprehensive EU legal framework for ADR in public-law matters.[582] Another possibility is that mediation proceedings would form part of the specific agreement or pact governing the IRBD. In this regard, the river commission would play an important role. Indeed, the question arises which entity would be best suited as a neutral party to tackle disputes or issues arising between the States sharing the international district: the EU Commission or the entity designated as the competent authority vis-à-vis the specific basin.

2.6.2. The exclusive jurisdiction of the CJEU: a challenge

170. The multi-level governance reality is particularly challenging in the context of dispute resolution. As mentioned above in the section "institutional mechanisms", there is a certain tension between the international legal landscape related to transboundary waters and the European one. Indeed, in the absence of specific European instruments for cooperation in these types of waters, Member States primarily rely on international structures and mechanisms. This constellation not only has an impact on the relationship between Member States in pursuing joint implementation of the relevant Directives, but also on the resolution of conflicts when cooperation fails. In many multilateral agreements governing transboundary waters construed on the basis of international law, States have provided for dispute-resolution mechanisms. However, the

579 Article 12(2) WFD and Article 8(5) FD.
580 Green and others (n 573).
581 Andrea Keessen, Jasper Van Kempen and Marleen van Rijswick, 'Transboundary River Basin Management in Europe Legal Instruments to Comply with European Water Management Obligations in Case of Transboundary Water Pollution and Floods' (2008) 4 Utrecht Law Review 35, 46.
582 Whereas the Mediation Directive 2008/52/EC currently expressly excludes public-law matters from its scope of application.

European Court of Justice has established exclusive jurisdiction over disputes between EU Member States with regard to the interpretation or application of EU law.[583] Therefore, whilst Member States make use of international agreements for the implementation of the WFD and the FD, as stipulated by these Directives, conflicts arising between them should be submitted to the CJEU, and not solved through the dispute-resolution mechanisms included in their respective agreements or through other mechanisms ensuing from international law.

171. In the context of the pillar of dispute settlement, differentiation should be made between transboundary waters shared exclusively by EU Member States, and such waters shared between Member States and non-Member States, for example the Rhine. In the latter case, the issue of concurrent jurisdiction is moot, as the CJEU does not have competence to settle those disputes.

172. The growing problem of concurrent jurisdiction of the European Court of Justice and the international tribunals and arbitration bodies has been widely discussed in international doctrine. In general, a tendency can be discerned where the CJEU increasingly tackles disputes touching on international law.[584]

173. Taking into account the specifics of the situation, namely a joint implementation by EU Member States of European secondary law, through international structures primarily governed by international law, theoretically speaking, disputes in this context may be adjudicated both by international courts and tribunals and by the CJEU. However, a central issue in the debate with regard to judiciary powers related to transboundary waters in the European Union, is Article 344 TFEU, which states "Member States undertake not to submit a dispute concerning the interpretation or application of the Treaties to any method of settlement other than those provided for therein".[585]

The Court has confirmed this principle by considering that to confer jurisdiction with regard to the interpretation or the application of the Treaty on another court, would be likely to adversely affect the allocation of responsibilities defined in the Treaties and the autonomy of the legal order of the European Union.[586] This provision relates to the primacy of EU law. Relevant is Declaration

---

583 Green and others (n 573).

584 Nikolaos Lavranos, 'The MOX Plant and IJzeren Rijn Disputes: Which Court Is the Supreme Arbiter?' (2006) 19 Leiden Journal of International Law 223. *Iron Rhine Arbitration, Belgium v Netherlands*, Award, ICGJ 373 (PCA 2005). Case C-459/03, *Commission of the European Communities v Ireland* [2006] ECLI:EU:C:2006:345.

585 Article 344 TFEU.

586 Nikolaos Lavranos, 'Freedom of member states to bring disputes before another court or tribunal: Ireland condemned for bringing the MOX plant dispute before an arbitral tribunal. Grand Chamber decision of 30 May 2006, Case C-459/03, Commission v. Ireland' (2006) European Constitutional Law Review 456, 456.

17 concerning primacy, which in turn refers to the Opinion of the Council Legal Service on the primacy of EU law. The Council Legal Service firstly stipulated that the primacy of EU law (then EC law) is a cornerstone principle of the European legal framework. This followed from established case law, e.g. *Costa/Enel*.[587] Therefore, on the basis of Article 344 TFEU, disputes between Member States related to the interpretation or implementation of EU law, must be exclusively tackled by the CJEU. In several cases, beyond the scope of governance of transboundary waters, Member States have brought disputes related to European law to international courts and tribunals.

174. On the other hand, in the IJzeren Rijn case, which concerned a dispute between Belgium and the Netherlands about the division of costs for the reopening of a railway line, the arbitration tribunal did not consider EU law as an obstacle for adjudication, even though EU-law aspects came into play.[588]

175. It is clear that the primacy of EU law over national law supports the exclusive jurisdiction of the EU. However, it can be questioned to what extent this primacy applies to disputes in the international water law sphere. Indeed, the TFEU and preceding treaties have been created by Treaties and are therefore originally based on international law. Moreover, the Water Framework Directive specifically aims to implement the UNECE Water Convention. However, considering the *sui generis* character of EU law, with its own standards and hierarchical norms, the EU legal order can be considered as having supremacy over international law, to the extent that it applies to Member States.[589] Indeed, the UNECE Water Convention is a mixed Treaty, in the sense that both the EU and the Member States are parties to the Convention. The Treaty is therefore part of the EU legal order[590], resulting in exclusive jurisdiction of the Court of Justice.[591] However, International adjudicating bodies are free to disregard assumptions of exclusive jurisdiction put forward by the CJEU.[592]

176. As to the initiation of proceedings before the CJEU, the initiative can either be taken by the Member States in question or by the Commission. The former option has a higher threshold, due to the political sensitivities related to one Member State initiating proceedings before the Court against another Member State.

---

587 Declaration 17 concerning primacy. Case C-6-64, *Flaminio Costa v E.N.E.L.* [1964] ECLI:EU:C:1964:66.

588 Nikolaos Lavranos, 'The MOX Plant and IJzeren Rijn Disputes: Which Court is the Supreme Arbiter' (2006) 19 Leiden Journal of International Law 221, 232.

589 Ibid 223.

590 See Chapter I.

591 Keessen, Van Kempen and Van Rijswick (n 581) 45.

592 E.g. OSPAR, Arbitral Tribunal, FinalAward, MOX Plant (2003) available atwww.pca-cpa.org.

2.6.3. Between mediation and exclusive jurisdiction: No middle ground

177. Enforcement and dispute resolution are especially challenging in the context of International River Basin Districts. First, there is clear tension between on the one hand, the exclusive jurisdiction of the CJEU and on the other hand, the fact that the Directives provide that Member States may resort to existing international mechanisms for the implementation of their provisions. Second, and as explained in pillars two, three and four, the exact obligations with regard to transboundary cooperation and coordination in IRBDs remain vague and weak. Therefore, as a result, enforcement is vague and weak as well. Third, it seems that the classical EU instrument for dispute resolution, i.e. the recourse to the CJEU proceedings, does not fit the shoe. Indeed, considering the fact that the main unit of governance is based on ecological boundaries as opposed to the traditional administrative, nationalistic borders, the threshold for dispute resolution should be set lower. On the other hand, the mediation procedure included in the WFD and the FD seems too open-ended and the legal status of the outcome of the procedure is not clear at all. The only certainty on the basis of the Directives is that the Commission should issue its reply within six months following application.

3. CONCLUSION

178. This Chapter has conducted a 360° analysis of the EU legal framework for water quantity management in International River Basin Districts. The two aspects most prominent in this analysis have been the red threads throughout the Chapter, i.e. (a) institutional mechanisms for cooperation in these transboundary waters and (b) the position of water quantity in the EU legal landscape as well as its relationship to water quality. The "five pillar"-research has been valuable in structuring this comprehensive analysis.

179. Table 4 provides a concise overview of these five pillars as applied to the water quantity management in the European Union.

Table 4. Pillars of the EU Legal Framework for Transboundary Water Management

The five pillars of the legal framework for transboundary waters applied to EU water quantity management		
Pillars & Directives	*Water Framework Directive*	*Floods Directive*
1. Scope	- River Basin Districts[593] - Surface waters - Associated groundwater - Aquifers - International RBDs ≠ unit of governance - Transboundary confined aquifers? - Water quality - Water quantity: accessory	- River Basin Districts - Surface waters - Associated groundwater - Aquifers - International RBDs ≠ unit of governance - Transboundary confined aquifers? - Floods - Groundwater included - Possibility to exclude sewerage
2. Substantive requirements	- Good quantitative status groundwater - Cost recovery - Programme of measures (not at IRBD level)	- Member States should set objectives (own discretion)
3. Procedural requirements	- General duty to cooperate - Obligation of joint implementation - RBMPs: no enforceable duty to produce joint plan - No explicit cross-border public participation - Cross-border notification in the context of the SEA and EIA Directives	- General duty to cooperate - No explicit obligation of joint implementation - FRMPs: no enforceable duty to produce joint plan - No cross-border public participation - Cross-border notification in the context of the SEA and EIA Directives
4. Institutional mechanisms	- Common Implementation Strategy - Non-binding recourse to EC - Existing international treaties	- Common Implementation Strategy - Non-binding recourse to EC - Existing international treaties
5. Dispute settlement	- CJEU - Mediation procedure on the basis of TFEU - Non-binding recourse to the Commission	- CJEU - Mediation procedure on the basis of TFEU - Non-binding recourse to the Commission

180. The most important bottleneck in transboundary water management in the EU on the basis of the Water Framework Directive and Floods Directive is found in a combination of the first pillar, Scope, and the fourth pillar, Institutional Mechanisms. More specifically, the International River Basin District is not defined as unit of governance. This is problematic, as the unit of governance forms the basis for the compliance with all subsequent substantive and procedural obligations defined by the Directives. As the provisions with regard

593 Including land systems associated with water, see *infra* for definitions.

to the "Scope" are currently construed, the RBDs and the "parts of the IRBDs located on the territories of the Member States" are loosely tied together through vague cooperation requirements. These requirements are purely procedural and – in the end – optional. The latter adjective is used consciously, as it is clear from the interpretation of the Directives that a lack of promulgating joint River Basin Management Plans and Flood Risk Management Plans for the entire IRBD does not have legal repercussions for the Member States in question. If Member States do not reach an agreement when trying to come up with an international RBMP or FRMP, it is sufficient for them to draft a plan for the part of the IRBD in their territories.

Related to this bottleneck is that there is a certain lack of clarity in many of the provisions in the Directives on some important questions. The legal value of cooperation and coordination requirements for the implementation of the Directives is not always clear. As explained above, it is not clear to what extent legal grounds can be found in the Directives to demand transboundary public participation from the Member States. The purposive approach to interpretation of the relevant public participation may warrant it, but the consideration that IRBDs are not defined as a unit of governance makes it a questionable mission. The same vagueness exists with regard to notification requirements, which are anchored in international customary law for transboundary waters, but not explicitly defined by the Directives, although this vagueness is partly cleared up and covered by the EIA Directive, as several measures falling in the realm of quantitative water management are covered by either Annex I or Annex I of this Directive.

181. The second issue also relates to Scope, and is the discrepancy in legal basis between water quality management and water quantity management in the TFEU framework. From a conceptual point of view, it is hard to imagine that integrated water resources management will be reached at some point in the European Union, when the quantitative aspects of water management are subject to the unanimity procedure and the qualitative aspects to the qualified majority procedure. Indeed, this is an artificial distinction. For surface water, substantive obligations for quantitative water management only exist to the extent necessary to achieve quality-related objectives and thus merely in an ancillary manner. The unanimity requirement has proven problematic in the run-up toward the adoption of the WFD, as the Commission had to thread carefully in order to keep the focus on water quality management. Environmental flow requirements only partly and implicitly form part of the analysis of the status of the surface-water bodies. For groundwater, the good quantitative status does form part of the requirement of the overall good status for groundwater bodies. There is no explicit legal framework yet for water scarcity and droughts. Although the Floods Directive deals with the issue of an abundance of water, and therefore rather the quantitative aspects of water management than the qualitative aspects, it has also circumvented Article 192(2) TFEU. It can be argued that because of the fact that the Floods Directive only focusses on procedural requirements, it does not substantively influence the territories of the Member States. By extension, the cooperation requirements embedded in the Floods Directive are even less prominent than in the WFD. For example, Article 3(4) WFD provides for an obligation of result for Member States sharing an IRBD to ensure coordination in the achievement of the environmental objectives of Article 4 of the Directive as well as for the drafting of the programmes of measures in IRBDs. The FD does not contain a similar obligation. A comprehensive approach to water management would require re-evaluating the applicable decision-making procedures with the goal of levelling the playing field in terms of decision-making procedures.

182. The third issue is that the fifth pillar is drenched in tensions between the EU level and the international level. Indeed, the Directives provide for utilisation of existing mechanisms, which have been created on the basis of international law. However, the tension lies in the exclusive jurisdiction of the CJEU over issues between the Member States concerning the implementation of the Directives. The problem might lie in the fact that the EU Directives only include a limited dispute-resolution mechanism, as the procedure included in Article 12 WFD merely stipulates that the Commission should provide feedback within six months, but that the mechanisms included in the existing international river basin mechanisms would be void if viewed in conjunction with the competence sphere of the Court of Justice. Indeed, proper dispute resolution mechanisms defined in agreements between EU Member States sharing International River Basin Districts would likely collide with the exclusive jurisdiction of the CJEU on the basis of Article 344 TFEU. There is no middle ground between the mediation procedure and the exclusive jurisdiction reality of the Court of Justice.

183. The following Chapter describes a case study of flood risk management in the Scheldt IRBD. This case study aims to shed light on how the EU cooperation requirements have been implemented at the international district level, both from the viewpoint of bottom-up implementation of cooperation provisions in the national legal frameworks, and from the viewpoint of the organisation of joint-governance mechanisms at the district level, i.e. the 2002 Scheldt Agreement and the International Scheldt Commission. This Chapter will also evaluate the resilience of the Scheldt river management regime and the manner in which the EU legal framework facilitates cooperation regarding this river.

CHAPTER III

'COME HELL AND HIGH WATER' – THE LAW OF THE SCHELDT DISTRICT

1. INTRODUCTION AND APPROACH

1.1. GENERAL

184. As explained in Chapter II, the Water Framework Directive and Floods Directive have steered requirements with regard to water management applicable to Member States along hydrological boundaries. Member States sharing the International River Basin District of the Scheldt are thus subject to these requirements. Chapter II, however, also revealed vulnerabilities in the legal framework embodying this river basin management approach.

This Chapter will examine the institutional mechanism for water management in the Scheldt, which supplements the analysis carried out in Chapter II. The feedback loop in this regard will be explained in the section 1.2 below. The knowledge gained through the "five pillars"-structure adopted in the preceding Chapter is highly relevant to the data presented in this Chapter III. Chapter II has mapped the substantive and procedural obligations applicable to Member States, and indicated which of these are a matter of best effort and a matter of result. The Chapter has shown the "ins and outs" of the institutional mechanisms for transboundary water management, which avail themselves to the Member States. A full picture emerges through a combination of theory and practice, which is why the legal knowledge gained in Chapter II needs to be supplemented by an analysis of how these requirements have been translated into the reality of a specific river basin.

For the sake of analysis, this Chapter will tease out one aspect of the umbrella concept of water quantity management, i.e. flood risk management, for reasons explained in detail in section 1.3.

1.2. APPROACH TO FINDING THE SCHELDT PUZZLE PIECES

185. The International River Basin District Scheldt (IRBD Scheldt) is defined as the area of land and sea, designated by the Parties to the 2002 Scheldt Treaty in accordance with the Water Framework Directive, which consists of the river basin of the Scheldt River, the related basins and the related groundwaters and coastal waters.[594]

Figure 6. International River Basin District Scheldt

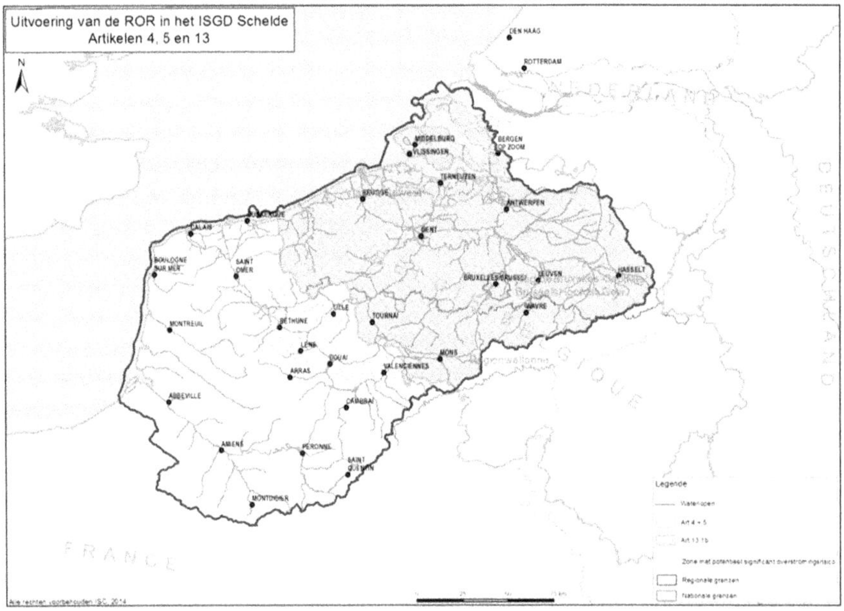

The IRBD Scheldt has a surface of approximately 36,500 km^2, and comprises France, Belgium (the Flemish Region, Walloon Region and Brussels Capital Region) and the Netherlands. The largest part of the Scheldt River, as pictured, runs through the Flemish Region, 32.9%, and the Walloon Region, 10.3%, in Belgium, followed by France and the Netherlands. Only a marginal proportion of the river flows through the Brussels Capital Region.[595]

The IRBD Scheldt is thus shared between these EU Member States, which use the relevant legal instrument based on international law, i.e. the Scheldt Treaty and its International Scheldt Commission, to implement the requirements of the

594 Article 1(d) of the International Scheldt Treaty (2002 Scheldt Treaty) (3 December 2002) <www.isc-cie.org/images/Documents/ACC_GENT_SCHELDT_AGREEMENT.pdf> accessed 12 July 2017.

595 More specifically, 0.4%.

WFD and FD. The Scheldt River is entangled in a web of legal and governance echelons, as depicted in figure 6. The main focus in this chapter is on the national and the regional/local dimensions in relation to the international level.

186. This chapter is structured around several components, pictured in figure 6. These will be described sequentially with the goal of cumulatively contributing to the disentanglement of the complicated web of actors, legal instruments, layers of governance and cooperation mechanisms that are in place in the IRBD Scheldt.

187. This chapter aims to realise this disentanglement by taking the following steps:

(a) The legal landscape *and toolbox for flood risk management* in the different jurisdictions of the IRBD Scheldt will be described, with the goal of pointing out the main similarities and differences between these jurisdictions and providing insight on the coherence in the implementation of the Floods Directive in the District.[596] This national-level analysis is useful to pinpoint and provide examples of scenarios where cooperation in the IRBD could have added value. Moreover, in contrast to the United States Delaware case presented in Chapter IV, the actual implementation of EU requirements for water management, in casu, flood risk management, in the individual States is necessary. Indeed, the focus of this book is on the EU legal framework and the manner in which it steers Member States toward good governance in International River Basin Districts. In order to formulate valuable recommendations, the national-level implementation needs to be looked at on top of the basin-level mechanism.

(b) Once the various aspects of the national legal frameworks related to flood risk management have been described, the research will move onto the question *which hydrological scales exist* in the three Scheldt countries. This is necessary because the IRBD Scheldt in its totality consists of various sub-basins and sub-sub-basins crossing the borders of the jurisdictions.

(c) The question then arises *which competent authorities* are active at these scales. Are they functionally decentralised? Or do they stem from existing administrations? Do they have legal personality and decision-making power? This is a pivotal element in identifying regional cooperation opportunities and inter-scalar links with the broader, more international scale of the IRBD.

(d) This fourth step will specifically address *cooperation* on water quantity management at all relevant levels and aims to provide answers to the question

596 The Brussels-Capital Region does not form part of the national-level analysis, as only 0.4% of the Scheldt runs through its territory. Moreover, the national analyses of the Walloon Region, Flemish Region, France and the Netherlands sufficiently demonstrate the complex multi-level governance challenges existing in the basin. Adding the Brussels-Capital Region, and its rather specific situation, to the analysis would detract from the readibility of this case study, taken into account the already overly complex interplay of actors and rules.

how the duty to cooperate stemming from international and European water law has been operationalised in the Scheldt River.[597] This section will move beyond flood risk management, as cooperation mechanisms are relevant to water (quantity) management in all its facets. It will be analysed how the national, regional and international echelons promote cooperation in the IRBD and to what extent this cooperation is coherent with the hydrological scale narrative promoted by the EU.

i. The national dimension will examine how the "duty to cooperate" is enshrined in the legal frameworks of the Flemish and Walloon Regions in Belgium, France and the Netherlands.
ii. The regional level of cooperation relates to the available EU, bilateral and transnational instruments for cooperation between the stakeholders in the sub-basins and sub-sub-basins of the Scheldt River and the analysis of actual cooperation practice.
iii. The international level of cooperation entails an impact assessment of the mechanism at the level of the International River Basin District, i.e. the Scheldt Treaty and the International Scheldt Commission. How is the international mechanism, which is used to implement the WFD and FD, 'sculpted' and does it promote actual river basin management?

188. This chapter takes a "bottom to top" approach, by starting from the national perspectives, through the regional, to the international. This sequence is justified because of the fact that water resources in the Scheldt District are mainly governed in the specific Member States in question, with (spoiler alert) only a marginal role for the entity active at IRBD level. Because of the more limited role of this entity on the one hand, and the relevance of the analysis of regional cooperation mechanisms, on the other hand, for the normative recommendations of Chapter VI, it is necessary to first understand the national legal frameworks before looking at the international level.

189. The countries sharing the Scheldt IRBD differ in the way their legal and administrative framework is structured, and where the power centres are situated.[598] France is a centralised state, where local authorities play an increasingly significant role in water management.[599] The Netherlands is a decentralised unitary state, where functional decentralisation is quintessential in the water

597 See Chapter I for an explanation of the international level, and Chapter II for an explanation of this principle at EU level.

598 Willemijn Van Doorn-Hoekveld, 'Transboundary Flood Risk Management Management: Compatibilities of the Legal Systems of Flood Risk Management in the Netherlands, Flanders and France' (2017) 26 European Energy and Environmental Law Review 81.

599 Corrine Larrue and others, 'Analysing and Evaluating Flood Risk Governance in France: From State Policy to Local Strategies' (STAR-FLOOD Consortium 2016) 13.

management sphere.[600] Belgium is a federal country, where the government at state level only has limited competences for water management, and where the three Regions, i.e. the Flemish Region, the Walloon Region and the Brussels-Capital Region have developed their own water and flood risk management legal frameworks as these regions have the constitutional competences to do so.[601] In the context of water management, with the exception of certain aspects, which pertain to the federal level[602], these three regions should be seen as three different "countries". Flood risk management in the international Scheldt District should be viewed in light of this institutional background.

2. LEGAL LANDSCAPE AND TOOLBOX IN THE SCHELDT DISTRICT

190. This section covers the first piece of the Scheldt puzzle, and looks into the legal framework and the instruments for flood risk management in the different jurisdictions sharing the Scheldt, namely Belgium, France and the Netherlands.

2.1. FLOOD RISK MANAGEMENT STRATEGIES IN DIFFERENT INSTITUTIONAL SETTINGS

191. Considering the institutional constellation of the European Union, i.e. the fact that its Member States are sovereign States with separate flood risk management strategies, it is pertinent to examine the legislative and policy frameworks applicable in each of the countries and regions sharing the Scheldt River Basin District, confronting the geopolitical and legal situation with the hydrological reality. Considering the procedural character of the Floods Directive, and the related absence of substantive requirements applicable to States sharing basins[603], analysing the way these different Member States have approached flood risk management in the Scheldt is the first step in determining opportunities and bottlenecks for cooperation in this basin. Indeed, as the FD does not stipulate the measures or instruments that should be included in the applicable legal framework or the Flood Risk Management Plans, these may

600 M Kaufmann, HK Van Doorn-Hoekveld, W., Gilissen and M Van Rijswick, 'Analysing and Evaluating Flood Risk Governance in the Netherlands. Drowning in Safety?' (STAR-FLOOD Consortium 2016).

601 The Brussels Capital Region will not be studied here, as only 0.4% of the Scheldt River runs through its territory.

602 E.g. insurance.

603 Ibid.

vary tremendously in the respective countries.[604] In this comparison, the five Flood Risk Management Strategies identified in the STAR-FLOOD project are a useful tool.[605] These are, more specifically, the strategies for *flood risk prevention*, which aims at keeping people away from the water through e.g. proactive spatial planning policies; *flood defence*, which is the most traditional strategy and which aims to keep water away from people through hard infrastructure such as dikes; *flood mitigation*, which relates to accommodating floods through mitigating its consequences e.g. green urban infrastructure and adaptive building; *flood preparation*, which tackles the preparedness of the population for floods; and *flood recovery*, which aims at helping society to recover as quickly as possible following a flood event through e.g. ex-post compensation mechanisms.[606] There are opportunities for cooperation in each of these strategies.

These degrees of cooperation may vary depending on the strategy in question. For example, cooperation requirements should be more apparent in the defence strategy than in the flood-recovery strategy. Indeed, measures adopted in the context of the former strategy by, for example, the upstream State, have a direct impact on the waters of the downstream State.[607] The development of an ex-post compensation scheme for damages resulting from floods, e.g. the question whether private insurance mechanisms are dominant or public compensation schemes, on the other hand, has a less apparent impact on the other States in the basin. Indeed, floods themselves are transboundary in nature, whereas their effects are more local. Thus, with regard to these types of issues, information exchange and exchange of best practices would be more appropriate than coordination on the measures as such.

The analysis of the countries and regions sharing the Scheldt IRBD below will be guided by these five FRMPs in order to provide a bird's-eye view on the similarities and differences between the different countries and regions on the one hand, and the opportunities and bottlenecks for cooperation on the other hand. As is clear from the above, flood risk management encompasses an incredibly broad spectrum of types of measures and regulations, from prevention to recovery.

---

604 Sally Priest and others, 'The European Union Approach to Flood Risk Management and Improving Societal Resilience: Lessons from the Implementation of the Floods Directive in Six European Countries' (2016) 21 Ecology and Society 50.

605 STAR-FLOOD has published a special issue in Ecology and Society, in which papers have been devoted to each of the strategies discussed in this book: Peter Driessen and others, 'Toward More Resilient Flood Risk Governance' (2016) 21 Ecology and Society 53.

606 For example, the CAT-NAT scheme in France, the Calamities Compensation Act in the Netherlands.

607 For example, infrastructure to drain the water away upstream will lead to water flowing downstream more quickly.

Figure 7. Flood Risk Management Strategies

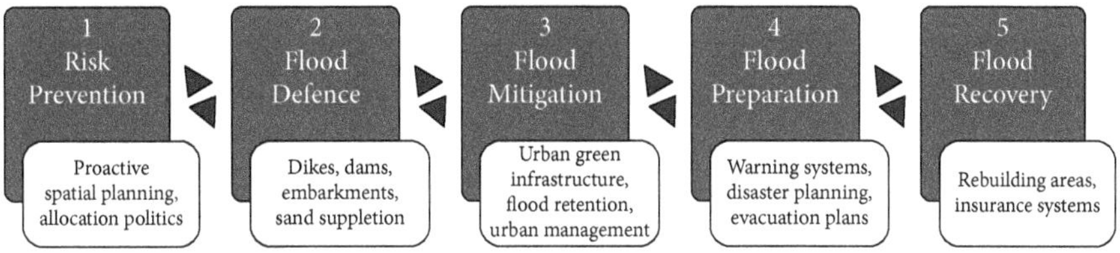

Source: Hegger and others, 2013

An in-depth legal comparative analysis of each of these types of measures, e.g. the question whether non-fault liability for flood risk management instruments exists in the relevant jurisdictions and to what extent[608], is beyond the scope of this book. The intention of the comparative analysis is to draw bottom-up lessons and conclusions on the manner in which cooperation mechanisms at the hydrological, transboundary scale can be improved at EU level.

2.2. THE FLOOD RISK MANAGEMENT REALM

192. The "legal framework and toolbox" section will be guided by several broad themes that have been identified when researching the applicable frameworks in the countries and regions sharing the Scheldt District:
(a) the physical vulnerability of the country or region in question to flood risks;
(b) the general implementation of the Floods Directive in the country or region;
(c) the legal instruments to operationalise preventive flood risk management;
(d) the application of plans and programmes;
(e) the existence, or lack, of (legally enshrined) safety standards;
(f) the framework for preparing for and recovering from floods;

2.2.1. The physical vulnerability to flood risks in the Scheldt States

193. All Scheldt countries are vulnerable to flood risks, although to a varying extent and degree of exposure. 13.5 million inhabitants populate the Scheldt IRBD, with a 5% rise in 2011 in comparison to the year 2000.[609] The density and population growth differs across the basin: Brussels saw a rise of 18.7% between

[608] See Willemijn van Doorn-Hoekveld, 'Compensation in Flood Risk Management with a Focus on Shifts in Compensation Regimes Regarding Prevention, Mitigation and Disaster Management' (2014) 10 Utrecht Law Review 216. Willemijn J Van Doorn-Hoekveld and others, 'Distributional Effects of Flood Risk Management – a Cross-Country Comparison of Preflood Compensation' (2016) 21 Ecology and Society 26.

[609] International Scheldt Commission, 'Roof Report 2016-2021' (2015) <www.isc-cie.org/images/Documents/ODB2-PFPG2_RAPPORT_NL-FR_VDEF.pdf> accessed 17 July 2017, 12.

the years 2002 and 2011 whereas this is only 0.8% in Zeeland.[610] In the Flemish Region, approximately 220,000 people are living in flood-prone areas.[611] 99% of these potentially affected people are located in the River Basin District of the Scheldt River.[612] An important aspect of flood vulnerability in the Flemish Region relates to the high degree of land sealing. Approximately 26% of the territories of the Flemish Region and the Brussels Capital Region is already made up of built-up area.[613] By 2050, this percentage of sealed, built-up land is expected to further increase to 30-50% of land use in a business-as-usual scenario.[614] In terms of the impact of climate change, a report by the Flemish government predicts a primary increase of floods in tidal rivers and in summer due to heavy rainfall.[615] A solid flood risk management framework that takes into account issues such as the special relationship between spatial planning and water management, is therefore quintessential in the Flemish Region.

194. In the Walloon Region, the glooming issue of land sealing and the degree of built-up area is not as apparent as it is in the Flemish Region. However, the region is vulnerable to flood risks. In the period between 1969 and 2014, 74% of Walloon municipalities in the Scheldt District suffered floods that were recognised as a general disaster through Royal Decree.[616]

195. As a low-lying delta, approximately 59% of the territory of the Netherlands can be considered as a flood-prone area.[617] As a result, flood risk management is high on the political agenda, and takes centre stage, for example, in the national adaptation to climate change programme. An important aspect of flood risk

---

610 Ibid 12.

611 CIW, 'Annex III to the Flemish River Basin Management Plan for the Scheldt 2016-2021' (2015), www.integraalwaterbeleid.be/nl/stroomgebiedbeheerplannen/overwegingsdocumenten/overwegingsdocument-bij-beheerplannen-vlaamse-delen-schelde/at_download/file accessed 14 April 2017.

612 CIW, 'Flemish River Basin Management Plan for the Scheldt 2016-2021' (2015) <www.integraalwaterbeleid.be/nl/stroomgebiedbeheerplannen/stroomgebiedbeheerplannen-2016-2021/documenten/Vlaams_deel_stroomgebied_Schelde.pdf> 108.

613 Lien Poelmans and Anton Van Rompaey, 'Detecting and modelling spatial patterns of urban sprawl in highly fragmented areas: A case study in the Flanders–Brussels region' (2009) 93 *Landscape and Urban Planning* 10-19.

614 Lien Poelmans, 'Modelling Urban Expansion and its Hydrological Impacts' (Ph.D. Thesis, KU Leuven 2010) <http://statbel.fgov.be/nl/binaries/MODELLING%20URBAN%20EXPANSION%20AND%20ITS%20HYDROLOGICAL%20IMPACTS_tcm325-116170.pdf> accessed 12 July 2017, 17.

615 Flemish Government, 'Vlaams Adaptatieplan 2013-2020' (2013) <https://www.lne.be/sites/default/files/atoms/files/2013-06-28_VAP.pdf> accessed 12 July 2017, 19.

616 Stratec, Environmental Impact Assessment of the Flood Risk Management Plan of the International River Basin District of the Scheldt (2015) 26, <http://environnement.wallonie.be/inondations/files/rie/RIE_PGRIEscaut_vf_NL.pdf>, accessed 31 March 2016.

617 Susceptible to floods from the sea, fluvial and, increasingly so, fluvial flooding. See Kaufmann (n 600) 7.

management in the Netherlands, is the right of citizens to be protected from floods. The explicit government responsibility in this regard, is therefore deeply rooted into the Dutch Constitution. In framing water management in general, and flood risk management in particular, this public responsibility aspect is paramount. In terms of its location in the Scheldt basin district, approximately 3,200 km^2 of the Scheldt District is located in the Netherlands, with 470,000 inhabitants occupying the Dutch part of the basin.

196. Finally, approximately 15% of France is situated in flood-prone areas. In recent years, especially the Xynthia storm had a significant effect on the country, with 53 casualties, major power outages, damaged dikes, and insured damages amounting to 2.5 billion EUR.[618] In the IRBD Scheldt, France is the upstream country, with 4.5 million inhabitants in the district. The flow of the basin starts in the North of France, before descending to Wallonia, Flanders and the Netherlands. The surface of the part of the district running through France amounts to 18,500 km^2.[619] The Scheldt District is included in the Artois-Picardie district, along with the Meuse.

2.2.2. Implementation of the Floods Directive: General

197. Although the Scheldt is shared by three countries – Belgium, France and the Netherlands – four separate legal frameworks need to be discussed. More particularly, the legal framework applicable in Belgium should roughly be divided into two parts, as the bulk of competences touching on flood risk management are governed by the regions[620], as opposed to the federal level, which necessitates a separate review of the Flemish and Walloon Region.[621] Certain topics, such as recovery following flood, i.e. insurance or public ex-post compensation mechanisms, and the preparedness for floods, i.e. warning systems and crisis management, pertain to the federal level.

198. In the context of water management, the Flemish Region can be characterised as a frontrunner regarding the topic of integrated water

---

618 European Parliament, 'Draft Report on the proposal for a decision of the European Parliament and of the Council on mobilisation of the European Union Solidarity Fund, in accordance with point 26 of the Interinstitutional Agreement of 17 May 2006 between the European Parliament, the Council and the Commission on budgetary discipline and sound financial management' (2010) 2010/2237 (BUD) <www.europarl.europa.eu/meetdocs/2009_2014/documents/budg/pr/835/835711/835711en.pdf> accessed 12 July 2017, 6.

619 International Scheldt Commission, 'Roof Report 2016-2021' (2015) <www.isc-cie.org/images/Documents/ODB2-PFPG2_RAPPORT_NL-FR_VDEF.pdf> accessed 17 July 2017.

620 Such as environmental matters, water law, and spatial planning.

621 As mentioned, the Brussels Capital Region does not form part of this analysis. This transfer of competences to the regions has occurred since the state reforms in the 1980s. Article 6, Para 1 of the Special Act on Institutional Reform of 8 August 1980, Belgian Official Journal 15 August 1980.

management, but a laggard in terms of coherence and efficiency of competent authorities, given the large degree of fragmentation.[622] The latter characteristic significantly hinders effective international and regional cooperation at the different basin scales in the IRBD Scheldt. This will be discussed in the following paragraphs.

Similarly to the European legal framework in the "pre-Water Framework Directive" era, the legal framework for water-related issues in the Flemish Region was fragmented, and regulated in various sectoral instruments, e.g. surface-water requirements were included in the Surface Water Act of 1971, and groundwater quantity management was included in the Decree related to the management of groundwater of 1984. Since the entry into force of the Decree Integrated Water Policy (DIWP), the provisions related to water management in the Flemish Region have been consolidated and bundled into one main legal instrument.

The 2003 DIWP therefore constitutes the main legal framework for flood risk management in the Flemish Region, and it implements both the Water Framework Directive and the Floods Directive.[623] This means that integrated water resources management, i.e. approaching the quantitative and qualitative management of water resources in an integrated manner, promoting multi-functional use, is one of the cornerstones of Flemish water management.[624] In line with the WFD, the main goals of the DIWP relate to achieving a good status of the water systems. The DIWP expressly emphasises the link between water quality and water quantity by including the requirement to determine environmental quantity norms.[625] Indeed, and in contrast with the WFD, the DIWP does not consider the quantity norms as completely accessory to the qualitative aspects.[626] Therefore, the legislator has provided a way to regulate the quantitative aspects, so that it not only serves to achieve a good ecological status of the surface waters, but can also have other functions. In this respect, the Flemish legislator, like the Dutch legislator, has exceeded what is required by the WFD. Another essential aspect of this integrative approach relates to the creation of the Coordination Committee on Integrated Water Policy through the 2003 DIWP (CIW), an entity that focuses both on quality and on quantity and brings together the relevant stakeholders into one governing body, under the auspices of the Flemish Environment Agency.[627]

The Floods Directive was implemented into the DIWP in 2010.[628] From a longitudinal point of view, since the turn of the century, the flood risk management

---

622 Mees, Suykens and others (n 525).

623 Decree of 18 July 2003 on integrated water policy (DIWP), Belgian Official Journal 14 November 2003.

624 See Article 4 DIWP.

625 Article 5 DIWP.

626 See Part II of this book, where the accessory character of water quantity management is explained. Still, the concept of "environmental quality norms" is sufficiently broad to cover the quantitative aspects as well.

627 Their acronym is the "VMM".

628 Decree of 16 July 2010 modifying the Decree of 18 July 2003 on integrated water policy, Belgian Official Journal 19 August 2010.

legal framework has gone through three periods. In the period prior to the implementation of the Floods Directive, i.e. in the pre-2010 period, measures related to flood risk management already existed, but did not form part of a comprehensive legal framework for flood risk management. This in contrast with the Netherlands, where a strong legislative framework for flood risk management had already been developed by the time the Floods Directive entered into force. In the second period, from 2010-2013, the Floods Directive was implemented into the DIWP, and the DIWP therefore provided for a structural framework to carry out flood risk management related measures by competent authorities such as the water authorities. In the third period, post-2013, a substantial reform of the DIWP took place, in which planning processes and planning structures were overhauled. This reform has had a major impact on flood risk management. In this reform, the planning levels for the river basin and sub-basin levels were integrated, water boards were abolished as separate governing entities, and so forth.[629]

In the three periods of flood risk management mentioned above, shifts in flood risk management strategies can be discerned. Especially in the flood risk prevention strategy, there has been a great increase of instruments at the disposal of water managers to keep people away from the water. The flood risk prevention strategy therefore gained significant importance in the past ten years. For example, the water assessment, which also exists in the Netherlands and will be explained below, was subject to a substantial reform in 2011, because it was evaluated as inadequate to serve its purpose.[630] Also, new instruments saw the light, e.g. signal areas, which were subject to a more stringent water assessment. The mitigation strategy also gained more importance in Flemish flood risk governance, e.g. through the Rainwater Regulations, which promote the installation of green roofs. This means that, although the defence strategy remains important in Flemish flood risk governance, the applicable legal framework strongly focuses on the risk prevention strategy as well.

199. In terms of the three phases of the Directive, i.e. the preliminary risk assessment phase, the cartography phase and the risk management plan phase, the first phase was skipped, in accordance with this possibility provided by Article 13 FD.[631] The flood risk and hazard maps that were developed prior to and following the entry into force of the Directive are a crucial element in the implementation of the instruments developed by the DIWP and its executive orders, e.g. the

629 Decree amending various provisions of the Decree of 18 July 2003 on the Integrated Water Policy, Belgian Official Journal 1 October 2013.

630 The bottlenecks will be discussed below.

631 Because detailed data and models had already been produced at that time (e.g. in the context of the Sigma Plan). Moreover, the Flemish Region indicated before 22 December 2010 that it was drafting flood hazard and flood risk maps.

water assessment and signal areas.[632] In accordance with the integrated water management approach adopted, the Flood Risk Management Plans (FRMPs) were fully integrated into the WFD River Basin Management Plans (RBMPs).[633] The RBMPs identify the watercourses in which a potentially significant risk exists, expand on the methodology that forms the basis for the cartography, sets forth the prioritisation of measures, and so forth.[634]

The timeline for water management and flood risk management related legislation in the Walloon Region is similar to that of the Flemish Region, with the Water Framework Directive constituting a turning point in the new millennium and the implementation of the Floods Directive in 2010. Like the Flemish Region, the Walloon Region embraced the concept of integrated water resources management. The importance of including the entire water cycle in the framework for the management of water was acknowledged by the Walloon legislator in 1999.[635] The first provision of the 1999 Decree stated that the water cycle is managed in a comprehensive and integrated manner with the continuing goal to safeguard its existence in the context of sustainable development. With the implementation of the Water Framework Directive, the 1999 Decree was superseded by the 2004 Water Code, which was included in the Environment Code.[636] Similar to the Flemish DIWP, the Walloon Water Code represented a consolidation of water-related legislation in the Region. Whereas the integrated approach to water quantity and water quality management has also been translated at planning level in the Flemish Region, the Walloon Region chose not to integrate the Flood Risk Management Plans with the River Basin Management Plans.[637] This was justified in the parliamentary proceedings by reasons of readability.[638] Up until the entry into force of the Decree of 4 February 2010 (which implemented the Floods Directive), the provisions with regard to flood risk management were scattered in this Water Code and in the legislative framework for spatial planning.

200. Like in the Flemish Region, the preliminary flood risk assessment in the sense of Article 13 FD has not been carried out. As reflected in the Water Code, the whole territory of the Walloon Region is marked as an area of potentially significant flood risk.[639] As for the second phase of the FD, the cartography,

632 These are mainly available through: www.waterinfo.be; it should be noted that different maps are being used in the prevention and the recovery strategy.
633 This means that the first-cycle FRMPs were integrated into the second-cycle RBMPs, as these had already been issued once in 2009.
634 CIW (n 612).
635 Namely through the Decree of 15 April 1999 regarding the water cycle and the creation of a "Public Company for Water". Belgian Official Journal 22 June 1999.
636 Belgian Official Journal 23 September 2004.
637 But their timelines have been coordinated.
638 Walloon Parliament (Session 2009-2010), no. 129/1, 11.
639 Article D.53-1 Water Code.

there are three types of maps in the Walloon Region, whilst the FD requires the promulgation of two types of maps. The three types of maps are flood-hazard maps[640] and flood risk maps within the meaning of the FD and flood-danger maps developed in the context of Plan PLUIES.[641] This last map was developed prior to the entry into force of the FD, and constitutes the basic document for the implementation of specific legal flood risk related instruments, such as the Walloon equivalent of the water assessment and of determining the high-risk zones relevant to the risk-differentiation process in the context of insurance. The third phase of the FD was finalised in March 2016, i.e. the FRMPs for the Walloon parts of the International River Basin Districts.

The Netherlands has a long history of flood risk management and the right to be protected from floods is enshrined in the Constitution, with Article 21 stating that "it shall be the concern of the authorities to keep the country habitable and to protect and improve the environment", after which it goes on to stipulate the responsibilities of the public competent authorities in this regard. As in the Flemish Region, Walloon Region, and also in conformity with the development in European water law, the Dutch legal landscape for water management evolved from a package of fragmented legal competences toward an integrated water management approach. A consolidation of existing legal instruments with regard to water resources forms an intricate part of this development. In the Netherlands, this development took place in the 1980s, when the concept of integrated water management found its way into the flood risk management realm.[642] The pinnacle of this consolidation surge was the Water Act.[643] This Act takes an integrative approach to water management, as its goals relate both to water quantity management, i.e. the prevention and limitation of floods and water scarcity, and water quality management, i.e. the protection and improvement of the ecological and chemical quality of the water system. It also focuses on the fulfilment of societal functions by water systems.[644] A distinction is made between water quality management, water quantity management and water-safety management. The last one refers to the management of surface water to the extent that it is not covered by water quality or quantity management.[645] The object of governance on the basis of the Water Act is the water system, which is the coherent whole of surface-water bodies and groundwater bodies, including related water-storage areas, dikes and supporting infrastructure.[646] It is significant that this flood-

640 Return period of (1) 25 years, (2) 50 years, (3) 100 years and (4) extreme circumstances.
641 The final sets of maps were published in December 2013, see Belgian Official Journal, 9 January 2014. The maps are available through "Géoportail de la Wallonie".
642 Marlon Boeve and Frank Groothuijse, *Omgevingsrecht* (Europa Law Publishing 2013).
643 Act of 29 January 2009 setting forth rules related to the management and use of the water systems, Dutch Official Journal 12 March 2009.
644 Article 2(1) Water Act.
645 Article 3(2)(3) Water Regulations with regard to the management and use of water resources (Water Regulations).
646 Article 1.1 Water Act.

protection infrastructure is considered as part of the water system, which is not the case in the Flemish and the Walloon Region.

At the turn of the century, the idea grew that the defence strategy alone, i.e. the hard infrastructure to prevent floods from occurring, will not result in a resilient flood risk management realm. Spatial planning policies to create more room for rivers and more space for water in general moved to the forefront. As mentioned above, the water assessment in the Flemish Region constitutes an important bridging mechanism between the domain of water management and spatial planning. In the Netherlands, a similar bridging mechanism exists and it is equally referred to as the water assessment. The scope of the Dutch water assessment, however, is more limited in comparison to its Flemish counterpart. Whereas the Flemish water assessment applies to permits, plans and programmes, the Dutch water assessment only relates to spatial plans and decisions. In these decisions, the competent authority should only motivate how the water system has been taken into account.[647]

In 2012, the flood risk management legal framework was further strengthened through an amendment to the Water Act, namely when the Delta Act related to water safety and freshwater supply entered into force. This amendment created a Delta Commissioner, Programme and Fund. The Delta Programme provides both for water quantity and water quality, and, in the Dutch interpretation, for water-safety management provisions. Indeed, for a period covering six years, it indicates specific measures to tackle floods and water shortages, with indications of the financial means to support these measures.[648]

201. The Floods Directive has been implemented in the Netherlands in the Water Act and the Water Decree.[649] In comparison with other countries, e.g. Poland, where the Directive has caused a landmark shift in the applicable legal and policy framework, the Directive has not had a huge impact on the flood risk management policy framework applicable in the Netherlands. Indeed, the Dutch risk-based approach and implementing policies served as a source of inspiration in the development of the Directive.[650] Like the Flemish Region, the Netherlands has not carried out a preliminary flood risk assessment in the sense of Article 13 FD, because flood-hazard maps and flood risks maps had already been drawn up for the whole territory of the country.[651] In 2003, the National Administrative

647 Order of 21 April 2008 implementing the Spatial Planning Act, Dutch Official Journal 29 April 2008.

648 Article 4.9 Water Act.

649 Bulletin of Acts, Orders and Decrees 2009, 490 and Bulletin of Acts, Orders and Decrees 2009, 548. Other Acts are important, e.g. the Delta Act, which constitutes the legal basis for the engineering works that have been carried out in the south west of the country. Marleen van Rijswick and Herman Havekes, *European and Dutch Water Law* (Europa Law Publishing 2012) 97.

650 Ibid.

651 Kaufmann and others (n 600).

Agreement on Water Affairs was promulgated for the first time.[652] This Agreement covers the entire water cycle, and designates the responsibilities of the respective actors in this regard.

202. As in Belgium, the French Constitution does not provide for an explicit constitutional right for citizens to be protected from floods, although generally the government needs to ensure the safety of its citizens.[653] The WFD was implemented into French legislation through an amendment to the Environmental Code in 2004.[654] The Floods Directive was transposed into French law through the 2010 Grenelle II Act.[655] The implementation of the Floods Directive took place at three levels: (i) the national level, (ii) the level of the hydrological district and the (iii) High-Risk Areas.[656] At the level of each hydrological district, the Coordinating Basin Prefect is tasked with carrying out a preliminary flood risk assessment, identifying the High Flood Risk Areas (TRI), elaborating flood risk cartography, and a Flood Risk Management Plan. In contrast to the Netherlands and the Flemish and Walloon Regions, France did carry out the preliminary flood risk assessment and identified these High Flood Risk Areas (TRI). The 1982 Act related to the compensation of victims of natural disasters also plays a pivotal role in French flood risk management, as will be further explained below.

2.2.3. Legal instruments to operationalise flood risk management

203. In the Flemish Region, the past decade has seen a tremendous rise in legal instruments embodying the risk-prevention strategy. In the 2003 DIWP, several instruments were introduced with the aim of mitigating the issue of land sealing, especially in flood-prone areas, and focussing more strongly on preventing flood risks and mitigating their consequences. In addition to classical instruments such as expropriation and compulsory purchase, the above-mentioned water assessment, signal areas and the duty to inform are the most prominent instruments bridging the domains of water management and spatial planning.[657] Many of these instruments show great similarities to instruments

652 Updated in 2008 and 2011.

653 Willemijn Van Doorn-Hoekveld (n 598).

654 Act no. 2004-338 of 21 April 2004 implementing Directive 2000/60/EC, French Official Journal 22 April 2004.

655 Act no. 2010-788 of 12 July 2010 related to the environment 2010, French Official Journal 13 July 2010.

656 Comité de Bassin Artois-Picardie, 'Schéma Directeur d'Aménagement et de Gestion des Eaux du Bassin Artois-Picardie pour la période 2016 à 2021' (2016) <www.artois-picardie.eaufrance.fr/IMG/pdf/sdage_2016-2021.pdf> accessed 12 July 2017, 28.

657 For more information on bridging mechanisms, see Herman-Kasper Gilissen and others, 'Bridges over Troubled Waters: An Interdisciplinary Framework for Evaluating the Interconnectedness within Fragmented Domestic Flood Risk Management Systems' (2016) 25 Journal of Water Law 12.

used in the other Scheldt countries and regions. As part of the water assessment, the permit-issuing authority should ask advice from the relevant water manager before granting a permit to build or adopt a spatial implementation plan or programme.[658] This way, the impact of this permit, plan or programme on the water system is assessed.[659] The water assessment constitutes a clear application of the prevention principle well known in environmental law, which is also one of the backbone principles of the DIWP.[660] In order to mitigate the impact on the water system, the DIWP includes a hierarchy of measures, which roughly emphasises preventing any impact whatsoever, and, subsequently, when prevention is not possible, compensatory measures *in natura* should be taken to mitigate the impact. In the Netherlands, the water assessment also exists, but has a narrower scope. Indeed, whereas the water assessment in Flanders pertains to permits, plans and programmes, the instrument in the Netherlands relates only to spatial plans.

The water assessment has gone through several stages of development since its conception in 2003. The main reform took place following the 2010 floods that heavily struck the Flemish Region. In the context of an evaluation that took place in the wake of these floods, it was concluded that the water assessment did not function adequately in practice, and was therefore too much of a "paper tiger". The main bottlenecks related to the lack of information predicting the outcome of the water assessment, resulting in excessive administrative burdens, meaning that it should be clear to the developer or applicant which actions should be taken to avoid a negative water assessment resulting in a re-submission of the application. Furthermore, a screening of 290 permit applications carried out by the CIW showed that for 27% of the applications, no advice had been requested, even though this was necessary on the basis of the applicable legislation.[661] Finally, the lack of follow-up to check compliance with the conditions set as a result of the water assessment was considered a significant problem. An inquiry conducted by the CIW showed that only 2.4% of the files referred to actual follow-up in the field. Through a 2011 Order, the water assessment was revised, among other things, to render the sourcing of advice obligatory, and the thresholds above which advice needs to be sought more straightforward.[662] The Order also sets stricter motivation requirements in the water paragraph, i.e. the

658 Ann Carette and Peter De Smedt, 'Het Vernieuwde Decreet Integraal Waterbeleid: Sneller En Beter?' (2013) Tijdschrift voor Milieurecht 576.

659 Article 8 DIWP.

660 See Explanatory Memorandum to the DIWP, Flemish Parliament (Session 2002-2003), no. 1730.

661 Order of the Flemish Government of 20 July 2006 determining the rules for the application of the water assessment, Belgian Official Journal of 31 October 2006. See also John Toury, and Martin Denys, 'Vernieuwde watertoets – vergoedingsregeling bij bouw- en verkavelingsverbond' (2012) Nieuw Juridisch Weekblad 82. See the Order of the Flemish Government of 14 October 2011, Belgian Official Journal 14 November 2011. This Order entered into force on 1 March 2012.

662 For example, the schemes on the basis of which the competent authority should determine whether or not to seek advice from the water managers were very unclear and complex prior to the reform.

paragraph included in the permit which refers to the impact on the water system and the compliance with the goals of the DIWP. Indeed, the Court of Cassation has since annulled permits for insufficiently precise motivation with regard to the water assessment, stating that the decision to grant a building permit should include a formal motivation, which must explain that the activities pertaining to the permit do not have harmful effects on the water system, or that these effects will be mitigated.[663]

Signal areas are another important element of the Flemish flood risk management realm, and aim to control the development of undeveloped parcels that have a hard destination and are situated in flood areas, to avoid an increase of risks.[664] Signal areas are areas that are important for the water system in the sense that they would serve as a sponge inducing water infiltration, but where the interest of this water system can come into conflict with the hard spatial destination of the area. The Flemish government has drawn up plans of action for each of the identified signal areas. For example, in areas where the spatial destination can be considered as compatible with their water-storage capacity, special conditions can be imposed, e.g. through a more stringent water assessment. In areas where there is an incompatibility, either a standstill policy can be applied, i.e. the prohibition to build on or further develop the land, or a proactive spatial planning policy, i.e. rezoning measures. An important question in the context of signal areas relates to finances. Through the Order of 20 June 2014, the Flemish government re-oriented the Rubicon Fund in order to provide municipalities with the necessary resources to implement signal areas.[665] Municipalities and provinces may ask for 60% of the compensation for loss resulting from government planning decisions paid to the owner in the context of the signal areas from the Flemish government.

Finally, the duty to inform is another bridging mechanism between the sectors of water management and spatial planning and entails that every buyer or lessor, for contracts exceeding a period of nine years, should widely disperse information on the flood-prone character of a building, i.e. in advertisements on the Internet, brochures, and so forth.[666] This instrument can be considered as a versatile instrument that could easily be applied in other countries and regions as well, because it is not as context-specific as the water assessment and signal areas for example. Indeed, whereas the water assessment and, especially, signal areas stem from the delicate spatial planning situation in the Flemish Region, considering the high degree of land sealing, the duty to inform is relevant for any country seeking to raise awareness for flood risks at a practical level.

---

663 *De Win and De Boeck v the Flemish Region* [2011] Council of State n 212.266 of 28 March 2011, para. 21.

664 Peter De Smedt, 'Towards a New Policy for Climate Adaptive Water Management in Flanders: The Concept of Signal Areas' (2014) 10 Utrecht Law Review 107.

665 Belgian Official Journal 25 September 2014.

666 Article 17*bis* DIWP.

204. As in the Flemish Region, the legislative framework in the Walloon Region includes certain instruments with the aim of creating more space for water, e.g. expropriation of immobile goods for flood risk management.[667] The Walloon Region has an instrument similar to the Flemish water assessment on the basis of the spatial planning code, more specifically through Article 136 CWATUPE, which will be replaced by D.IV.57 Urban Planning Code (CoDT) in the future[668], which stipulates that actions, works and permits that are exposed to flood risks may be prohibited or be subject to special conditions.[669] Whereas the Flemish water assessment forms part of the DIWP, and thus the water management framework, the Walloon equivalent is governed separately in the spatial planning legal framework – a similar situation as that in the Netherlands. Similarly to the Flemish water assessment, the permit-issuing authority is supposed to submit a request for advice to the relevant water manager, which in turn may lead to conditions being set or to refusal of the permit. There is no legal obligation to source this advice from the water manager. Once the advice is sourced, the permit-issuing authority is not obligated to follow the advice, as it is non-binding.[670] Further details on how the provision should be applied by the competent authorities are not embedded in the legal framework, but are set forth in a Circular of the competent Minister, a 2003 Circular that was drawn up following floods that occurred in the beginning of that year.[671] The Circular sets out guidelines, aimed at the competent authorities in the municipalities, related to granting permits in flood-prone areas. The Circular states that each permit application within one of the parameters set in the flood-hazard map ("cartographie d'aléa d'inondation") should meet the conditions determined by the government in a regional planning regulation.

This makes the case weaker for the municipality who wishes to base its refusal of the permit on the advice. Moreover, the cartography on the basis of which the advice is issued does not have legal value, as has been confirmed in case law and the parliamentary proceedings preceding the adoption of Article 136, or Article D.IV.57 CoDT.[672] The cartography is considered as a document contributing

---

667 Article D.53-11, Para 1 Water Code.

668 This has been applicable since 1997, i.e. five years prior to the Flemish water assessment. Until recently, this instrument was referred to as "Article 136 CWATUPE". The "CWATUPE", the previously applicable spatial planning legal framework has been replaced by the "Code du Développement Territorial", commonly referred to as the "CoDT".

669 Article 136 CWATUPE.

670 Further non-binding guidance for competent authorities is included in the so-called "Circulaire Foret" of 9 January 2003, Circulaire relative à la délivrance de permis dans les zones exposées à des inondations et à la lutte contre l'imperméabilisation des espaces, Belgian Official Journal 4 March 2003.

671 Ibid.

672 For example, in its Judgment of 29 October 2008, the Council of State stated that a particular Order of the Walloon Government of 3 May 2007 which adopted cartography of flood risks of the undercurrent stream of the Amel (the Amblève), does not, on its own, necessitate a building prohibition. *Henriette Peeters v the deputy of the provincial Council of Limburg* [2008]

to the decision-making process of the competent authority with regard to the identification of flood-prone areas, without it having an autonomous legal value. The Walloon equivalent to the water assessment is less elaborately and prescriptively developed in the legal framework: whereas the Flemish DIWP sets the thresholds for advice, the Walloon instrument does not. One of the main issues with regard to spatial planning and flood risk management, is that the designated high-risk areas are not legally binding in the sectoral plans. It is therefore possible that a sectoral plan allows construction in a zone that has been designated as a "red zone" in flood risk cartography. With regard to this aspect, this means that the water managers lack the legal basis to prevent further construction in high-risk areas.[673] This situation differs from the French flood risk management realm, where sectoral plans do prohibit construction in these designated high risk areas and are more comparable with the Dutch approach.

205. An interesting legal instrument exists in the Dutch water management realm to encourage cooperation and coordination between regional water authorities in the same River Basin District. In these so-called "water agreements", the regional water authorities, with the goal of achieving a coherent water management policy, should make arrangements for the issues that exceed their respective package of competences. The water authorities have the right to take the initiative to conclude these water agreements when they see fit: their conclusion does not depend on certain situations stipulated by law. In principle, these water agreements are to be concluded between the competent water authorities, but other public bodies may be invited to join the discussions, on the condition that they have responsibilities with respect to the water system.[674] Whereas traditionally, these water agreements were instruments to conclude agreements on the quantitative aspects of water management, they have developed toward dealing with a more integrated scope, including water quality management as well.[675] According to Gilissen, this instrument could play a role in facilitating transboundary water management, i.e. by concluding transboundary water agreements that would carry out a function similar to the Dutch water agreement within the meaning of the Water Act, but then between a Dutch water authority and another competent authority within the same IRBD across the border, to regulate water-related issues that exceed their respective geographical areas of management.[676] This instrument will be interesting for Chapter V of this book.

Council of State n 187.450 of 29 October 2008. Also, see Flemish Parliament (1996-1997) no. 233/1, 17.

673 Interview with a civil servant of the department of non-navigable watercourses of the Walloon Region ("DG03"), 4 November 2014.

674 Article 3(7)(2) Water Act.

675 Herman Kasper Gilissen (n 548) 120.

676 Ibid 130. This option will be discussed in the following parts of this book.

2.2.4. *The application of plans and programmes*

206. All Scheldt countries and regions have issued management plans for their respective parts of the Scheldt District for the reporting period 2016-2021, which is the second reporting period for the WFD (the River Basin Management Plans) and the first reporting period for the FD (the Flood Risk Management Plans).

207. The Flemish RBMPs mirror the measures included in the DIWP, and reflect the three-step strategy of capture, storage and drainage.[677] By building further on the legal framework of the DIWP, existing rules and structures were called on: the Coordination Commission on Integrated Water Policy (CIW), which had already been appointed as competent authority for the WFD, was made responsible for the implementation and execution of the FD. The CIW, in which all Flemish water managers are represented, thus became competent for drafting both the RBMPs and the FRMPs. Integration was also pursued at the planning level: hence, the FRMPs have been integrated into the second generation of RBMPs. Administratively, this creates the advantage that various procedural steps for drafting RBMPs can be combined with the similar phases for drafting FRMPs, e.g. public consultation and the gathering of various pieces of advice. Content-wise, the FRMPs are fully integrated in the RBMPs, e.g. the proposed measures are brought together into one set of measures to reach the goals of both Directives. The integrated content is also clearly illustrated by the fact that it was no longer found necessary or even desirable to use the term FRMP.[678]

208. In the Walloon Region, the RBMP and FRMP were issued separately. The development of the FRMPs has taken place in a number of phases. In a first phase every sub-basin is provided with a Technical Sub-basin Committee.[679] These Committees are composed of the watercourse managers, representatives from the Walloon administrative departments, provinces, municipalities and wateringues. Through a series of roundtables, suggestions are made for measures against flooding, ranging from flood prevention, protection, preparation, recovery and post-crisis evaluation. In the subsequent phase, the Committee proposals are discussed in a new round of gatherings, this time involving the River Contracts, which are the relevant entities at the hydrological scale and which will be discussed in the section on hydrological scale governance below.[680]

---

677 *Infra.*

678 The DIWP uses the concept of 'flood risk management provisions' to define the FRMP as part of the RBMP. However, FRMP is used in the DIWP for the FRMP of before 22 December 2010 that corresponds to the conditions of the FD. Also the administrations still use the term FRMPs.

679 Free translation of: Comité Technique de Sous-Bassin Hydrographique, CTSBH.

680 These roundtables are organised for 4 different themes: spatial planning, river overflow, run-off & erosion, and crisis management.

209. Before the FD was introduced, the Walloon Region had already issued a non-binding plan which is a relevant addition to its legal framework: The Plan PLUIES, short for "Global Plan for the Prevention and the Fight against Floods and their Effects on the Victims", plays an important role in flood risk management in the Region. The Plan PLUIES was called into life following floods on the territory of the Walloon Region in August 2002 (in the valley of Mehaigne) and January 2003 (in the Meuse and Scheldt basins). This plan was adopted in 2003 and establishes the policy lines of the region with respect to preventing and dealing with floods and flood risks. The plan promulgates five specific objectives of management and reduction of risks: (i) improving knowledge of flood risk, (ii) reducing and slowing down surface run-off of water in catchment areas, (iii) improving riverbeds and alluvial plains, (iv) reducing vulnerability in flooding areas, and (v) improving crisis management of disasters.[681] There is a coherent relationship between Plan PLUIES and the FRMPs for the Walloon Region, as the measures included in the FRMP are also included in the Plan PLUIES.[682]

Figure 8. PARISs as a Bridge between FRMPs and RBMPs

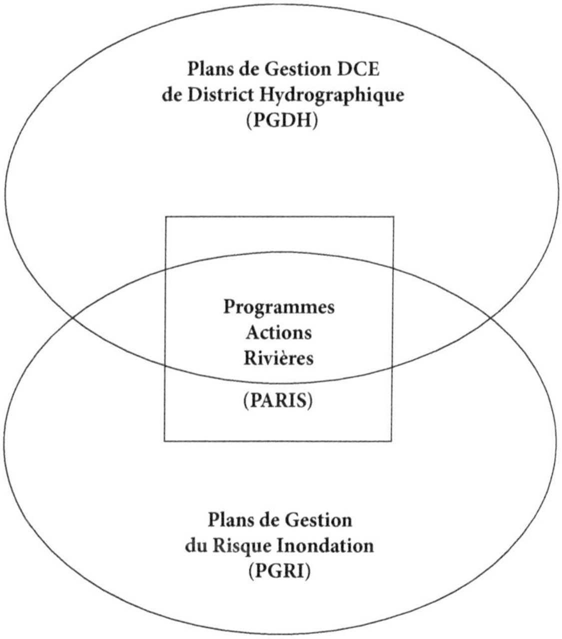

Source: SPW DG03

[681] These have been translated into 30 specific actions such as the adaptation of a regional-planning regulation.

[682] There are four FRMPs in the Walloon Region: for the Walloon parts of the Rhine, Scheldt, Meuse and Seine.

210. Not only the Plan PLUIES is relevant in terms of plans and programmes, but also the so-called "Integrated and Sectorised Action Programmes for Watercourses" or "PARISs", which are plans especially targeted toward the integrated and sustainable use of water resources and which combine the RBMP and FRMP at local level.[683] These should be seen as local counterparts of the EU-steered plans. The PARISs, developed by water managers, do not have a legal basis in the Water Code or other legislative framework. In light of the realisation of the PARISs, watercourses in the Walloon Region were divided into 6000 sectors, and to these sectors management objectives apply: flood protection, nature conservation, fluvial transport or recreation. Only one PARIS has since been actually developed.[684] The bodies specifically active at the hydrological scale, the Walloon river contracts, are not involved in the elaboration of the plans, which is a clear bottleneck in light of integrated water management.[685]

211. In the Netherlands, the National Water Plan outlines the general water policy applicable in the Netherlands, and also includes the respective River Basin Management Plans in the context of the EU Directives.[686] The main goals of the Dutch FRMP for the Scheldt relate to the avoidance or decrease of flood risks in the Scheldt River Basin District, whilst also improving the quality of water and the provision of societal functions stemming from the water resources. As has been mentioned above, the Netherlands adopts a risk-based approach – as does the Floods Directive. The flood risk within the meaning of the Scheldt FRMP is therefore also defined as probability times consequences. Spatial zoning plans are drafted by the municipalities and define the functions of the areas in question and the associated rules.

212. In France especially, there is a significant number of plans and programmes with relevance to water management and spatial planning that should be issued by the various administrative layers. There are three types of strategies and plans, one of which ensues from the implementation of the Floods Directive: the Flood Risk Management Plans. The other two levels are the national strategy and the local strategies. As mentioned above, substantial competences for flood risk management have been transferred to the municipalities in 2014.[687]

---

683 Programme d'Actions pour Rivières au sens large par une approche Integrée et Sectorisée.
684 Mees, Suykens and others (n 525).
685 To date, only one PARIS has been established, i.e. for the Lesse sub-basin.
686 I.e. the Dutch part of the four International River Basin Districts.
687 Before that, in 1983, competences for spatial planning were transferred to the municipal authorities. Decentralisation Act, 7 January 1983. An important aspect of this transfer is the question of financial resources. Several options can be envisaged in this regard, e.g. a vertical allocation of a fixed /variable budget from the level of the State to the level of the municipalities, or a mechanism in which these local entities can raise their own funds, such as the regional water authorities in the Netherlands.

Following a series of floods, the French legislator issued legislation with respect to the prevention of natural disasters. Through the 1995 Act related to the reinforcement of the protection of the environment, the instrument of plans for the prevention of foreseeable natural disasters (PPRN) was created.[688] Flood events are one of the natural phenomena included in the scope of this instrument. The Environmental Code further specifies the desired content and aim of such plans, i.e. to designate the zones in which danger exists and to subsequently prohibit construction or industrial operation or stipulate further conditions for construction and operation in these zones.[689] Another type of zone is the "precaution zone", which is not directly impacted by the risk, but with regard to which activities could aggravate the risk.[690] These plans have binding force, as non-compliance with provisions stipulated in the plan triggers penalties set in the Urban Planning Code. The PPRNs in question are considered as a public easement, and should be attached as an annex to the land-use plan.[691] The liability of local authorities that have granted building permits, despite the area having been designated as a "danger zone" and the awareness of said authority of the risk in question, has been confirmed in case law.[692] Specific to floods is the so-called "PPRI", which is the Flood Risk Prevention Plan, created in 1995 and drawn up at the level of the State. This means these are not based on the Floods Directive, but on pre-existing domestic law. The PPRI sets zoning requirements etc. that should be taken into account by local authorities.

The National Flood Risk Management Strategy (SNGRI)[693], which is different from the Flood Risk Management Plans[694], outlines the guiding principles and priorities of the national policy. The SNGRI is approached as the step that needs to be taken prior to the promulgation of the Flood Risk Management Plans. It has been drawn up by the Joint Flood Committee, which is an entity at national level representing a whole range of stakeholders.[695] Three main goals have been set in the SNGRI: to increase the safety of the exposed population, to stabilise and reduce the costs of damages, and to shorten the "return to normal" of the areas

---

688 Act no. 95-101 of 2 February 1995 related to the enforcement of environmental protection, French Official Journal 3 February 1995.

689 Article L562-1, 1° Environmental Code, Ordinance n 2000-914 of 18 September 2000 related to the legislative part of the Environmental Code, French Official Journal 21 September 2000.

690 E.g. forestry, farming, and so forth. See Article L562-1, 2° of the Environmental Code.

691 The Urban Planning Code states that the local urban plans should include as annexes all public easements that influence the territory in question. See Article L126-1 of the French Urban Planning Code.

692 E.g. *Min. Equip. c/Ste SMAC* [2003] Council of State n 213991 of 13 June 2003.

693 The National Flood Risk Management Strategy or "Stratégie Nationale de Gestion des Risques d'Inondation", also referred to as the SNGRI.

694 The Flood Risk Management Plan or "Plan de Gestion des Risques D'Inondation", also referred to as the PGRI.

695 OECD 'Seine Basin, Île-de-France, 2014: Resilience to Major Floods' (2014) OECD <www.oecd.org/gov/risk/Flood-risk-management-seine-river-executive-summary.pdf> accessed 12 July 2017.

that have been hit. The subsidiarity principle constitutes an important aspect of the strategy. The Flood Committee acts as an entity similar to the Coordination Committee (CIW) in Flanders, in that it represents relevant stakeholders. The Action Programmes for Flood Prevention (PAPI), launched in 2001, have promoted integrated flood risk management at the sub-basin level.[696]

As required by the FD, France issued, on top of its amalgam of plans and programmes, a Flood Risk Management Plan for the period 2016-2021 for the part of the Scheldt on its territory. These FRMPs mirror the River Basin Management Plans, referred to as SDAGEs in France. There are five main goals around which the Plan is centred: (i) flood risk prevention, i.e. to avoid or reduce the exposure to risk, (ii) flood defence and mitigation, i.e. keep water away from people or manage the water flow, (iii) awareness raising, i.e. improving knowledge and sharing information, (iv) preparation and recovery, through crisis management and insurance systems, and (v) river basin management and enhancing downstream-upstream solidarity. The Plan adopts the risk-management approach also prevalent in the Floods Directive, i.e. risk is a combination of the flood hazard and the consequences in terms of human health, economic activity, the environment and cultural heritage. Indeed, this is in conformity with French implementing legislation, i.e. the Grenelle II and the Decree of 2 March 2011 related to the management of flood risks. As mentioned above, the TRI or High-Risk Areas have been identified throughout the country. For the Scheldt River, eleven TRI have been identified.[697]

FRMPs are adopted by the State (Prefect of the basin) in the River Basin Committee (RBC).[698] The binding force of the FRMP regarding the administration in its decision-making process is explicitly mentioned, e.g. in terms of town planning and water management. The RBC represents several stakeholders in the field of water management, but, as explained in the above section, the approval of FRMPs depends on the Basin Coordinator Prefect. In principle, strategies for flood risk management are adopted by municipalities, or groupings thereof. The municipalities may join forces at the sub-basin level, but this does not constitute a top-down requirement. Future plans and decisions must be compatible with the FRMP: Flood Risk Prevention Plans (PPRI), Local Water Management Plans, town-planning documents etc. However, the PPRI remains the main instrument, as the FRMP remains rather vague in the setting of its objectives.

The implementation of the Water Framework Directive has strengthened certain water management plans. The River Basin Committee revises the so-called "Schéma Directeur d'Aménagement et de Gestion des Eaux" (SDAGE) on a six-year basis.[699] The basin-coordinating prefect subsequently approves this scheme.

696 Corinne Larrue and others (n 599) xiv.
697 Inter alia: Calais, Saint-Omer, Béthune-Armentières, Lille, Lens, Doai, Valenciennes.
698 Article L.566-3 of the Environmental Code.
699 See *infra*.

The SDAGE has binding legal force, as subsequent administrative decisions, e.g. in the urban planning realm, should comply with the provisions of the SDAGE. The SAGE, or the water management plan, constitutes an operationalisation of the SDAGE with the goal of implementing the provisions included in the SDAGE, and should be developed at the level of the sub-basin, i.e. by the Local Water Committee.[700] Whereas the SDAGE must obligatorily be drafted and issued, the SAGE has a more optional character. Efforts have been made, in the context of integrated water management, to link the FRMP (PGRI) and the SDAGE. For example, the chapters dealing with floods that are included in the SDAGE are absorbed by the PGRI, and there are mutual compatibility requirements between the respective planning documents.

213. In addition to the water management plans, in the context of land-use planning, there are several relevant plans, i.e. the Local Urbanistic Plan (PLU), which is the spatial plan issued at the level of the municipalities defining building zones, and the Plan for Territorial Coherence (SCOT), which is an inter-municipal spatial planning tool with the goal of creating coherence with respect to several policy domains, such as mobility and the environment.[701] The interdependencies between the water management plans on the one hand and the urban-planning documents on the other, have been provided for in the legislative framework. The SCOT should be compatible with both the SDAGE and the SAGE, and the PLU should be compatible with the SCOT. Only when there is no SCOT, should the PLU be compatible with the SDAGE and the SAGE. The SCOT and PLU should be compatible with the FRMP issued in the context of the Floods Directive.[702]

2.2.5. The existence (or lack thereof) of legally enshrined safety standards

214. An important difference between some of the countries and regions sharing the Scheldt River relates to the existence or absence of legal safety norms for flood risks. There is a legal philosophical angle to this question, as one may wonder whether it is equitable that citizens in a certain country cannot invoke legal safety standards, whereas citizens across the administrative border within the same hydrological basin do have this possibility. Approaching the question from the perspective of the authorities, safety norms also provide them with greater legitimacy to carry out, sometimes contentious, flood risk management measures that can be based on solid legal requirements. Within the realm of the flood-protection strategy, in contrast to the Netherlands, the Flemish Region has not adopted legal safety standards. Suggestions have been made by water managers in the Flemish Region to introduce safety standards for flood risks, e.g. linked to the

---

700 "Schéma d'Aménagement et de Gestion des Eaux".
701 Article L151-1 and following of the Urbanistic Code.
702 L111-1-1 CU. Article L566-7 CE.

type of activity on the land, or expressed in terms of flood damages. Respondents in the Flemish Region stress that the safety norms should be sufficiently refined in order to be useful.[703] There are certain area-specific safety norms in the Flemish Region as well, albeit not legally enshrined, as they are in the Netherlands. For example, the Sigma Plan is a project of the Flemish Region that aims to protect the Scheldt Region from floods, whilst also respecting nature, economy and recreation, involving 250 km of the Scheldt River.[704] In the Sigma Plan, various safety norms have been established. These safety norms result from a cost-benefit analysis, in which several possibilities related to the protection against floods have been weighed.[705] Flanders has made agreements with the Netherlands regarding certain areas, to better align the safety norms across the basin, for example, at the border with the Netherlands.[706]

215. One of the main differences between Flemish and Walloon flood risk management on the one hand, and Dutch management on the other, is that in the context of the latter, legal safety standards have been developed.[707] In this regard, a distinction should be made between the areas that are protected by primary flood-defence structures and the areas protected by secondary or regional flood-defence structures. Since January 2017, new safety norms have been included in the legal framework. These are based upon the consideration of the risks of a flood in combination with its consequences. A safety level of 10-5 is applicable to all persons living behind dykes or dunes, i.e. their chance of death as a result of a flood cannot exceed 1:100,000 per year.[708] All primary flood-defence structures need to be compliant with these safety norms by 2050. The Netherlands is the only country in the Scheldt basin in which such legally enshrined safety norms are applicable. The question arises whether it is problematic that one country has legal protection norms, whilst in the same basin, across the international administrative border, no such norms are applicable. It can be said that these legal safety norms grant the Dutch competent water authorities more legitimacy to carry out works in comparison to the competent authorities in the Flemish Region, which operate without such legal norms supporting their actions. On the other hand, an advantage of the lack of

703 Interview with former Member of Parliament Flemish Region, 7 July 2014.

704 Flemish Parliament, debate Committee on public works, mobility and energy, meeting (15 October 2002) <https://docs.vlaamsparlement.be/website/htm-vrg/327696.html> accessed 12 July 2017.

705 Whereas in the Netherlands, safety standards have been determined before in the legal framework, and have to be complied with.

706 Interview with civil servant navigable watercourses Waterwegen en Zeekanaal, 13 January 2014.

707 Which are included in the Water Act.

708 Regeling van de Minister van Infrastructuur en Milieu, van 2 december 2016, nr. IENM/BSK-2016/283517, ter uitvoering van de artikelen 2.3, eerste lid, en 2.12, vierde lid, van de Waterwet, houdende regels voor het bepalen van de hydraulische belasting en de sterkte en procedurele regels voor de beoordeling van de veiligheid van primaire waterkeringen, Dutch Official Journal 20 December 2016.

legally enshrined safety standards would be that Flemish water managers have more flexibility to respond to ad-hoc situations.[709] Like the Flemish and Walloon Region, and unlike the Netherlands, France has not developed legal safety standards for flood risks for the entire country. However, the Environment Code sets forth certain safety levels in the context of permit applications for the instalment of flood-defence infrastructure.[710]

2.2.6. Preparing for and recovering from floods

216. In Belgium, the competences with regard to the preparation and recovery strategy are mainly situated at the federal level in Belgium, and therefore apply both to the Walloon and the Flemish Region. The Royal Decree of 16 February 2006 makes a distinction between 3 different types of plans: multi-disciplinary Emergency and Intervention Plans, mono-disciplinary Intervention Plans and Internal Emergency Plans.[711] These plans are activated whenever an emergency situation or an imminent emergency situation emerges.[712] The multi-disciplinary Emergency and Intervention Plans (EIP) consist of a General Emergency and Intervention Plan (GEIP) and a Specific Emergency and Intervention Plan (SEIP). The GEIP contains the necessary information and general guidelines to ensure the management of an emergency situation. The SEIP supplements the GEIP with additional specific guidelines for particular risks.[713] EIPs are drawn up at federal, provincial and municipal level.[714] The 2006 Royal Decree and related Ministerial Circulars contain provisions related to the minimum content and the method of drawing up these plans.[715] Crisis response is divided into a municipal, provincial and federal phase.[716] Factors such as the nature of the emergency and the means available determine which phase should be launched. For example, the municipal

709 Interview with civil servant Flemish department of navigable watercourses, 26 March 2014.

710 Article R214-119-3 of the Environment Code. Three types of standards exist: 1/200 for Class A, 1/100 for Class B and 1/50 for Class C. This applies to permits introduced after 1 January 2020.

711 Article 2 of the Royal Decree of 16 February 2006, Belgian Official Journal 15 March 2016.

712 Ibid, Article 6.
An emergency situation is defined here as "any event that causes or can cause damaging effects for the social life, such as a serious disturbance of the public safety, a serious threat to the live or health of persons and/or to important material interests, and which requires the coordination of the disciplines to take the threat away or to limit the harmful consequences".

713 Article 3 of the Royal Decree of 16 February 2006.

714 Ibid, Article 3.

715 The EIP should include, among other things, information about: the modalities of the different alarm phases, the procedures concerning alarming the population, the communication means to employ, the means of transport and the reception and accommodation of the victims in case of an evacuation. The SEIP should contain provisions about: a description of the risk involved, the data regarding persons who are specifically involved in the risk, the accident scenarios and intervention procedures for each scenario, and so forth. Article 26 and Article 27 of the Royal Decree of 16 February 2006; Ministerial Circular of 26 October 2006 concerning the emergency and intervention plans.

716 Article 7 of the Royal Decree of 16 February 2006.

phase is launched and is scaled up to a higher phase when several municipalities are involved.[717]

In the Netherlands, a prominent feature in the flood risk management realm in the Netherlands are the so-called "safety regions", which are especially important in terms of preparedness for floods, i.e. for crisis management and evacuations.[718] These safety regions are regulated on the basis of the Safety Regions Act[719], which requires municipalities to jointly establish public bodies responsible for emergency management, and to then delegate their competences with regard to this topic to these bodies.[720] The Safety Regions Act determines the division of the territory of the Netherlands into safety regions, and the municipalities in question are then required to make the proper arrangements in accordance to the Act.[721] There is alignment between the activities of the regional water authorities and those of the safety regions.[722] For example, the safety regions are responsible for emergency management and have to draw up specific plans, i.e. policy plans and crisis plans.[723] The latter plans must be aligned to the calamity plans issued by the regional water authorities.[724] As is true for crisis management regarding flood risks in Belgium, several plans based on specific risks need to be drafted in order to prepare for flood events.[725] Moreover, the chairman of the regional water authority of the territory of the safety region may participate in meetings conducted by the boards of the safety regions, but does not have a formal legal role.[726] As is the case in Belgium and the Netherlands, several calamity plans are issued in France as well, e.g. the Plan ORSEC is drafted by the department prefecture and municipal rescue plans are drawn up by municipalities.[727]

217. In terms of the recovery strategy, especially the French and the Belgian system are similar, and the Dutch scheme is rather different. In the Netherlands, the recovery strategy is only considered as a marginal aspect of the flood risk management realm of the country, as the focus is on preventing floods from

---

717 Ministerial Circular of 26 October 2006.

718 Willemijn van Doorn-Hoekveld, 'Compensation in Flood Risk Management with a Focus on Shifts in Compensation Regimes Regarding Prevention, Mitigation and Disaster Management' (2014) 10 Utrecht Law Review 216.

719 Act of 11 February 2010 on Safety Regions.

720 Maria Kaufmann and others, *Analysing and Evaluating Flood Risk Governance in the Netherlands. Drowning in Safety?* (STAR-FLOOD Consortium 2016) <www.starflood.eu/documents/2016/03/wp3-nl-final-webversion.pdf>, accessed 15 April 2016.

721 These are included in the annex of the Act, see Article 8 of the Safety Regions Act.

722 Regional water authorities will be discussed below.

723 Article 14: these policy plans need to be established every four years, and outline the policy with regard to the tasks of the safety region.

724 Ibid.

725 Policy plans, crisis management plans, calamity plans. See Ibid.

726 Article 12 Safety Regions Act.

727 Herman-Kasper Gilissen and others, 'A Framework for Evaluating the Effectiveness of Flood Emergency Management Systems in Europe' (2016) 21 Ecology and Society 27.

happening in the first place. In France and Belgium, the mechanisms for financial ex-post compensation are rather well developed and have been deeply rooted in the legislative frameworks for several decades. Needless to say, the consideration that these countries in one and the same River Basin District, with regard to this aspect of flood risk management, are polar opposites is relevant in this context.

The Belgian ex-post compensation scheme, like the French scheme, revolves around an automatic and compulsory inclusion of the flood coverage in the first-party fire insurance, with a public fall-back mechanism.[728] Since the entry into force of the 2005 Insurance Act, an individual that takes out a property-level fire insurance, is automatically covered for flood risks.[729] In theory, fire insurance is not mandatory, but approximately 95% of the owners and 89% of the renters have fire insurance, as it is generally a condition to obtain a mortgage and to conclude a rental contract. Prior to this 2005 Insurance Act, floods were not covered by insurance, and compensation was granted through a public fund, for which a cumbersome procedure had to be followed and only after the acknowledgment of the flood event as a natural disaster through Royal Decree. In that case, the National Calamities Fund partly compensated the victims in question. Following a series of natural disasters, the Belgian legislator considered the necessity for insurance related to natural disaster coverage.[730] The Belgian scheme includes a link between the recovery strategy and the flood risk prevention strategy by advocating a risk-differentiation approach. Buildings constructed in high-risk areas do not benefit from the government's cap on insurance fees, and can also be refused coverage by the insurer. High-risk zones in the insurance context, in general, are the zones that are subject to recurrent and serious floods. The recurrence period is defined as the "inverse of the annual probability of occurrence or exceeding of a particular event".[731] For example, for an event with a recurrence period of 25 years, there would be a chance of 1 in 25 that an event of the same intensity would occur the following year. As for the importance of a flood, the amount of overflowing water, and thus the depth of the water is decisive. These high-risk zones are designated through the 2007 Royal Decree on High-Risk

728 Cathy Suykens and others, 'Dealing with Flood Damages: Will Prevention, Mitigation and Ex-Post Compensation Provide for a Resilient Triangle?' (2016) 21 Ecology and Society 1. Michael Faure and Véronique Bruggeman, 'Catastrophic Risks and First-Party Insurance' (2008) 15 Connecticut Insurance Law Journal 1.

729 See Act of 17 September 2005 modifying, as regards the Insurance for Natural Catastrophes, the Act of 25 June 1992 on the Land Insurance Agreement and the Act of 12 July 1976 on the Repair of Certain Damage Caused to Private Goods by Natural Disasters, Belgian Official Journal 11 October 2005.

730 E.g. the windstorms in January 1990, the earthquakes in 1983. See Véronique Bruggeman, *Compensating Catastrophe Victims: A Comparative Law and Economics Approach* (Wolters Kluwer 2010).

731 Own translation.

Areas.[732] The Regions are responsible for designating these high-risk areas.[733] The above-mentioned stricter regime for buildings in the high-risk areas, i.e. the fact that these do not benefit from the cap and suffer the possibility of refusal, only applies to those built 10 months following the entry into force of the 2007 Royal Decree.[734] As mentioned above, there is a public fall-back mechanism. Indeed, the Disaster Fund intervenes when certain thresholds of the insurers are exceeded, or for those damages not covered by the insurance policy. A major State Reform in Belgium changed the institutional settings of the country, and transferred the competences for the Disaster Fund from the federal to the regional levels, i.e. the Flemish Region, Walloon Region and the Brussels-Capital Region.[735]

In 2010, Belgium was hit by serious floods, and subsequently carried out an evaluation of the flood risk governance realm. An important conclusion relates to the criticism that water managers and insurance actors should enhance their cooperation.[736] An important aspect of flood risk management that could benefit significantly from an improved synergy between public and private actors in this regard, transcending the recovery strategy, relates to adaptive building. Indeed, it is paramount that measures not only focus on avoiding construction in flood-prone areas, but also on flood-proofing buildings that already exist, especially taking into account the degree of land sealing in the Flemish Region and the population density.[737] Insurers are hesitant to take into account these adaptive building measures at property level, e.g. in flood-prone areas, in their calculation of the fees on a structural basis[738], and the legislator and executive bodies have not provided guidance or regulations in this regard.

218. In France, the CAT-NAT scheme constitutes a cooperation mechanism between the State and private parties, i.e. the insurance companies.[739] National

732 Belgian Official Journal 23 March 2007.

733 The criteria on the basis of which the Regions delineate the risk zones are determined by the Royal Decree of 12 October 2005, Belgian Official Journal 21 November 2005.

734 Which is 23 September 2008. In these zones, insurers can refuse coverage. Also, if the insurer does insure, the maximum tariffs set forth by the Tariff Office do not apply. This means that the insurer can choose the height of the premium.

735 See the new Article 6, Para 1, II, 5° of the Special Act of 1980. X, Government Agreement of 1 December 2011.

736 E.g. the suggestion was raised that both parties should conclude a cooperation agreement with the goal of mutually sharing natural disaster data. For example, interview with civil servant Flemish Environment Agency 19 April 2014.

737 In other words, the lack of available space.

738 E.g. doubts relating to the effectiveness of the measures, the influence of human intervention, etc. Interview with stakeholder in the insurance sector, 5 March 2014.

739 For more information, see Suykens and others (n 728); Jennifer K Poussin, WJ Wouter Botzen and Jeroen CJH Aerts, 'Stimulating Flood Damage Mitigation through Insurance: An Assessment of the French CatNat System' (2013) 12 Environmental Hazards 258; Jennifer K Poussin, Wouter Botzen and Jeroen CJH Aerts, 'Effectiveness of Flood Damage Mitigation Measures: Empirical Evidence from French Flood Disasters' (2015) 31 Global Environmental Change 74; Olivier Moréteau, 'Policing the Compensation of Victims of Catastrophes:

solidarity is the backbone of French flood risk management, and this is especially apparent in its ex-post compensation scheme. Natural disasters are automatically included in first-party insurance policies for buildings and moveable property.[740] In contrast to the Belgian insurance scheme, French citizens pay a fixed rate as a trade-off for the natural disasters coverage, regardless of the degree of risk to which they are exposed based on the location of their building. The solidarity principle is thus taken to the next level, in comparison to the way the system works in Belgium, where people living in high-risk areas do not benefit from the government's cap on insurance fees for flood risks and can be confronted with a refusal by the insurer to cover their building.[741] The Central Reinsurance Fund (CCR) is a public fund that acts as a reinsurer. In France, there is a link between prevention and recovery through the interaction between the CAT-NAT and the Barnier Fund. There is a redistribution mechanism as 12% of the proceeds of the premiums flow into the Barnier Fund, which is a fund used to support measures for the prevention of natural phenomena, including floods.[742] In contrast to the Belgian scheme, insurance only steps in after acknowledgment of the flood event as a natural disaster through Royal Decree.

219. The recovery strategy is defined in a very different manner in the Netherlands compared to the other countries and regions in the Scheldt basin.[743] The Netherlands prioritises the prevention of floods and flood risks, e.g. through spatial planning policies, defence infrastructure, and so forth. This stems from the geophysical circumstances of the country, as a low-lying Delta, and more specifically the consideration that the damages resulting from floods, should a flood event take place, would be insurmountably high, taking into account, among other things, the fact that 59% of the country is a flood-prone area and approximately 26% of the country is below sea level. The preparation and recovery strategies are therefore less prominent in comparison to the risk-prevention, mitigation and defence strategies.

220. Even more so than the preparation strategy, the recovery strategy, in comparison to the other flood risk management strategies such as risk prevention and defence, takes up a rather minor place in the Dutch flood risk management realm.[744] Following the 1953 floods that hit several European countries, Dutch

---

Combining Solidarity and Self-Responsibility' in Willem Van Boom and Michael Faure (eds), *Shifts in Compensation between Private and Public Systems* (Springer 2007).

740 Bruggeman (n 730).

741 Of course, it should be noted that the natural disaster coverage is included in the fire policy and constitutes just one of the elements of coverage.

742 Suykens and others (n 728).

743 Ibid.

744 Bruggeman (n 730); Véronique Bruggeman, 'A Critical Comparison of the Main Compensation Mechanism for Victims of Natural Catastrophes in Belgium and the Netherlands. With a Law and Economics Twist' (2011) 8 Journal for European Environmental & Planning Law 46.

insurance companies declared that flood risks would be uninsurable in the Netherlands.[745] Ex-post compensation in the Netherlands takes place through the Calamities Compensation Act (CCA), where it should be noted that the compensation fund is exclusively managed by public authorities. This stems, among other things, from the above-mentioned consideration that flood risk management is very much seen as a government responsibility on the basis of the Constitution. A pilot insurance project, the so-called "Neerlandse", was established in 2011[746], which mainly focuses on fluvial flooding. However, the Neerlandse insurance is only taken out by a marginal number of citizens.[747] Flood risk management in the Netherlands in general is characterized by a high degree of solidarity between citizens, as all citizens contribute to the implementation of the government responsibility to protect its people from floods. However, the main distinction that is relevant in this context is between those living outside the defence structures and those living within. If damages from floods occur outside these structures, the victims will not be able to benefit from CCA compensation.[748] Inside the defence structures, full compensation is not guaranteed, as the calamities fund is not equipped to compensate in case of a major flood event.[749]

2.2.7. Public participation in the basin countries

221. Public participation is seen as one of the cornerstones of the WFD and FD.[750] Moreover, the Aarhus Convention and implementing EU Directives require EU Member States to involve the public in decision-making processes. On the basis of Articles 6 and 7 of the Aarhus Convention, each party must make appropriate provisions to enable public participation during the preparation of plans and programmes relating to the environment (Article 7 of the Aarhus Convention). To this end, the public should be informed by public notice (or individually), early on in the environmental decision-making process, and in an adequate, timely and effective manner (Article 6 of the Aarhus Convention). All Scheldt countries have public-participation processes and transparency requirements in place, enshrined in the legal frameworks. Dissemination and consultation of draft River Basin Management Plans and Flood Risk Management Plans form an integral part of flood risk management frameworks of the Scheldt parties.[751]

745 Suykens and others (n 728).
746 Insurance for pluvial flooding is available since 1998 and is part of property insurance.
747 Ibid (n 728).
748 Ibid.
749 Kaufmann and others (n 720).
750 See Chapter II of this book.
751 Larrue and others (n 599) 43. Mees, Suykens and others (n 525). Kaufmann and others (n 720) 36.

For example, in the Flemish Region, public participation with respect to water management is one of the key principles embedded in the Decree Integrated Water Policy.[752] The DIWP requires the relevant authorities to actively involve the citizens in the preparation and determination of the integrated water policy, as well as in the specific execution of the policy in practice.[753] Citizens and stakeholders should be consulted for a 6-month period, written remarks may be submitted and should subsequently be taken into account. Each individual may also request background documentation and information used in the course of the preparation of the draft RBMP.[754] Parallel to the public consultation procedure and during the same period of time, the plans are submitted to advisory institutions.[755]

222. However, in practice, participation is rather limited and takes place at the end of the decision-making process, e.g. in France and the Flemish Region.[756] One of the reasons of this limited participation in practice, relates to a lack of awareness of flood risks, which is the case e.g. in Belgium and the Netherlands.[757] Public debates on this issue are rare.[758] The requirement to take into account the feedback from the population in the context of the consultation procedure, is only legally enshrined in the Netherlands.[759]

This means that public consultation takes place rather late in the decision-making process, which is a more general issue in flood risk management, common in several other Member States.[760] For example, the Flemish DIWP provides

752 See Article 6, 8° of the Decree Integrated Water Policy. This is the first time that participation is included as a principle in environmental legislation in the Flemish Region. The participation principle within the meaning of the DIWP entails that citizens should, at an early stage and in an adequate, timely and effective way, have a voice in the preparation, the determination, the implementation, the follow-up and the evaluation of the integrated water policy. The inclusion of this principle is therefore an explicit translation of the relevant provisions included in the Aarhus Convention. The Explanatory Memorandum to the DIWP explicitly refers to the Aarhus Convention. See Explanatory Memorandum to the draft Decree Integrated Water Policy, Parl. St., Vlaams Parlement (2002-2003) no. 1730/1, p. 21.

753 Certificate for the implementation report of the Aarhus Convention (25 November 2010) <www.unece.org/fileadmin/DAM/env/pp/reporting/NIRs%202011/rapport_VL_2010_Ned_met_wijzigingen_def.pdf>, accessed 12 February 2014.

754 Article 37 Para 2 DIWP.

755 See Article 37 Para 3 DIWP and Article 47 Para 1 DIWP. E.g. the environmental council, the socio-economic council of Flanders, the water boards, etc.

756 Maria Pettersson and others, 'How Legitimate Is Flood Risk Governance in Europe? Insights from Intra-Country Assessments' (In review with Water International). Mees, Suykens and others (n 525).

757 Mees, Suykens and others (n 525) 19; Kaufmann, and others (n 720) 36.

758 Larrue and others (n 599) 43.

759 Ibid.

760 Maria Pettersson and others, 'How Legitimate Is Flood Risk Governance in Europe? Insights from Intra-Country Assessments' (In review with Water International); Kristina Ek and others, 'Strengthening and Redesigning European Flood Risk Practices Towards Appropriate and Resilient Flood Risk Governance Arrangements Design Principles for Resilient, Efficient

that, following publication of the draft RBMP, and at the latest 90 days before expiration of the period within which written remarks can be submitted, the CIW should, in cooperation with the sub-basin secretariats, organise an information and participation meeting.[761]

2.3. PRELIMINARY CONCLUSION

223. The above sections have shown that flood risk management approaches differ considerably between the countries and regions in the international Scheldt basin, from prevention to recovery. Flood risk management clearly does not embody the "one size fits all" slogan, important in assessing subsidiarity, e.g. at the EU level. However, it cannot be denied that inappropriate legal frameworks and policies in one of the Scheldt countries, either in the short or in the long term, have an impact on the existence and extent of flood risks in the other countries. Taking into account the differences in institutional and legal frameworks as well as in the geographical realities, the existence of cooperation and coordination mechanisms and practices is the way to go forward, rather than a complete harmonisation of specific flood risk management measures and instruments.

224. In the Flemish Region, the legislature has carried out a substantial reform of the applicable legal framework following the evaluation of the first-generation RBMPs for the period 2009-2015 and the occurrence of a heavy flood event in 2010. The goal of this reform was to simplify planning processes and structures with a view to increasing their effectiveness. This reform has encouraged positive changes in the Flemish legal landscape, for example, the public participation processes for the different plans have been streamlined. Moreover, in the past ten years, legal instruments have been adopted to prevent flood risks and mitigate floods. These include the water assessment, signal areas and the duty to inform. Characteristic of the Flemish legal framework is the integrated approach to water management, as the Floods Directive and Water Framework Directive

and Legitimate Flood Risk Governance; Lessons from Cross-Country Comparisons' (STAR-FLOOD Consortium, 2016).

761 See Article 37 § 6 DIWP and Article 47 § 4 DIWP. Following the finalisation of the public consultation, the public institutions submit all their written remarks to the advisory institutions responsible for drawing up the water management plans (i.e. the CIW and the sub-basin secretariats). Once the management plans have been finalised, they are submitted to the Flemish Government, who notifies the relevant authorities (for example, at municipal and provincial level). Finally, the plans are published in the Belgian Official Gazette and are made available for inspection at the offices of the provinces, municipalities, secretariats and water boards. See Article 37 § 8 DIWP and Article 47 § 6, and Article 48 DIWP. Certificate for the implementation report of the Aarhus Convention (25 November 2010) <www.unece.org/fileadmin/DAM/env/pp/reporting/NIRs%202011/rapport_VL_2010_Ned_met_wijzigingen_def.pdf>, accessed 12 February 2014.

have been integrated into a single legal framework, the RBMPs and FRMPs have been merged into one single plan, and the Coordination Commission Integrated Water Policy develops policies for both water quality and water quantity.

225. In the Walloon Region, the Floods Directive was implemented in the Water Code, which also serves as the implementing framework for the Water Framework Directive, in 2010. The Plan PLUIES constituted the initial step toward integrating different flood risk management strategies, by which the Floods Directive has further induced comprehensive flood risk management in the region. The Walloon legal framework has not integrated the Flood Risk Management Plans with the River Basin Management Plans. The approach to flood risk governance of the Walloon region has developed from a focus on flood defence toward more mitigation-based measures, e.g. natural flood-control zones and re-meandering. Instruments to prevent construction in flood-prone areas, such as the water assessment, also exist in the Walloon Region.

226. Protecting the population against floods is a constitutional duty of the government, which is quintessential in the legal framework for flood risk management in the Netherlands. Functional authorities established at the hydrological scales are subject to this constitutional requirement. From a legal point of view, the Netherlands approaches water management in an integrated manner, with the Water Act covering both water quality and quantity. In addition to traditional measures stemming from the defence strategy, spatial planning instruments are used to create more space for water. An example is the water assessment, which relates to the impact of spatial plans and decisions on the water system. An important component in the Dutch legal framework is the Delta Programme, which aims to enhance climate resilience and the robustness of fresh-water supply and water safety. As the Netherlands already had an extensive legal framework in place for water management and flood risk management, the Floods Directive did not have a substantial impact on its legal landscape.

227. From an institutional perspective, the framework for flood risk governance in France can be described as centralised with strong local competences. There is a multitude of plans relevant to flood risk governance in France, e.g. the PPRN, PPRI, SDAGE, SAGE, PLU, and SCOT. An important development in the French legal landscape relates to the transfer of competences to the local level through the 2014 MAPAM Act, and the increasing possibilities for municipalities to bundle their competences, e.g. in EPTBs. This development and the inter-municipal cooperation vehicles are also relevant to the regional level. The recovery strategy plays an important role in the French legal framework, with a clear legal link to the prevention strategy. A good illustration of this link is the partial transfer of proceeds of the premiums in the context of the CAT-NAT schemes into the Barnier Fund, which in turn promotes measures to prevent floods.

228. There are opportunities to cooperate with respect to all flood risk management strategies identified, although clearly in varying degrees, from (i) exchange of lessons learnt, to (ii) exchange and coordination of information, to (iii) coordination of measures, as illustrated in the table below.[762] What has become apparent in the overview of the flood risk management frameworks of the countries and regions sharing the Scheldt, is that similar legal instruments exist throughout the basin, although slightly different in every country.[763] For example, the water assessment is used in the Flemish Region, Walloon Region and the Netherlands. The main necessity for cooperation regards the defence, risk-mitigation and preparation strategies, e.g. with regard to hard infrastructure to protect the floods, alarm systems, and so forth. Spatial planning and water management decisions as part of the risk prevention strategy have a more implicit transboundary character. For example, when the Flemish water assessment is applied incorrectly, and a permit is granted for an activity which influences the drainage regime of the watercourse, this will have consequences further downstream.

Table 5. Flood Risk Management Strategies and Examples of Cooperation

Risk prevention	Risk mitigation	Defence	Preparation	Recovery
- Information exchange on FRM instruments (e.g. Dutch & Flemish water assessment, Flemish duty to inform)	- Information exchange on procedures green roofs, flood-control areas, etc.	- Coordination on measures: dredging, dike elevations - Cooperation on determining safety standards for the basin - Coordinating the meaning of "significant risk"	- Cooperation on calamity plans, flood forecasting, warning systems	- Information exchange on risk differentiation - Information exchange on link with prevention (e.g. French Barnier Fund)

229. Several pieces of the puzzle are still missing before river basin management in the Scheldt District can be evaluated. First, there needs to be clarity on which authorities operate at which scales. Then, the manner in which the legal and policy frameworks of the Scheldt parties provide for inter-basin interaction, e.g. the question whether the national legal instruments require cooperation and with regard to which measures, how the parties interpret the "significant increase of risk" component of the Floods Directive, and so forth. Moreover, it needs to be clear which instruments exist to facilitate these interactions.

---

762 With regard to substantive and procedural aspects, i.e. Pillars II and III from the Chapter II pillars.

763 Encouraged by, among other things, the FD and WFD.

3. THE NUTS AND BOLTS OF (HYDROLOGICAL SCALE) GOVERNANCE AND THE ACTOR LANDSCAPE IN THE SCHELDT DISTRICT

3.1. THE DESIGNATED HYDROLOGICAL SCALES

Figure 9. Sub-basins in the Flemish Region

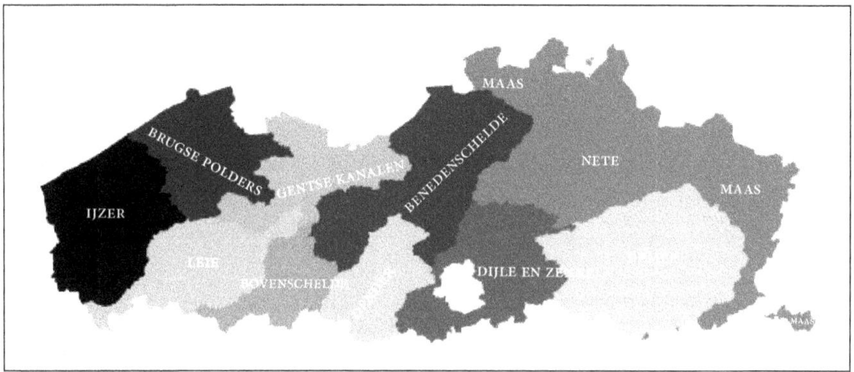

Source: www.integraalwaterbeleid.be

230. Now that the legal frameworks and instruments have been clarified, it is necessary to assess the question of the governance at hydrological scale. Indeed, Chapter II has teased out the precise meaning of "river basin management" or hydrological scale management from the EU perspective, which now needs to be operationalised at the basin level. Which hydrological scales have been designated in the respective basin countries? Which authorities operate at these scales? What are their characteristics in terms of type, legal personality, decision-making power, and so forth? An analysis of these river basin management bodies at local scale is interesting, because it might provide a mirror to river basin management at the higher, more international level. Moreover, in an ideal situation, links across hydrological scales should exist in the EU.[764] This is indeed one of the criteria to determine whether the governance mechanism as a whole can be considered resilient or not.

231. In terms of planning in light of the WFD and the FD, four main levels of hydrological scales can be distinguished. The highest level is the River Basin District that forms part of the international River Basin District within the meaning of the WFD and FD. In the Flemish Region, the Scheldt and the Meuse basins correspond to the level of the River Basin District and are governed by

[764] Priest and others (n 604).

the Flemish government. Furthermore, there are four river basins in the Flemish Region, i.e. the Meuse, Scheldt, Brugse Polders and the IJzer.[765] The third relevant level is the level of the sub-basin, of which there are eleven in the Flemish Region. Finally, there are 103 sub-sub-basins, which play a role at the most local level of governance.[766]

232. Four IRBDs are situated in the Walloon Region, i.e. the Scheldt, the Meuse, the Rhine and the Senne. Fifteen sub-basins have been identified, five of which are located in the Scheldt basin, as pictured in figure 11.[767]

233. The Netherlands also has four IRBDs running through its territory, i.e. the Ems, Meuse, Rhine and Scheldt. These River Basin Districts have been split up further into seven sub-basins, as pictured in figure 10.[768] Many important governance decisions related to water management take place at this sub-basin level.

Figure 10. Sub-basins in the Netherlands

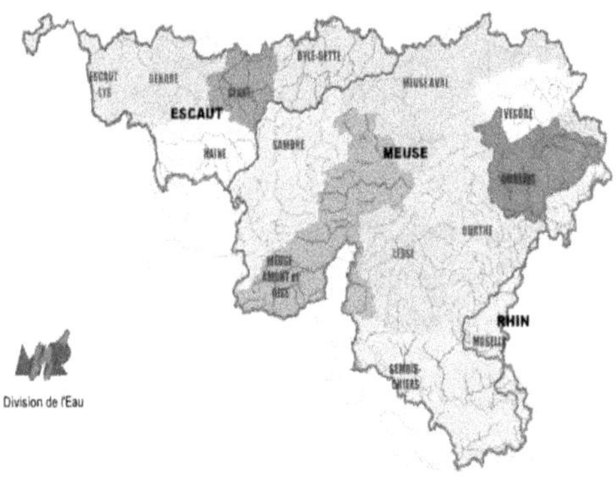

Source: helpdesk water.nl

Figure 11. Sub-basins in the Walloon Region

Source: Union of Cities and Municipalities Wallonia

765 These are primarily governed by the Region and more specifically by the Coordination Commission Integrated Water Policy under the auspices of the Flemish Environment Agency. Moreover, the level of the Flemish Region is relevant, because of legislation and policies established at this level, and because of policy notes that are integrated into the planning output toward the EU. This will be explained in the section on competent authorities.

766 The sub-basin boards are the relevant governance bodies for the sub-basins and the sub-sub-basins, which will be discussed below.

767 I.e. Dendre (Dender in Dutch), Dyle-Gette, Escaut-Lys, Haine and Senne.

768 These sub-basins are the Ems, Scheldt, Meuse, Nether Ems, Rhine-North, Rhine-East and Rhine-West.

234. The main IRBDs in France are the Rhine, Meuse, Scheldt and Rhone, which is shared with Switzerland.[769] Thirteen sub-basins have been identified in France, of which four are overseas territories and six are transboundary, as illustrated in figure 12.[770] These are often combined under the auspices of one authority. This is also the case for the Scheldt RBD, which constitutes, together with the Samber RBD, the basin "Artois-Picardie", for which the RBMP and FRMP have been adopted.

Figure 12. Sub-basins in France

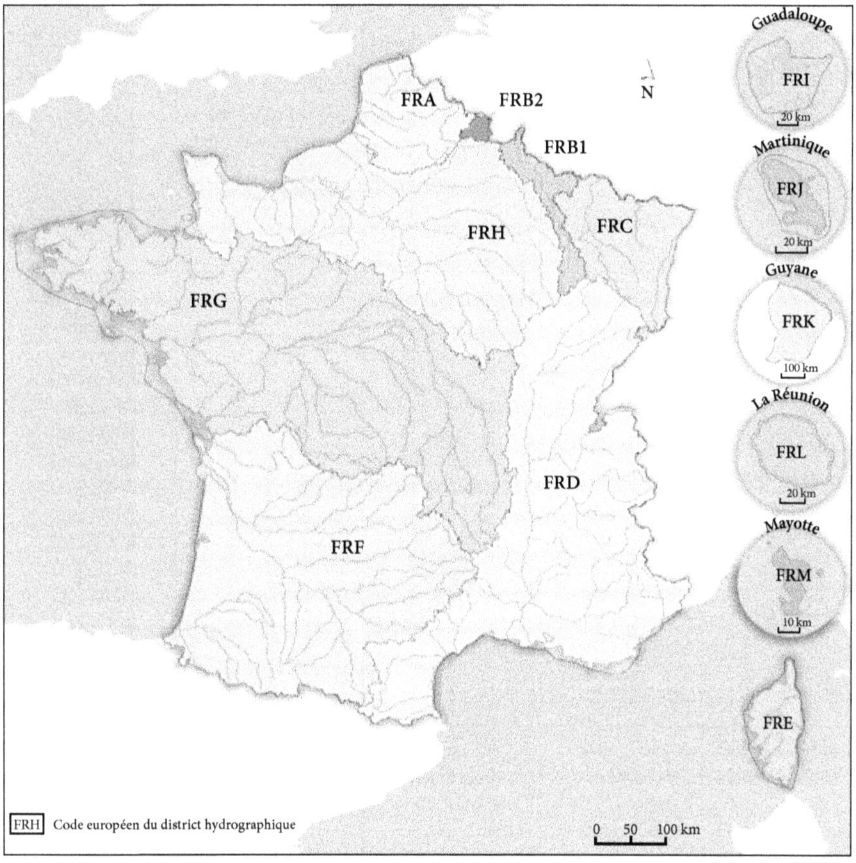

Source: French Ministry of the Environment

769 Other transboundary river basins are Garonne, Ebro and Po, of which small stretches cross borders.

770 Rhône, Adour Garonne, Rhin-Meuse, Artois Picardie, Seine and Normandie. Commission, 'Commission Staff Working Document Accompanying the Report from the Commission to the European Parliament and the Council on the Implementation of the Water Framework Directive (200/60/EC) River Basin Management Plans' (2012) SWD(2012) 379 final.

3.2. THE COMPETENT AUTHORITIES AT THE RELEVANT (HYDROLOGICAL) SCALE

Table 6. Watercourses and Competent Authorities[771]

Watercourses	Flemish Region	Walloon Region	The Netherlands	France
Different types of categorisation	**Navigable:** – Department of Mobility and Public Works – The Flemish Waterway – Agency of Maritime Services and Coast – Maritime Access	**Navigable:** Department of navigable watercourses (DG02)	**National water systems and primary flood defences:** – Ministry of Infrastructure and the Environment – Rijkswaterstaat – Delta Commission	**State:** Regional prefect Departmental prefect Mayor
	Non-navigable: Flemish Environment Agency (VMM) and Coordination Commission Integrated Water Policy (CIW) Provinces and Municipalities Provinces and Municipalities 02. Polders and wateringues	**Non-navigable:** – Department of non-navigable watercourses (DG03) – Provinces – Municipalities – Wateringues		**Regional:** – Regional Council – Departmental Council Municipality + Intermunicipal Body / Association (e.g. EPCI (public entity for intertownship cooperation))
Sub-basin and sub-sub-basin	– All actors mentioned above – Sub-basin boards (but only advisory role)	– All actors mentioned above – River contracts	– Provinces – Regional water authorities (hydrological scale entities) – Municipalities (spatial planning)	– Water Agency (water boards) – Basin regional prefect – River Basin Committee & Local Water Committee (CLE)

771 The specific water management authorities are included in this table, other relevant entities such as those responsible for spatial planning, forests and land have not been included.

				- River contracts - EPTB and EPAGE (public river (sub-) sub-basin territorial authority) - River or catchment basin syndicates

235. The Scheldt parties make different categorisations to designate responsibilities for their parts of the IRBD. The Flemish Region and the Walloon Region make a distinction between navigable and non-navigable watercourses, and within the latter category, three further sub-categories which are governed by different administrative layers. France and the Netherlands, in turn, make distinctions mainly based on national versus regional water systems. The Netherlands has the most prominent hydrological scale governance with functional authorities having responsibilities for the designated catchments.

236. In the Flemish Region and the Walloon Regions, separate authorities have responsibilities for navigable and non-navigable watercourses, respectively. In the Flemish Region, the Department of Mobility and Public Works (MOW) is the manager of navigable watercourses, but the management is in the hands of three authorities, namely the Flemish Waterway[772], the Agency of Maritime Services and Coast and Maritime Access. In the Walloon Region, the department of navigable watercourses DG02 is the competent authority. In both regions, three categories of non-navigable watercourses exist, i.e. the first, second and third.

Before 2014, in both regions these were governed respectively by the level of the region, the provinces and the municipalities. However, an internal state reform in the Flemish Region reshuffled the competences with respect to non-navigable watercourses, mainly with the goal of administrative simplification and efficiency increase.[773] The most important shuffle in this regard has been the rescaling of

---

772 Waterways and Sea Canal and NV De Scheepvaart have merged into "the Flemish Waterway" since 1 January 2018.

773 Decree of 28 February 2014 modifying the Act of 28 December 1967 related to unnavigable watercourses, Belgian Official Journal 11 April 2014. See Advice of the SERV, 'Ontwerp van decreet tot wijziging van de Wet van 28 december 1967 betreffende de onbevaarbare waterlopen, meer bepaald de wijziging van de classificatie en andere diverse wijzigingen' (12 September 2013) n 2013/039 <https://www.minaraad.be/themas/hinder/ontwerpdecreet-onbevaarbare-waterlopen-adviesvraag/13-039%20advies%20onbevaarbare%20waterlopen.pdf/download> accessed 12 July 2017.

the third category of watercourses to the second category of watercourses, i.e. a vertical upward governance shift from the level of the municipalities to the level of the provinces. The rescaling was not made mandatory, and several options have been provided to the municipalities, i.e. a cooperation agreement between the municipality and province in question, the status quo and the rescaling from the third to the second category.[774] As mentioned above, the *ratio legis* of these rescaling efforts related to the wish for efficiency gains in watercourses management in the sense that more efficient management should lead to an improved allocation of the available resources. For example, as a result of the transfer from the third to the second category, approximately 50,000 to 100,000 EUR per year is freed up for municipalities, which should be used to better manage their canals and for improved flood risk management.[775]

In the Walloon Region, this rescaling exercise was not carried out. A different type of rescaling has been suggested in the Walloon Region in 2004, i.e. a transfer from the provinces to the region.[776] This suggestion has, to date, not been carried out. In 2012, the Union of Cities and Municipalities of Wallonia issued a recommendation arguing for the transfer of the third to the second category, mainly based on the consideration that municipalities do not have sufficient financial resources to tackle such water resources management, similarly to the Flemish Region. This 3rd to 2nd category transfer has not yet been performed in the Walloon Region, as will be discussed in the section on the Walloon Region below.

237. In the Walloon Region, the first category is managed by the department of non-navigable watercourses DG03, the second category by the provinces and the third by municipalities and wateringues. Various provinces and municipalities can therefore be involved in one single sub-basin. In the Flemish Region, the first-category watercourses are governed by the Flemish Environment Agency. Polders and Wateringues, consisting of riparian land owners, have some competences in a few 2nd and 3rd level watercourses.[777]

238. This means that the Flemish water management scenery is very fragmented, and to improve coordination within this scenery, the Coordination Commission Integrated Water Policy (CIW) was created in 2003. The CIW prepares policies and brings together delegates from several relevant organisations active in water

774 Also, non-classified watercourses were either abolished or upgraded to the second or third category.

775 See press release of the Minister for the Environment (9 May 2014), available on the Internet at: <www.cdenv.be/actua/herschikking-onbevaarbare-waterlopen-stelt-beheer-op-punt> accessed 16 September 2016.

776 Through the Environmental Code, Decree of 27 May 2004 including Book I of the Environmental Code, Belgian Official Journal 9 July 2004.

777 In the provinces of Antwerp, Limburg and at the coast.

management, thus tackling both water quantity and water quality management. It is also under the auspices of the CIW that the RBMPs and FRMPs are developed.

239. These authorities responsible for water management, unlike in the Netherlands, are not functionally decentralised but stem from existing administrative arrangements, and therefore do not correspond to the hydrological scales. In addition to these water managers, the Flemish and Walloon Regions have introduced hydrological scale entities.

In the Flemish Region, the sub-basin boards were introduced to coordinate water management issues at the sub-basin scales. These are established under the auspices and supervision of the Flemish Region. Their geographical scope corresponds to the hydrological delineation of sub-basins, and therefore runs across the region, provinces and municipalities, which means that they are different from the threefold categorisation mentioned in the section above. Prior to the reform of the DIWP in 2013, water boards existed in the Flemish Region. Although in Dutch, these water boards' names, i.e. "waterschappen", are synonyms of the Dutch regional water authorities, they are very different in terms of competences. Indeed, the Flemish water boards did not possess legal personality or any sort of decision-making power. Water boards were abolished as a separate deliberation platform in the 2013 reform of the DIWP and water-board secretariats were integrated into the secretariats of the sub-basin board structures.[778]

These sub-basin boards were created through the DIWP[779] to coordinate, in line with the integrated water management approach, the water quantity and quality issues at local scale, and to help draft the RBMPs in light of the EU WFD requirements.[780] There are three levels of organisation within each sub-basin constellation: (i) the Sub-Basin Secretariat carries out the day-to-day managerial functions of the sub-basin and is, for example, responsible for drafting the sub-basin-specific parts of the River Basin Management Plans, (ii) the Sub-Basin Management, which is further split up into a General Assembly, in which representatives of different departments, e.g. the Flemish spatial planning department, and political representatives from local governments are brought together, and a Sub-Basin Board, and a (iii) Sub-Basin Council, in

778 See Explanatory Memorandum to the Decree modifying the Decree Integrated Water Policy, Parl. St. (2012-2013) 2072 (1) 5.

779 Although sub-basin committees already existed in the 1990s.

780 Mees, Suykens and Crabbé (n 493). Further details with regard to these sub-basin boards were defined in 2005: Order of the Flemish Government of 9 September 2005 related to the geographical categorisation of water systems and the organisation of the integrated water policy, Belgian Official Journal 2 December 2005.

which civil-society actors from relevant sectors are represented. The respective provincial governors preside over the meetings of the sub-basin boards.

The sub-basin boards are, in comparison for example with the Dutch regional water authorities, not autonomous and financially independent and do not possess legal personality. Indeed, they fully depend on the level of the Flemish Region, more specifically the Flemish Environment Agency, for financing, approval of decisions, and so forth. The sub-basin boards do not have binding decision-making power, and can only provide advice. One of the issues related to the sub-basin boards relates to the existing resource deficiency, as a lack of personnel hampers the roll-out of an effective joint governance body.[781] There is no direct hierarchical link or authority between the sub-basin coordinator and the personnel of the sub-basin board. Sub-basin boards do not have legal personality, and therefore do not have the competences, e.g. to apply for funding from EU-level projects. In order to be able to participate in research or other projects, the sub-basins have to be linked to an existing administration. As a consequence, participation of the sub-basin board in said project is linked to the interests of the administration, and not necessarily pursuant to the sub-basin-wide perspective.[782] From a comparative analysis of the Walloon river contracts on the basis of comparative legal research and qualitative empirical research, which will be discussed below, it has been concluded that the sub-basin board constellation needs improvements to face the challenges of flood risk management in the future in order to effectively represent the quintessential sub-basin hydrological scale. Comparative analysis and exchange of lessons learnt with the Walloon river contract model is relevant in this context.[783] The sub-basin secretariats are not directly represented in the Coordination Commission Integrated Water Policy[784], under the auspices of which water quality and quantity management policies are developed.[785] This is problematic from the viewpoint of inter-scalar links, one of the criteria for resilient transboundary water management as adopted in this book.

240. Cooperation at the level of the sub-basin remains ad hoc and devoid of any structural shape. The DIWP does not provide for any sort of representation of stakeholders of the sub-basins across the border, e.g. from the Netherlands or France. The sub-basin-specific parts in the RBMPs and FRMPs therefore only refer to policies and projects of the part of the sub-basin on Flemish territory. The sub-basin-specific parts for those sub-basins where bilateral contacts with authorities across the border, even the regional border with the Walloon Region,

781 Number of fulltime staff in the sub-basin boards investigated varies between 1 and 2.5. Ibid.
782 Ibid.
783 Mees, Suykens and Crabbé (n 493) 12.
784 They are only represented in the CIW through the provincial governor.
785 Representatives of sub-basin boards are represented in certain working groups of the Coordination Commission Integrated Water Policy.

are relevant and existing, do mention succinctly through what kind of structure this interaction takes place, and with regard to which part of the sub-basin. One of the issues of course relates to the mismatch in administrative form between the authorities on both sides of the respective borders, e.g. the Dutch Regional Water Authorities would be confronted in the Flemish Region with the amalgam of the provinces, the Flemish Environment Agency, the managers of the navigable watercourses, and the Flemish waste-water treatment company. Cooperation between the different water managers in the Flemish Region is often inadequate and the scenery in general is particularly fragmented. This of course raises the threshold for cooperation across the border, as the question arises, which organisation should be addressed. In the Flemish Region, the degree of built-up area is significantly higher than in the Walloon Region, making it particularly interesting for both regions to work together in the management of watercourses, e.g. the creation of buffer areas on the territory of the Walloon Region benefits the water drainage in the Flemish Region.

241. In the Walloon Region, hydrological scale entities are the river contracts ("contrats de rivières"), which have been developed to coordinate issues such as flood risk management, bring together relevant actors at sub-basin level and sensitise the population to, in this case, the existence of flood risks.[786] River contracts in the Walloon Region are considered as an "apolitical device".[787] These instruments are inspired on the French model and were launched for the first time in 1993.[788] Even though they have a voluntary character, they are perceived as more effective bodies operating at the hydrological scale than the Flemish sub-basin boards.[789] At the time of writing, with the exception of the sub-basin Moselle, all Walloon sub-basins are covered by river contracts.

242. River contracts are bodies with legal personality[790] that bring together all relevant public and private actors in the context of sustainable water management, contributing to an integrated approach to water quantity and quality management.[791] The river contracts aim to facilitate and coordinate, not take binding decisions.[792] The participation of these actors occurs on a voluntary

786 Although many other issues can be covered by the river contracts, e.g. water quantity and quality management, spatial planning, nature protection, and so forth.
787 Interview with civil servant of the department of non-navigable watercourses, DG03, of the Walloon Region, 26 May 2014.
788 A first circular in this regard was issued in this year. Mees, Suykens and others (n 525).
789 Mees, Suykens and Crabbé (n 493).
790 Although these entities are referred to as "contracts", these should be seen as actors, not as documents. More specifically, non-profit organisations.
791 Article R.45, 4° Water Code. Interview with civil servant of the department of unnavigable watercourses of the Walloon Region, 26 May 2014.
792 Similar to the Flemish sub-basin boards and in contrast to the French SAGE for example.

basis. The river contracts are much more of a bottom-up initiative than the Flemish sub-basin boards. Indeed, whereas the Flemish sub-basin boards are created and developed at the level of the Flemish legislator and government, the Walloon river contracts are developed more locally, with financial and other support from the higher Walloon Government.[793] The local initiator is responsible for drafting a protocol agreement, after having discussed the goals and measures to be taken with the participants based on the specific needs of the sub-basin, with a horizon of three years. After the consent of several municipalities and at least one province, the protocol agreement is submitted to the level of the Walloon Region, i.e. to the competent Minister. Following the final approval by the competent Minister, the initiator should take the necessary steps to create the legal personality for the river contract in question, where the statutes should obligatorily set forth the goals of the contract, which are also outlined in the Water Code.[794] The protocol agreement is therefore the main legal basis for further cooperation at the sub-basin level, as it defines the goals of the river contract[795], the responsibilities of the actors and the different sources of funding from the layers involved.

Once the river contract is created, four bodies are important. First, a coordination unit is responsible for the day-to-day management of the river contract, funded through the public actors involved in the constellation.[796] The coordination unit, for example, has the task to raise public awareness, coordinate and follow up on the actions set out in the protocol agreement, and so forth. Second, the directorate tackles the administrative aspects of the organisation, e.g. personnel recruitment and the management of budgets.[797] Third, the General Assembly, also referred to as the river committee, is organised twice a year, and represents the meeting place for the representatives of the partners, e.g. in the river contract for the downstream Meuse ("Meuse Aval"), the General Assembly is composed of three groups – the local actors (associations), the local entities (municipalities and provinces), and the regional entities (e.g. DG03, which is the department for non-navigable watercourses of the Walloon Region).[798] Fourth, local river committees act as brainstorm platforms for the local actors.[799] These meetings of the river committees are usually open to all public and private actors

---

793 See Order of the Walloon Government of 13 November 2008 amending Book II of the Environmental Code containing the Water Code on river contracts, Belgian Official Journal 22 December 2008 (The 2008 River Contract Order).

794 See Article R.48 in conjunction with Article R.51 Water Code.

795 Which should be in line with the goals of the Water Code.

796 Mees, Suykens and Crabbé (n 493).

797 See also Article R.46 of the 2008 River Contract Order.

798 More information is available on the website of the river contract "Meuse Aval": <www.meuseaval.be/index.php/le-contrat-de-riviere-meuse-aval/definition-et-cadre-legislatif/80-le-contrat-de-riviere-meuse-aval> accessed 12 July 2017.

799 Francis Rosillon and Jérôme Lobet, 'Transboundary River Contract Semois-Semoy between Belgium (Wallonia) and France' in Patrick Meire and others. (ed), *Integrated Water Management: Practical Experiences and Case Studies* (2008).

interested in participating. Several local river committees can be created for each river contract.[800] As mentioned above, the participation in river contracts is voluntary, but the participation rate is high, e.g. in 2016, 215 of 262 municipalities on the territory of the Walloon Region were involved in a river contract.[801] The clear voluntary character of participation does not equal a lack of necessity to achieve goals in the case of river contracts. Indeed, as a legal entity, the stakeholders involved in the river contract are legally obliged to set goals and report back on how and when these are achieved.

The coordinator has three years to develop a protocol agreement, in which a field inventory should be included. This field inventory should contain the conditions that should be improved in the sub-basin, a list of priority data on the basis of which the inventory has been drawn up, and the dissemination of the results of the inventory to the public.[802] The protocol agreement should include, among other things, information on the aquatic environment and how it can be maintained, participation methods, a detailed budget related to the functioning of the entity, and so forth. A yearly activity report should be drawn up and approved by the river committee, where all river contracts pertaining to the same sub-basin are obligated to draw up a coordinated activity report.[803] The protocol agreement has a legal validity period of three years. In the third year, the coordinator should evaluate and revise the actions announced in the agreement, as well as the field inventory, and submit the documents to the board, which, in turn, submits them to the river committee. The Water Code stipulates the criteria for the evaluation of the project, which, for example, relate to the specific results of the river contract on water quality and the dynamics of the working groups. Ultimately, the competent Minister decides on the extension of the protocol agreement. In terms of financial resources, the river contracts are financed through the region, which finances approximately 70% of the activities, and the provinces and the municipalities, which finance up to 30%. The reporting requirements and requirements with regard to the fulfilment of sensitisation activities, mainly relate to the 70% funding portion, i.e. within the relationship between the river contract and the Walloon Region. The other groups involved in the operation of the river contract, such as the stakeholders from agriculture, NGOs, and so forth, are not required to provide financial support.[804] The financial allocation mechanism is defined in the legislative framework, where the amount of financial allocation depends on the size of the sub-basin and the number of inhabitants. As river contracts have legal personality, they may apply for funding from EU projects as well, e.g. in the context of INTERREG.

800 For example, the river contract for the downstream Meuse, has 5 local river committees.

801 Interview with civil servant of the department of non-navigable watercourses, DG03, of the Walloon Region, 26 May 2014.

802 Article R.52 Water Code.

803 Article R.54 Para 1 Water Code.

804 Although they can on the basis of Article R.55 Para 2 Water Code.

243. River contracts are generally positively perceived by actors active in water-related issues in the Walloon Region.[805] Not all river contracts in the Walloon Region are as active and successful, and one of the factors determining the active involvement of the river contract in the sub-basin relates to the dynamics of the contract coordinator. For example, the river contract for the sub-basin Dender was one of the first contracts created in the region, in 1992. However, between 2003 and 2010, the river contract was put on hold and no activities were carried out. Key characteristics of the river-contract mechanisms are the fact that these are governed bottom-up at the local level and supported by a flexible legal framework at the level of the region: thus, a bottom-up enterprise with a support mechanism sprouting from the higher layers.[806] The river-contract bodies, in contrast to the Flemish sub-basin boards, have legal personality as NGOs, and independence in terms of personnel and the allocation of their financial resources. As discussed in the previous chapter, the Flemish sub-basin boards form part of the Flemish government, and are very strongly linked to the level of the region for personnel and financial resources. Furthermore, the process of the creation of the river contracts in itself is voluntary, and the effective functioning of the mechanism depends considerably on the dynamics of the coordinator and leading parties involved.[807]

244. In the Netherlands, there are broadly four layers of governance with respect to water management: the State (the National Water Authority), which governs the main River Basin Districts, the regional water authorities, which govern the regional water bodies, and the provinces and municipalities, which also have specific tasks in water management.[808] At the national level, the Ministry of Infrastructure and the Environment is the competent authority for water management and spatial planning, and, for example, is the entity supervising primary flood-defence structures.[809] Rijkswaterstaat is part of this Ministry and is the executive agency responsible for the management of the main water systems. Of these four levels, the regional water authorities are based on functional democracy principles, whilst the others are based on general democracy principles.[810] Indeed, one of the characteristics of Dutch water law is the mechanism of functional

805 Mees, Suykens and Crabbé (n 493).

806 Mees, Suykens and Crabbé (n 493).

807 Interview with civil servant of the department of unnavigable watercourses of the Walloon Region, 26 May 2014.

808 Although the Water Act only refers to the State and the regional water authorities as competent authorities in the realm of water management. For example, the municipalities are responsible for processing excess groundwater. A similar issue as in the Flemish and Walloon Region is that the borders of provinces and municipalities do not follow these hydrological boundaries, and are therefore different from the sub-basins. Van Rijswick and Havekes (n 649) 129.

809 Willemijn Van Doorn-Hoekveld, 'Transboundary Flood Risk Management Management: Compatibilities of the Legal Systems of Flood Risk Management in the Netherlands, Flanders and France' (2017) 26 European Energy and Environmental Law Review 81.

810 Alfred Van Hall, 'Water Stuurt Ruimtelijke Inrichting. Nieuw Denken over Oud Water.' in Frank Maes and Luc Lavrysen (eds), *Integraal Waterbeleid in Vlaanderen en Nederland* (die Keure 2003) 206.

decentralisation, on the basis of which the water boards operate.[811] These regional water authorities are an intricate part of the Dutch institutional water landscape and their importance cannot be over-emphasised. These are very different from the regional water authorities (water boards) that used to exist in the Flemish Region. However, the situation is wholly different in the Netherlands, where the regional water authorities form an intrinsic part of the institutional organisation of the country, dating back to the Middle Ages.[812]

As mentioned above, regional water authorities are functional bodies that operate to reach specific goals and carry out specific responsibilities in the context of water management.[813] They are public bodies, with all related consequences in terms of their competences to issue permits, raise taxes, and so forth. The role of the regional water authorities is rooted in the Constitution, which provides for a primary role of provinces in this regard in terms of supervision and the organisation of the authorities, and the legislative power.[814]

There have been significant developments in the scale on which these water authorities operate. Whereas in 1950, 2650 water authorities were active[815], in 2017 only 22 water authorities operate on Dutch territory, thus implying a significant scale increase. For the Dutch part of the Scheldt IRBD, the relevant regional water authorities are the Scheldt Streams and Brabantse Delta. The water authorities are autonomous in their financing schemes, on the basis of taxes levied for their services. The geographical areas in which the regional water authorities operate, are based on hydrological boundaries, i.e. they do not coincide with the territories of municipalities and provinces, but do not necessarily coincide with the level of the sub-basins in the sense of the WFD either. There are other factors that determine the catchment scale at which the water authorities are active, e.g. the geographical coverage of protection by dikes.[816] As mentioned above, there are 22 regional water authorities for seven WFD sub-basins.

Water authorities and provinces are closely connected on the basis of the Constitution. Provinces have the right to define the organisational ins and outs of the water authorities, the composition of their administration, and so forth. This provincial competence, however, is limited by the fact that the law determines the scope of these provincial orders, and that they can also be issued insofar as the law has not provided differently.[817] For example, the Water Authorities Act determines the procedure to be followed when two or more provinces do not reach agreement on the creation or abolishment of a water authority on their territories.[818] The

---

811 Ibid.
812 Van Rijswick and Havekes (n 649).
813 See Article 1 of the Water Authorities Act.
814 Article 133 of the Dutch Constitution. Van Rijswick and Havekes, no. 262.
815 Van Rijswick and Havekes, no. 262.
816 Ibid.
817 Article 133 (1) Constitution.
818 Article 7 Water Authorities Act.

water authorities have the power to regulate on the basis of the Water Authorities Act. Indeed, whenever a law, a general measure of administration or a provincial order is provided for, the water authorities may issue orders, as long as they are compatible with these acts and regulations of the higher levels.[819] In terms of legal hierarchy, the orders issued by the water authorities have legal authority similar to that of regulations issued by municipalities.[820] In addition, as provided for by the Constitution, regulations issued by the water authorities can only be annulled if they are in violation with the law or the public interest.[821]

245. The organisational structure of the water authorities is rather straightforward with a clear legal hierarchy ensuing from the Water Authorities Act. There are three bodies active in the water authorities' context: the governing board, the executive committee and the chair. The governing board is responsible for issuing the regulations necessary to allow the water authority to adequately fulfil its tasks. It should also draw up a package of documents and maps, a so-called "ledger", that indicates protection zones, the water infrastructure, watercourses, etc. within the geographical scope in which the water authorities operate.[822] The ledger should also determine the maintenance duties e.g. with respect to the relevant infrastructure. In general, the governing board may delegate certain tasks to the executive committee, with the exception of an exhaustive list of tasks stipulated in the Water Authorities Act, for example, determining the budget and raising taxes.[823] The executive committee is responsible for the day-to-day tasks of the water authority in question, for the preparation of the issues that should be discussed by the governing board and for the implementation of the decisions of the governing board.[824] The chair, in turn, should promote the administrative integrity of the water authority and the adequate fulfilment of the authority's tasks.[825] The president is also the legal representative of the water authority, and has the power to intervene and take measures if justified by the urgency of circumstances.[826] All decisions adopted by the governing board of the water authorities may be annulled by the provincial executive.[827] As mentioned above, annulment is only possible if said decisions are in clear violation with either the law or the public interest.

---

819 Article 59(1) Water Authorities Act. Article 59(2) further provides that if a topic covered by the water authorities order is, at a later stage, covered by acts and regulations promulgated by these higher levels, the former orders of the water authorities are repealed.

820 JM de Meij and IC van der Vlies, *Inleiding tot het Staatsrecht en het Bestuursrecht* (Kluwer 2004) 110.

821 Article 133 Constitution.

822 This concerns two ledgers, which are often combined. Article 78 Water Authorities Act.

823 Article 83 Water Authorities Act.

824 Except when the president should implement said decisions. Article 84 Water Authorities Act.

825 Article 94 Water Authorities Act.

826 I.e. when there is no time to convene the executive committee or governing board. See Article 94 Water Authorities Act.

827 Also non-written decisions that have legal consequences. Article 156 Water Authorities Act. See Van Rijswick and Havekes, no. 262.

One of the indicators of a resilient joint body governing water resources relates to its financial independence.[828] One of the bottlenecks in the Flemish sub-basin board structure, is the dependence of the body on financial resources coming from the higher government (the Flemish Region). The Walloon river contracts, in contrast, have more room to manoeuvre in terms of allocation of funding. The Dutch water authorities are practically self-sufficient, as they have the power to raise taxes to support their activities. Indeed, water authorities have their proper taxation mechanism, and only rarely rely on funding from the state level.

Another important aspect determining the success of flood risk management measures is the issue of enforcement. The water authorities have an explicit legal basis to enforce their rules and regulations. In case of non-compliance, or risk of non-compliance, the executive committee may apply an "administrative enforcement order" in order to enforce its rules.[829] The administrative enforcement order is a measure available to government authorities with the goal of putting an end to an violation or to return to a lawful situation. The administrative enforcement order is not necessarily a fine that is imposed. In certain cases, it is not the executive committee, but the chair of the water authority that applies the administrative enforcement order.[830] In addition to these administrative sanctions, the water authorities also have penal competences, e.g. the civil servants of the water authority may track down the violations of the orders issued by the authority.[831]

246. France being a centralised State, with strong regional and local competences for flood risk management[832], there are roughly two types of public entities active in this context, the state authorities and the local authorities. Within the realm of the state authorities, there are three layers of governance, (a) the competent Ministries, i.e. the Ministry for the Environment and the Interior, (b) the prefects, who are representatives of the State at regional and departmental level, responsible e.g. of overseeing the implementation of state budgets, and (c) the Water Agencies, which play an important role in water management at the level of the basin districts.[833]

At national level, a water parliament has been established, i.e. the National Water Committee, whose aim is to provide advice at the level of the basins on water-related issues with a national angle. The National Water Committee provides advice on the draft River Basin Management Plans (the acronym for these is "SDAGE"). It consists of representatives of the State, e.g. basin prefects, parliamentarians, senators, presidents of basin committees and local water

828 Bruce Hooper, 'River Basin Organization Performance Indicators: Application to the Delaware River Basin Commission: Supplementary File' (2010) 12 Water Policy 10. This will be discussed in the evaluation section.

829 Article 61 Water Authorities Act.

830 Article 61(2) Water Authorities Act.

831 Article 85(1) Water Authorities Act.

832 And overall decentralisation tendencies, that have been developing since 1982.

833 Larrue and others (n 599) 9.

commissions. Another actor relevant at the state level, is the national agency for water and the aquatic environment, referred to as "ONEMA", which, among other things, collects relevant data and functions as the "water police".[834]

The local authorities include the Regional Councils, the Departmental Councils and the municipalities. Municipalities are gaining importance in French flood risk governance. On the basis of the 2014 Act, the competences for water management and flood risk prevention has been attributed to the municipalities, which are now obligated to maintain rivers, defence infrastructure, and so forth. A mechanism has also been provided on the basis of which municipalities may combine their competences and transfer their tasks in this regard to an Inter-Municipal Cooperation Organisation. It can therefore be noted that this development is the opposite of what has happened in the Flemish Region, where a 2014 internal state reform induced the transfer of competences for the management of third-category watercourses away from the levels of the municipalities to the levels of the provinces.

247. Watershed management in France has existed for decades, and more specifically since the enactment of the 1964 Water Act.[835] The 1966 Decree related to financial basin agencies then further established the basin authorities.[836]

These basins are often combined under the auspices of one authority. This is also the case for the Scheldt RBD, which constitutes, together with the Samber RBD, the basin "Artois-Picardie", for which the RBMP and FRMP were adopted. At the time, this choice was made in order to compromise between the reflection of hydrological realities and the pragmatic consideration not to install too many agencies.[837] As mentioned above, at first, these watershed authorities were marked as financial agencies, as the 1964 Act established financial basin agencies with legal personality and financial autonomy. They were further strengthened through the 1992 Water Act, which characterised water as *res communis* in France.[838]

The 1992 Act introduced the water management plans SDAGE and SAGE. The SDAGE is the compulsory River Basin Management Plan for each of the six main basins, which exists to report to EU level, although it existed prior to the Water

---

834 "Office national de l'eau et des milieux aquatiques".

835 Act no. 64-1245 of 16 December 1964 related to the regime and division of waters and the battle against pollution, French Official Journal 18 December 1964.

836 Further strengthened in 1992, 2006. Decree no. 66-700 of 14 September 1966 related to financial basin agencies created by Article 14 of the Act of 16 December 1964, French Official Journal 23 September 1966. Both the 1964 Act and the 1966 Decree were abolished by the Decree no. 2007-397 of 22 March 2007 related to the regulatory part of the Environmental Code.

837 E.g. keeping in mind that a minimum population threshold should be met before the creation of an agency could be justified. See Alexandre Brun, 'Les Contrats de Rivière En France: Enjeux, Acteurs et Territoires' (2010) 51 Les Cahiers de droit 679.

838 Water Act, Act no. 92-3 of 3 January 1992 related to water, French Official Journal 4 January 1992.

Framework Directive, and the SAGE is the more local and optional management instrument for the sub-basins.

The six water agencies are public entities with legal personality linked to the State, and are of an administrative nature.[839] Their goal is to facilitate, through financial resources, the realisation of useful works and measures for the whole hydrographic basin for which they are responsible.[840] The 1964 Water Act also installed a legal mechanism on the basis of which these entities raise fees from towns, farmers and the industry related to the use and pollution of water resources.[841] The water agencies are required to promulgate programmes stipulating actions to be taken in the years ahead, in the context of which the redistribution of the raised fees, for example through subsidies is described.[842] They implement the goals of the River Basin Management Plans (SDAGEs), as developed by the River Basin Committee. They are composed of several representatives from state and regional levels and water users.[843]

The Basin Coordinator Prefect[844] constitutes another form of operationalisation of the link between the State and the decentralised level, and coordinates state action in water management at the basin level.[845] The Basin Coordinator Prefect approves and enforces the River Basin Management Plan (SDAGE), as promulgated by the River Basin Committee. The Prefect also issues the Programme of Measures. The River Basin Committee operates at basin level or for a group of basins and advises on water-related projects of interest for the whole basin, and is often referred to as a water parliament. In sum, the River Basin Committee provides for the policy orientation for the basin or groups of basins, which the Water Agency implements. As mentioned above, the River Basin Committee develops and approves the SDAGE, which, among other things, marks the hydrographic boundaries of the sub-basins, and provides advice on the draft sub-water management plans (SAGEs). Decisions related to water resources must be compatible with the SDAGE, which indicates its binding force. The SDAGE is applied only at the scale of the river basin and promulgated under the legal tool of the "Arrêté", which may not contradict other laws.[846]

The River Basin Committees are composed as follows: 40% of local elected government officials, 40% representing water users and 20% representatives

839 Article L213-8-1 of the Environment Code.

840 Article 4 of the 1992 Water Act.

841 Alexandre Brun, 'France's Water Policy: The Interest and Limits of River Contracts' in Schneier-Madanes (ed), *Globalized Water: A Question of Governance* (2014) 140.

842 Ibid.

843 Article L213-8-1, § 2 of the Environment Code.

844 "Préfet Coordonnateur de Bassin ".

845 Article L213-7 Environment Code.

846 Article 3 of the 1992 Water Act. Commission, 'Report from the Commission to the European Parliament and the Council on the Implementation of the Water Framework Directive (2000/60/EC) River Basin Management Plans', COM(2012) 670 final.

of the State.[847] The Local Water Committees operate at the sub-basin level.[848] The Committees have decision-making powers and decisions are taken through majority voting.[849] The Local Water Committee develops the water development and management plan for the sub-basins (SAGE), which is approved by the prefect. The provisions of the SAGE should be compatible with those of the SDAGE. It is composed as follows: 50% of representatives of local authorities, 25% water users and 25% representing the State.[850] The Local Water Committees are therefore territorial commissions, where local representation is quintessential.[851]

The list of entities does not end here. Other relevant actors at the level of the basin are river syndicates and the so-called "EPTBs", which are the territorial public basin organisations.[852] These EPTBs represent a territorial collectivity cooperation mechanism, which combines regions, departments, municipalities, and so forth. Territorial collectivities or their groups may create an EPTB at the level of the (sub-)basin for the prevention of floods and the balanced management of water resources.[853] It has the legal character of a "syndicat mixte" or mixed syndicate, which is a French public cooperation mechanism.[854]

Furthermore, a so-called "EPAGE" may be created[855], which is a public institution for the planning and management of water at the level of a sub-basin and a territorial collectivity within the meaning of Article L. 5711-1 of the French Code on Territorial Collectivities.[856] The EPAGE is an entity operating below the level of the EPTB. The cooperation instruments such as EPTBs and EPAGEs should be seen in the context of the so-called "MAPAM"-Act of 2014.[857] This Act has created a "GEMAPI" competence which links water and flood risk management, thereby aiming to encourage integrated water management, and provides for a shift of competences in flood risk management to the local levels. Indeed, on the basis of the Act, competences in this regard have been transferred to the municipalities, which in turn can delegate these competences to an inter-municipal cooperation

847 Article L213-8 Environment Code.

848 Local Water Committees or "Commissions Locale de l'Eau".

849 Vera Zaporozhets, 'Power Distribution in French River Basin Committees' (2015) Toulouse School of Economics, Working Paper n TSE-558 <www.tse-fr.eu/sites/default/files/TSE/documents/doc/wp/2015/wp_tse_558.pdf> accessed 13 July 2017.

850 Article L212-4 Environment Code.

851 Maria Lauria Scaduto, *Comparative Analysis between France and Italy* (Springer International Publishing 2016) 34.

852 "Etablissement Public Territorial de Bassin". Arrêté du 7 février 2005 relatif à la délimitation du périmètre d'intervention de l'établissement public territorial de basin, French Official Journal 12 February 2005.

853 Article L213-12 Environment Code, which refers to the French General Code on Territorial Collectivities.

854 Article L5721-1 and following of the French Code on Territorial Collectivities.

855 "Etablissement public d'aménagement et de gestion de l'eau".

856 This will also be discussed in more detail in the chapter on regional cooperation below.

857 Act n° 2014-58 du 27 janvier 2014 de modernisation de l'action publique territoriale et d'affirmation des métropoles, French Official Journal 28 January 2014.

structure, the "EPCI". The EPCI is a territorial cooperation instrument in which several municipalities join forces and combine their competences into one public entity, which can raise local taxes for example. In turn, on the basis of the MAPAM-Act, the EPCI may delegate competences with regard to flood risk and water management to the entities operating at basin levels, such as the EPTBs, the EPAGE and the syndicates. Important competences in the realm of flood risk management have been transferred from the central state level to the local levels. However, these local levels are the cities and municipalities, not the entities at the hydrological, i.e. basin and sub-basin, scale, as flooding is considered as an urban issue for which these cities and municipalities are considered the appropriate layer of governance.[858] The intervention of basin organisations such as EPTB could therefore be relevant in this regard. However, the link and possible overlap with existing River Basin Committees and Local Water Commissions can be questioned. This is, however, beyond the scope of this PhD research.

248. The SDAGE Artois Picardie states that all territories of the Artois-Picardie River Basin District are covered by a SAGE.[859] As mentioned above, the process of developing a SAGE is optional, but, once the SAGE has been published, it is legally binding as decisions in the water realm should take into account the provisions of the SAGE. There are three types of committees in the governance structure of the Artois-Picardie basin and the elaboration of the FRMP, i.e. the supervising committee for floods (the so-called "comité de pilotage"), which is presided over by the coordinating Prefect of the basin, a technical committee, which is composed of representatives of the state services and a territorial strategy committee, which relates to local management structures. The River Basin Committee Artois-Picardie is an entity that combines both public and private partners, i.e. representatives of territorial groups, public department and state level administration, (non-) professional users and associations (e.g. nature protection), with an interest in the basin.

The above-mentioned three-tier approach to hydrological governance, i.e. the national water parliament, the water boards and the local water committee have, at times, been considered as structures that are too rigid at local level.[860] In the 1970s, at the time of the first-era water directives at EU level, the "clean river project" encouraged the creation of river contracts in France. The 1981 Circular created a legal basis for the establishment of river contracts for the realisation of the "clean rivers" initiative.[861] From the beginning, the goal of these river contracts was

[858] Larrue (n 599) 18.

[859] Comité de Bassin Artois-Picardie, 'Schéma Directeur d'Aménagement et de Gestion des Eaux du Bassin Artois-Picardie pour la période 2016 à 2021' (2016) <www.artois-picardie.eaufrance.fr/IMG/pdf/sdage_2016-2021.pdf> accessed 12 July 2017.

[860] Brun (n 837) 682.

[861] Circulaire 5 février 1981 relative aux contrats de rivières (1981) <www.gesteau.fr/sites/default/files/Circulaire_contratriviere_19810205.pdf> accessed 13 July 2017.

to bring together local stakeholders and users to achieve a collective goal.[862] As mentioned above, the Walloon river contracts were based on the French model.[863] The decentralisation movement that started in France in the 1980s, and thus the transfer of important competences to the local levels, was beneficial for the development of river contracts.[864] It should be noted that river contracts are not rooted in water-related legislation in France, but are governed by a series of non-binding circulars. In the French legal landscape, a river contract is essentially a contract between local authorities and the State, and may pertain to flood control, water quality management, public awareness to water pollution, and so forth.

Similarly to the river contracts in the Walloon Region, French river contracts are established on a voluntary basis. At a time where water quality on the one hand and flood risk management on the other hand were still very much tackled in separate legal frameworks, the river contracts integrated these two issues in one instrument. The procedure for approval of river contracts is quite similar to the one in the Walloon Region, in that the local actors involved first need to draw up a draft contract, which includes an assessment of the status of the basin and a list of measures to tackle the issues detected in this assessment. Up until 2004, this draft was to be drawn up by the Prefect and submitted to the Ministry for the Environment, in order to be approved by the National Approval Committee, which is composed of representatives of different interest groups.[865] If this Committee issued a positive advice, the draft river contract could be elaborated in more detail, i.e. related to the budget, entity responsible for the technical and administrative personnel, and the Local River Committee in question. Finally, the draft needed to be approved again by the National Approval Committee.[866] Since 2004, further decentralisation has taken place, as the role of the National Approval Committee has since been replaced by the water boards. This strengthened role of the basin authorities originated from the concern that local coherence was inadequate in the framework of approval by the National Approval Committee.[867]

River contracts cover a period of five years and are complementary to the instrument of the SAGE. Mechanisms have been put in place to tackle any overlap in planning and authorities with respect to river contracts. When the territorial scope of a river contract and a SAGE completely overlap, the Local Water Committee functions as the river committee.[868] If this is only partly the

862 Ibid.

863 The river contract also has a transboundary aspect, and there are a few examples of transboundary river contracts.

864 Francis Rosillon, 'Eau et Territoire À Travers L'expérience Des Contrats de Rivière En Wallonie (Belgique)' (2015) 25-26 Territoire en Mouvement Revue de Géographie et Aménagement 1.

865 Circulaire du 24 octobre 1994 relative au plan décennal de restauration et d'entretien des rivières (1994) <www.gesteau.fr/sites/default/files/circulaire_19941024_cr.pdf> accessed 13 July 2017.

866 Brun (n 841).

867 Circulaire ministérielle du 30 janvier 2004 relative aux Contrats de Rivière et de Baie (2004) <www.gesteau.fr/sites/default/files/Circulaire_300104.pdf> accessed 13 July 2017.

868 Maria Lauria Scaduto, *Comparative Analysis between France and Italy* (Springer International Publishing 2016) 52.

case, the river committee forms a sub-committee of the Local Water Committee. If the territorial scope of the river contract is larger than that of the SAGE, the members of the Local Water Committee with respect to the river contract area, become members of the river committee.

The river contracts have supported river basin management, as municipalities have been working together in an upstream-downstream synergy with the goal of producing drafts and projects that the State would approve of, and, ultimately, help finance.[869] Upstream-downstream cooperation in this regard makes sense, as measures taken individually by certain municipalities would not yield any benefits at basin scale as much as when measures are coordinated across the basin. The State is naturally more likely to approve the latter type of action. It should be noted that river contracts do not include binding measures or plans of actions, considering their voluntary nature, in contrast to the SAGE plans developed by the Local Water Committee. The lack of binding legal value of the contracts has been confirmed in case law: the contracts cannot produce legal consequences for third parties, for example.[870] In order to produce legal effects, the provisions in the river contracts may be included in the SAGEs; these two instruments are complementary.[871] This lack of legally binding effects *stricto sensu* does not mean that the entity created by virtue of the river contract is without clout. The steering committee, in contrast to the river committee under the auspices of a Walloon river contract, has project ownership.[872] Surveys for the period of 1990-2010 have showed that, from an environmental point of view, river basins in which river contracts have been implemented, did not significantly benefit in comparison to river basins in which no river contracts have been set up. The criticism of the French river contracts in this period of time related to the survey is that, with respect to water quality, the contracts were hardly used for the prevention of pollution, and with regard to flood risk management, the focus was on hard defence infrastructure instead of flood risk prevention.[873] Whereas in the Walloon Region, river contracts function as the sub-basin entities, in France, the contracts operate on top of the various types of sub-basin and basin water authorities in place.

249. The French and Walloon river contract also has a transboundary aspect. The decentralisation movement that started in the 1980s in France, and thus the transfer of important competences to the local levels, was beneficial for the development of river contracts.[874] A transboundary river contract was also concluded between France and the Walloon Region with regard to the Semoi/Semoy sub-basin. This

869 Brun (n 837) 694.
870 Tribunal Administratif de Bordeaux, 7 November 2006, Association des propriétaires et riverains exploitants agricoles de la Dordogne et autres.
871 Brun (n 837) 702.
872 Rosillon and Lobet (n 799) 202.
873 Brun (n 841) 146.
874 Rosillon (n 864) 6.

was done through a series of INTERREG programmes. The river contract presents specific measures that should be carried out on both sides of the border.[875] This type of transboundary river contract could constitute a valuable contribution to basin – sub-basin linkages in the context of cooperation in IRBDs. Considering especially the way in which river contracts in the Walloon Region are perceived positively, it could serve as a robust tool to improve communication, participation and stakeholder involvement in a transboundary setting. A similar suggestion was made by Gilissen, namely to introduce transboundary water agreements in the Dutch model.[876] These opportunities will be analysed in Chapter V of this book.

3.3. CONCLUSION

250. The following table provides an overview of the main entities created specifically at the hydrological scale.

Table 7. Entities at the Hydrological Scale and their Characteristics

Country or Region	Entities	Legally enshrined	Financial autonomy	Binding decision-making power
The Netherlands	Regional water authorities[877]	Yes	Yes	Yes
The Flemish Region	Sub-basin boards	Yes	No	No
The Walloon Region	River contracts	Yes	Partly	No
France	River Basin Committees / Local Water Commissions	Yes	No	Yes

251. In the Flemish Region, the management of watercourses is in the hands of the region, the provinces and the municipalities, depending on whether the watercourses are in the first, second or third category. Entities specifically operating at the hydrological scale, i.e. sub-basin scale, have been installed as well but merely have an advisory task and are strongly linked, in terms of legal, financial and human resources, to the level of the first-category water manager, namely the Flemish Environment Agency of the Flemish Region.

252. Water managers in the Walloon Region are the region, provinces and municipalities, depending on the category of watercourses.[878] The Walloon Region has also created river contracts at the hydrological – sub-basin – scales. These are created bottom-up by local actors with top-down support from the

875 Rosillon and Lobet (n 799) 205.
876 Herman Kasper Gilissen, (n 548).
877 Rijkswaterstaat is responsible for water management of the main stem rivers. However, this table focuses on the entities operating at the different hydrological sub-scales within the Scheldt River.
878 Similar to the Flemish Region.

applicable legal framework and in terms of financial resources. Participation in these river contracts is voluntary and one of the factors in the success or failure of the contracts relates to the dynamism of the coordinator and the actors involved.

253. In the Netherlands, the regional water authorities have been explicitly designated by law as water managers. Provinces and municipalities also have responsibilities with regard to water management, but they do not qualify as water managers *stricto sensu*. The regional water authorities have broad competences and financial autonomy, as well as legal authority to enforce their rules and regulations.

254. In France, a myriad of authorities operates at the hydrological scale, i.e. water agencies, basin prefects, river basin committees and local water committees and river contracts. Moreover, inter-municipal cooperation mechanisms are also relevant at the sub-basin levels, for example, catchment-basin syndicates. Overlaps may exist between these entities. For example, where cities and municipalities transfer competences to a territorial public basin organisation such as the EPTB in the context of the MAPAM Act, overlaps may exist with activities carried out by river basin committees and local water commissions. Tools that are relevant in the context of transboundary cooperation are inter-municipal cooperation mechanisms and transboundary river contracts. These instruments will be considered in the discussion of regional cooperation in the following sections and chapters.

255. In terms of hydrological scale governance, significant differences can be noted in the district. Where regional water authorities are concerned, the Dutch regional water authorities are entirely different from the Flemish sub-basin boards. Whereas the regional water authorities have broad and concrete policy and implementation competences and financial autonomy, the sub-basin boards are advisory bodies focusing on the fulfilment of procedural requirements stemming from the WFD and FD.[879] The competences for water management are at the level of authorities such as the region, provinces and municipalities, whose territorial scope does not correspond with the hydrological scale. Indeed, the sub-basin boards are not the water managers. The same goes for the Walloon Region, where different government departments are the designated water managers. The entities operating at the hydrological sub-basin scale are tasked with a coordinating and advisory role, to support the administration with sub-basin-wide perspectives. The river contracts in the Walloon Region have a completely different set-up in comparison to the Flemish sub-basin boards, having been created as a bottom-up process supported by a legal framework and some degree of financial autonomy. There again, competences are at the level of the region, provinces and municipalities, but whereas in the Flemish Region the competences for the third-category non-navigable watercourses have largely been transferred

879 Namely providing the sub-basin specific parts of the RBMPs and FRMPs.

to the provinces, this has not been done in the Walloon Region. The French legal framework has instituted actual river basin management, with the Water Agencies, River Basin Committees, Local Water Committees and the possibilities for Inter-Municipal Cooperation Associations to delegate actions to more basin-oriented entities such as the EPTB. However, the river contract in the Walloon Region is to communicate with entities such as the Local Water Committee, as the French river contract has a different role to play.

256. Now that the various legal regimes have been explained and the competent authorities have been described, both at the hydrological scale and beyond, the next section will turn to the three dimensions of cooperation in the District.

4. THE THREE DIMENSIONS OF COOPERATION IN THE SCHELDT DISTRICT

257. Sections 2 and 3 of this Chapter have described the national legal frameworks for flood risk management in the Scheldt District. This section presents the three dimensions of cooperation as applied to the Scheldt District, i.e. national, regional and international. The relevant primary and secondary EU law is interwoven in and will be discussed throughout the three dimensions. The three-dimensional approach to analysing flood risk management in the IRBD Scheldt will provide a good basis for the evaluation of its resilience.

4.1. NATIONAL LEVEL PROMOTION OF INTERNATIONAL COOPERATION IN THE SCHELDT

258. The WFD and FD require Member States sharing a River Basin District to coordinate their efforts as much as possible in the context of the implementation of the Directives. In Chapter II of this book, the enforceability of these coordination and cooperation requirements were already questioned. It is fitting to look into the legal frameworks of the States sharing the Scheldt River Basin District, in order to determine the extent to which cooperation with their counterparts across the border is promoted.

4.1.1. The duty to cooperate as enshrined in the national frameworks

259. The requirements included in the WFD and FD with regard to coordination in IRBDs have been implemented in the national legal frameworks to some extent.[880] In the Flemish Region, the DIWP stipulates that the Flemish government

880 Article 22 Para 2 DIWP.

should take the necessary measures so that the competent authority assigned to the International River Basin District Meuse and the International River Basin District Scheldt respectively draft a River Basin Management Plan for the entire IRBD, within the timeframes included in the DIWP.[881] At the same time, however, the DIWP provides that the Flemish government should determine River Basin Management Plans for each of the parts located on Flemish territory, if the adoption of an RBMP for the entire international River Basin District is not feasible within the timeframe included in Article 34 of the DIWP.[882] Cooperation is therefore made entirely optional. This is in line with the spirit of the Directives, which only include obligations of best efforts in this regard, as opposed to obligations of results.[883]

260. The Walloon legislative framework refers to international cooperation, although it takes a rather soft approach on the topic. The Water Code, and more specifically Article 2 7°, stipulates that a Walloon River Basin is defined as the part of each International River Basin District that is located on the territory of the Walloon Region. Furthermore, the second part of the Water Code pertains to the integrated management of the natural water cycle, and sets out the relevant provisions with regard to the creation of Walloon hydrological basins and sub-basins.[884] The Water Code stipulates that, within the limits of its competences, the Walloon Region negotiates the international and inter-regional agreements necessary for the creation and the organisation of International River Basin Districts.[885] Furthermore, the Water Code states that the Walloon Region promotes the international and interregional coordination necessary for the accomplishment of obligations imposed by the WFD and FD with the goal of preparing a single Flood Risk Management Plan or a collection of Flood Risk Management Plans coordinated at the level of the international River Basin District. Finally, on the basis of the Water Code, the Walloon Government adopts the rules and relevant decisions with the goal of ensuring the collection of information necessary to produce the documents to communicate to international bodies.[886]

261. In the Netherlands, most references to international cooperation for the transboundary aspects of water management are included in the relevant plans, i.e. the Water Plan with the Flood Risk Management Plan for the Scheldt as

---

881 Namely, in Article 34 of the DIWP.

882 Article 22 Para 2 of the DIWP. The Coordination Commission Integrated Water Policy has been appointed as the competent authority in this regard.

883 See also, Cathy Suykens, 'EU Water Quantity Management in International River Basin: Crystal Clear?' (2015) European Energy and Environmental Law Review 134; Jasper Van Kempen, *Europees Waterbeheer: Eerlijk Zullen We Alles Delen?* (Boom Juridische Uitgevers 2012) 362; Hey and van Rijswick (n 435) 242.

884 Part II of the Water Code, 'Gestion intégrée du cycle naturel de l'eau'.

885 Article D.12 Water Code.

886 As mentioned above, the competences for the recovery strategy are mainly situated at the federal level. Reference is therefore made to the explanation set forth in the Flemish section.

annexed thereto.[887] One section of the plan relates to international cooperation; however, this section is rather limited. Reference is made to the umbrella part of the FRMP drafted under the auspices of the International Scheldt Commission. For the Netherlands, the State was the main discussion partner in the international discussions.[888] For the smaller transboundary water bodies, bilateral international discussions took place, in the context of which the provinces and the regional water authorities took the lead.

262. As mentioned above, the WFD and FD have been implemented in French legislation through an amendment of the Environmental Code. As is the case in the Walloon and Flemish legal frameworks, the French Environmental Code and its implementing orders also refer to international cooperation. The Environmental Code stipulates that a ministerial order should delineate the basins or groups of basins in the country. When such basins extend beyond French borders, i.e. partly run through the territories of other EU Member States, the coordinating prefect of the basin is responsible, under the auspices of the foreign affairs minister, for ensuring coordination with the competent authorities of these other Member States in question.[889] This with the goal of designating the international district and developing a programme of measures that takes into account the entirety of the River Basin District. The ministerial order of 17 March 2006 further developed provisions with regard to the SDAGEs, and stipulated that, as for IRBDs, the summary of the plan should refer to the international coordinative commissions and the competent foreign authorities and which actions have been taken to ensure the coordination referred to in the Environmental Code.[890] The SDAGEs include the River Basin Management Plans within the meaning of the EU Water Framework Directive. There is a sanctioning mechanism applicable in case of non-compliance with the provisions of the SDAGEs. However, the sanctions section does not explicitly provide for a sanction in case of non-compliance with the coordination and cooperation requirement set with respect to the development of SDAGEs in IRBDs. In terms of international cooperation, the French FRMP Artois-Picardie refers to the coordinative activities taking place at the level of the International Scheldt Commission, and explicitly states that the plan only relates to the French parts of the IRBD Scheldt. The RBMP Artois Picardie, in the form of the SDAGE, does refer to international cooperation in the Scheldt District. It lists the important issues for the district, among other things, improving the good governance mechanisms for inter-regional and international cooperation. The reference to increased cooperation within the European Grouping for Territorial Cooperation is also mentioned.[891]

---

887 Adviescommissie Water, 'Bruggen Bouwen Nederlands Waterbeheer in Europees en Grensoverschrijdend Perspectief' (2007).

888 I.e. not the water authorities.

889 Article R212-2 Environmental Code. Article L212-2-3 Environmental Code.

890 Article 3 of the Order of 17 March 2006 related to the content of the SDAGEs, French Official Journal 15 April 2006.

891 See *infra*.

4.1.2. Cross-border consultation and public participation

263. As discussed in Chapter II, Article 14 WFD does not explicitly elaborate on the necessity for Member States to engage in transboundary public participation. In certain IRBDs, the parties provided for a mechanism to inform citizens throughout the district. For example, in the transboundary district of the River Eems, shared between Germany and the Netherlands, a document has been drawn up referring to Article 14 WFD, directed toward the citizens of the respective countries, explaining how they can actively participate in the plans and programmes issued for the international district. Such transboundary public participation mechanism is not present in the Scheldt District.

264. The Flemish RBMP Scheldt sets several policy goals, which includes a strong and coherent water policy that is based on consultation and coordination. In this regard, the plan states that it is the task of the Flemish Region to hold multilateral and bilateral consultations with the neighbouring countries and regions. In this regard, the RBMP acknowledges the importance of the sub-basin structures for the area-specific aspects of integrated water management.[892]

265. The Dutch Water Act provides that more specific rules should be stipulated with regard to the drafting of the national water plan, including the River Basin Management Plans for the districts that the Netherlands is involved in.[893] One of these specific rules should relate, according to the Water Act, to the consultation of competent authorities of the other States sharing these districts. Moreover, the Ministers in question should notify the competent authorities of the respective States of the input received on the draft international river basin management plan, amongst others from stakeholders, other departments, insofar as this input does not exclusively pertain to the Dutch part of the plan.[894] Other than that, the Water Act does not provide for more basic elements of transboundary cooperation with other competent authorities in the shared River Basin Districts.

The French SDAGE refers to the legal requirement of involving the competent authorities of countries affected by the plan in the same IRBD in the context of the public consultation procedure. The same holds true for the Walloon RBMP.[895] The French SDAGE mentions that the Flemish authorities requested an agreement related to the apportionment of the flow levels between the Scheldt and the Lys, which could, as mentioned in the SDAGE, prove to be a sensitive issue in light of climate change, i.e. in case of decreasing water levels.

---

892 As the Flood Risk Management Plan is absorbed by the River Basin Management Plan, the programme of measures also applies to flood risk management.

893 Article 4(3) Water Act.

894 Article 4(3)(2) Water Act.

895 Article 53(6) § 2 and § 4 of the Walloon Environment Code.

4.1.3. *The solidarity principle*

266. The solidarity principle stemming from the Floods Directive entails that Member States may not take measures that would lead to a significant increase of flood risks in another country within the international district.[896]

267. In the Flemish Region, the solidarity principle has been transposed in the Decree Integrated Water Policy, and its scope has been extended to the synergies between the three regions and between the various competent authorities for the sub-basins and sub-sub-basins. Indeed, the DIWP states that no measures can be taken that would lead to a significant increase of flood risks in areas in the same river basin, unless those measures are coordinated with the competent authorities of the other Member States or, within Belgium, regions and sub-basins and sub-sub-basins.[897] Neither the Decree, nor the Explanatory Memorandum, provide further specification on the interpretation of the solidarity principle. The Flemish RBMP for the IRBD Scheldt was published in the Belgian Official Journal through an Order of the Flemish Government in March 2016.[898] However, the plan does not refer to the solidarity principle.

268. With regard to flood risk management specifically, the French Environment Code mentions that the Flood Risk Management Plans must not include measures that would lead to a significant increase of flood risks upstream or downstream in other countries or regions sharing the basin.[899] However, substantial room for interpretation exists with respect to how "significant increase of risk" should be understood exactly, because the Environmental Code leaves this question open. A similar translation of the solidarity requirement exists in the Walloon Environment Code.[900]

269. The Dutch part of the Flood Risk Management Plan for the IRBD Scheldt was published in December 2015. In line with the flood risk management domain and the related division of responsibilities discussed above, the Ministries, provinces, municipalities, water authorities and safety regions cooperated in drawing up the plan.[901] The solidarity principle is explicitly mentioned in the Dutch part of the FRMP for the Scheldt. In the Netherlands, the solidarity

---

896 See Article 7(4) of the Floods Directive and Chapter II of this book. See generally, on solidarity in water management, Andrea Keessen and others, 'Solidarity in Water Management' (2016) 21 Ecology and Society.

897 Article 6(12) Decree Integrated Water Policy. See Explanatory Memorandum to the Decree modifying the Decree Integrated Water Policy of 18 July 2003 (2009-2010) 549, 1.

898 Belgian Official Journal 2 March 2016.

899 Article R566-10 Environmental Code.

900 Article D53-3 § 4.

901 As well as their umbrella organisations (e.g. the Union of Water Authorities (UvW), the Inter-Provincial Coordination Platform (IPO)).

principle is referred to as the principle of "not passing on". This principle, i.e. the fact that passing on flood risks to adjacent areas should be avoided, constitutes an important element in the choice of measures included in the FRMP. As for the regional water bodies, this principle is accommodated in the "capture – buffer – drain" three-step strategy discourse.[902]

4.2. REGIONAL COOPERATION IN THE SCHELDT DISTRICT

270. Although the main focus of this book is on cooperation at the level of the International River Basin District, and the synergies with the European legal framework, more regional and local cooperation is a topic that must not be neglected when analysing transboundary water management. Indeed, cooperation does not only take place at the level of the International Scheldt Commission. Within water bodies, several hydrological units can be identified. These hydrological units may be located in one country, or across one or more borders. For this reason, cooperation should (and does) take place at these lower scales as well. This more local type of cooperation is referred to as regional cooperation. The need for stronger regional cooperation in water management in general, has been acknowledged by stakeholders on several occasions.[903]

4.2.1. Tools for regional cooperation available to the Scheldt riparians

271. The table below provides an overview of the relevant regional cooperation tools available to the parties sharing the Scheldt. The instruments will be discussed in this section.

902 Ministerie van Infrastructuur en Milieu, 'Overstromingsrisicobeheerplan voor het Stroomgebied van de Schelde 2016-2020' (2015) https://www.rijksoverheid.nl/binaries/rijksoverheid/documenten/rapporten/2014/12/12/4-4-overstromingsricicobeheerplan-schelde/4-4-overstromingsricicobeheerplan-schelde.pdf accessed 13 July 2017. The Dutch FRMP also refers to transboundary watercourses that as such are not part of the Dutch plan because they do not represent a significant risk on Dutch soil, but where the risk does exist in the neighbouring country. In this case, the watercourse is included in the plan, in order to evaluate whether certain measures in said country have an impact on the risks in the Netherlands. See p. 31 of the plan.

903 Especially in the basin-specific parts of the Flemish River Basin Management Plans. Adviescommissie Water, 'Bruggen Bouwen Nederlands Waterbeheer in Europees En Grensoverschrijdend Perspectief' (2007) <www.adviescommissiewater.nl/Images/Advies%20over%20Nederlands%20Waterbeheer%20in%20Europsees%20perspectief%20-%2019%20november%202007_tcm308-268502.pdf> accessed 13 July 2017, 22. CIW, 'Tweede Waterbeleidsnota – Inclusief Waterbeheerskwesties' (2013) <www.integraalwaterbeleid.be/nl/publicaties/afbeeldingen/Tweede%20Waterbeleidsnota_LR.pdf/at_download/file> accessed 13 July 2017.

Table 8. EGTC v Benelux v Brussels Agreement

Characteristics	EGTC Regulation	Benelux Treaty	Brussels Agreement
Legal basis/ Framework	EU: Art. 175 TFEU (ex 159 EC Treaty)	– Council of Europe: Madrid Convention – Benelux Union Treaty	– Council of Europe: Madrid Convention
Goal	Achieving social, economic and territorial cohesion in the EU	Facilitating transboundary and inter-territorial cooperation	Facilitating transboundary cooperation
Duration	– Indefinitely	– Indefinitely – Can be denounced by one of the partes through written notification to the depository	– Indefinitely – Can be denounced by one of the parties through written notification[904]
Territorial scope of application	– EU MS[905] – Third countries neighbouring at least one Member State – OTCs[906]	– Benelux Union MS – Third Countries neighbouring one of the Benelux Union MS	– Belgium – France
Personal scope of application	– National authorities – Regional authorities – Local authorities – Public undertakings (Directive 2004/17/EC) – Entities carrying out SGEI – Any of the above from third countries	– Natinal authorities – All public authorities – All public institutions – Cooperation mechanisms between the above	In Belgium: – Provinces[907] – Municipalities – Inter-municipal cooperation groupings – Autonomous municipal corporation – Public Centre for Social Welfare or linked associations – Polders and wateringues (FL Region only)

904 Denunciation is effective by the end of the calendar year. It is important to note that a denunciation by one f the Belgian parties does not affect the participation of the other Belgian Parties.

905 Member States.

906 Overseas Countries and Territories.

907 In France: the region Champagne-Ardenne, the region Lorraine, the region Nord-Pas de Calais and the region Picardie, the municipalities, the departments, their cooperation agreements on the territories of said regions and their public institutions insofar as territorial communities participate in the transboundary cooperation.

Characteristics	EGTC Regulation	Benelux Treaty	Brussels Agreement
Material scope of application	Topics in the context of territorial, social and economic cohesion	All themes belonging to the competences of the participants	All themes belonging to the competences of the participants
Possible mechanisms	EGTC with legal personality	– BGTC with legal personality – Entity w/out legal personality – Administrative agreement	– Entity with legal personality or participation in foreign legal entity – local cooperation agreement for transboundary cooperation (LCTC) – Entity without legal personality
Initiative	Bottom-up and optional	Bottom-up and optional	Bottom-up and optional
Approval of national government	Yes Tacit approval after 6 months	No, unless required by internal law	Yes
Applicable law	– EGTC Regulation – Convention creating EGTC – National law for issues not covered by Regulation (determined by statutes of EGTC)	– The Convention – Internal law of one of the parties	– The Convention – Internal law of one of the parties (e.g. where entity has headquarters)
Transfer of competences	Yes, broad legal capacity. Transfer of governance competences from authorities to EGTC (no transfer of police and regulatory competences)	Only for the BGTC: Broad legal capacity + regulatory authority – binding decision-making power.	For the entities with legal personality: Legal personality, but no transfer of regulatory and police competences
Bodies and Personnel	EGTC: – Assembly + director – Additional bodies may be provided for in the statutes – Own personnel	BTGC: – Own personnel – Personnel made available by participants, but under the sole authority of the BTGC	LCTC: – Assembly + President & Vice-President – Statutes may create additional bodies – Brussels Agreement does not specify personnel arrangement

Characteristics	EGTC Regulation	Benelux Treaty	Brussels Agreement
Supervision and jurisdiction	Applicable administrative supervision determined in statutes	Administrative supervision procedures – internal law Parties can designate one common supervision authority, support by the Secretariat General of the Benelux Union Legal relationships governed by law that would be appicable if the participating authorities had executed competences themselves	Administrative supervision procedures – internal law[908] For the LCTC: internal law of country where headquarters are located

908 See Article 8 of the Brussels Agreement. The supervising authority should keep other supervising authorities informed; this is also in line with what is provided, e.g. in the Flemish Decree on Inter-Municipal Cooperation, more specifically Article 77.

272. With regard to the IRBD Scheldt, there are three main legal instruments through which regional, decentralised authorities can formalise cooperation with counterparts across the border, i.e. the European Grouping for Territorial Cooperation (EGTC), the Benelux Agreement and the Brussels Agreement. Decentralised authorities may base their formalised cooperation on these instruments and the tools they make available. Important to mention in this regard are the EU financing opportunities through the INTERREG projects, which are significant in the context of transboundary water management in the European Union, and which are used in practice as well.[909] Sub-basins in the Scheldt District have made use of the EGTC, but not specifically of the Benelux Agreement. The latter is not referred to as an instrument of cooperation in the sub-basin specific part for the Flemish part of the Scheldt district.[910]

– *European Union: The EGTC*

273. The EGTC was developed in the realm of EU law, whilst the Benelux Treaty is inspired by international law. Both legal tools are relevant in the context of the international Scheldt District.

The EGTC is a mechanism that was developed at EU level, i.e. through a Regulation, to promote transboundary cooperation at a more decentralised level and to remove some of the administrative barriers that complicate such cooperation initiatives.[911] The advantage of the Regulation as a legal instrument of choice is that, considering that it should be implemented 'As Is' and is directly binding in all its aspects, it creates a uniform mechanism throughout the European Union.[912] This helps remove some of the barriers inherent to transboundary cooperation, i.e. discrepancies in legal and institutional arrangements on different sides of the borders.[913] Although an EU Regulation is directly applicable in the Member States, the EGTC Regulation needs minimal implementation, mainly directed toward designating the competent authorities to approve the creation of the EGTCs in question. Originally, the EGTC was meant as a tool to facilitate the achievement of social and economic cohesion, and the EGTC Regulation explicitly restricts the tasks of the EGTC to achieving this goal.[914] More specifically, under the 2006

909 For example: French-Walloon transboundary river contract, as explained in the sections above.

910 The Flemish Region dominates the Scheldt in terms of geographical scope.

911 Regulation (EC) 1082/2006 of the European Parliament and of the Council of 5 July 2006 on a European grouping of territorial cooperation (2006 EGTC Regulation) [2006] OJ L 210/19.

912 In contrast to e.g. the Directive, which leaves substantial discretion to Member States in its implementation.

913 Elisabetta Nadalutti, 'Does the 'European Grouping of Territorial Cooperation' Promote Multi-level Governance within the European Union?' (2004) 51 Journal of Common Market Studies 756, 767.

914 Jirka Zapletal, 'The European Grouping of Territorial Cooperation (EGTC): A New Tool Facilitating Cross-Border Cooperation and Governance' (2010) Quaestiones Geographicae 29(4) 16.

Regulation, EGTCs were meant to implement cooperation projects co-financed by the EU through the Regional Development Fund, the Social Fund and/or the Cohesion Fund.[915] The 2013 amendment to the 2006 Regulation also significantly broadened the scope of the EGTC mechanism. In contrast to the 2006 version, the Regulation now allows EGTCs to be created with non-EU Member States as long as at least two EU Member States are involved and the third country neighbours at least one of the Member States, including its outermost regions and Overseas Countries or Territories (OCTs). The EGTC is governed by a Convention which is to be decided on unanimously by the founding members and which sets out the basic features of the mechanism. The amending 2013 Regulation deals with aspects of cross-border administrative mismatch, an issue that is especially relevant in the context of regional cooperation in water management. In the context of water management, different types of authorities at a different hierarchical level and of a different scale are confronted with one another in shared sub-basins. The 2013 amending Regulation allows national authorities to participate in the EGTC, with the *ratio legis* that competences that are national on one side of the border may be regional on the other, for example.[916] This is relevant for the IRBD Scheldt, e.g. for the Flemish Region, which is not viewed as a "decentralised authority" within the meaning of the 2006 Regulation and which was therefore excluded from such possibility to participate. The creation of the EGTC is a voluntary and bottom-up process. Neither the EU nor the state authorities can impose the establishment of such structure on decentralised authorities. The EGTC has legal personality and is liable for the acts of its bodies vis-à-vis third parties. It has to draw up and manage a budget and is liable for all its debts.

The amending 2013 Regulation brings the EGTC instrument into line with the Lisbon Treaty, which added a territorial angle to the EU cohesion policy. The scope of the EGTC is therefore no longer restricted to social and economic cooperation activities, but also covers issues enhancing territorial cohesion. By analogy, on the basis of the new Regulation, the EGTC is able to vie for EU funding other than the Union Cohesion Policy funding. The applicable legislation, in addition to the EGTC Regulation, that the EGTC is subject to, is the Convention creating the EGTC in question and national law for issues not covered by the Regulation. The regional authorities are not autonomous in their cooperation initiatives with their cross-border counterparties through the EGTC, as the national level government can intervene on several occasions. First, the authorities that intend to create the EGTC need to gain approval from the national governments. The latter may refuse approval for reasons of non-compliance of the Convention with the Regulation, or other EU and national law related to the competences of the

915 Ibid.

916 Recital 7 of the Regulation (EU) 1302/2013 of the European Parliament and of the Council of 17 December 2013 amending Regulation (EC) 1082/2006 on a European grouping of territorial cooperation (EGTC) as regards the clarification, simplification and improvement of the establishment and functioning of such groupings (2013 EGTC Regulation) [2013] OJ L 347.

members of the EGTC, or when the participation of the entity is not justified for reasons of public interest or policy of the country in question, or if the statutes of the EGTC are inconsistent with its convention. The national government needs to give its response within six months, where a lack of response implies tacit approval.[917] Second, Member States may restrict the powers and tasks that the EGTC may carry out without EU funding.[918]

- *CoE Instrument*

274. In 1980, the European Outline Convention on Transfrontier Co-operation between Territorial Communities or Authorities was adopted under the auspices of the Council of Europe.[919] This Convention is specifically aimed at structurally improving cooperation between territorial authorities or communities, which are defined as "communities, authorities or bodies exercising local and regional functions and regarded as such under the domestic law of each State".[920] The contracting parties may further specify which communities or authorities are covered through notification to the Secretary General of the Council of Europe. Indeed, the Convention encourages Parties to provide territorial cooperation projects, but leaves it up to the parties to further define the manner in which such cooperation should legally take place through special agreements.[921] Whereas the Madrid Convention was, at first, limited to regions bordering each other, the second protocol to the Convention broadened the scope to include non-neighbouring entities.[922] The scope of the Convention is very broad, and can cover a whole array of different sectors, including spatial planning and water management.[923] The third protocol further strengthened the legal framework by providing the possibility to create an entity for transboundary Euregional cooperation with legal capacity, the so-called "Euregional Co-operation Groupings". Through the entry into force of the third protocol to the Madrid Convention, the possibilities available for EU Member States under the auspices of the above-mentioned EGTC Regulation were made available to all Member

917 The approval of the State in which the EGTC would be located is necessary, however.

918 This is a result of the hesitant attitude of the Netherlands. See Explanatory Memorandum to the Act implementing the EGTC Regulation in the Netherlands (2008-2009) 31926 n 3.

919 European Outline Convention on Transboundary Co-operation between Territorial Communities or Authorities concerning Territorial Co-operation (Madrid Convention) (adopted 21 May 1980, entered into force 22 December 1981) ETS n 106.

920 See Article 2 Madrid Convention.

921 Jan Wouters and Maarten Vidal, 'Eindelijk een publiekrechtelijk kader voor gedecentraliseerde grensoverschrijdende samenwerking' (2005) De Franse Nederlanden 219-224.

922 Protocol No. 2 to the European Outline Convention on Transboundary Co-operation between Territorial Communities or Authorities concerning Territorial Co-operation (adopted 5 May 1998, entered into force 1 February 2001, ETS n 169.

923 Koen De Feyter and others (eds), *The Local Relevance of Human Rights* (Cambridge University Press 2011) 102.

States to the Council of Europe.[924] Considering the federal state structure of Belgium, it is relevant to note that the Madrid Convention and related protocols are together characterised as a mixed Treaty, where the relevant institutions are responsible for the implementation on their respective territories. Following the Madrid Convention, the Benelux Treaty was promulgated within the setting of the Benelux Union in 1986 and then replaced in 2014.[925] It allows decentralised authorities to formalise transboundary cooperation across the border between the Netherlands and Belgium.[926] Despite the existence of the EU-wide instrument of territorial cooperation, i.e. the EGTC Regulation and the Madrid Convention with its Third Protocol, the Parties to the 1986 Benelux Treaty still considered it relevant to revamp the latter and provide for an improved form of territorial cooperation in the Benelux Union.[927]

The Benelux cooperation instrument, in its updated form, was deemed to have added value, specifically over the EGTC Regulation. For example, it does not require prior approval of the central authorities for the creation of a joint entity by regional entities and provides the possibility to transfer regulatory competences to this entity, which is not possible in the context of the EGTC.[928] The Benelux Agreement originally applied to Belgian provinces, municipalities, and polders and wateringues and in the Netherlands between provinces, municipalities, regional water authorities and public entities. The additional Protocol to the 1986 Benelux Agreement widened the scope by allowing the Parties to allow other public-law legal persons to participate in the cooperation mechanisms, as long as one territorial authority of said country participates in said mechanism.[929] However, partly inspired by the EGTC Regulation and realising that the territorial scope of the Treaty was

---

924 See Strategische Adviesraad internationaal Vlaanderen en de Vlaamse Adviesraad voor Bestuurszaken, 'Advies over de instemming met het derde Protocol bij de Europese Kaderovereenkomst inzake grensoverschrijdende samenwerking betreffende euroregionale samenwerkingsverbanden' (20 July 2011) Parl. St. 1393 (2011-2012) <https://www.vlaanderen.be/nl/publicaties?uitgever=Strategische%20Adviesraad%20Internationaal%20Vlaanderen%20%20SAR%20iV%20/%20Flemish%20Foreign%20Affairs%20Council> accessed 13 July 2017.

925 In accordance with Article 30 of the Benelux Treaty on transboundary and inter-territorial cooperation, the Benelux Treaty no longer produces legal effects for a party to the Treaty on the date of entry into force of the new Treaty on its territory. The Flemish Region has implemented the Treaty through its Decree of 26 June 2015 related to the ratification of the Benelux Treaty on transboundary and inter-territorial cooperation ("Benelux Treaty"), as signed in 's-Gravenhage on 20 February 2014, Belgian Official Journal 16 July 2015.

926 It is relevant to note that the Madrid Convention and the three protocols have been signed by the Benelux countries as well as neighbouring countries Germany and France.

927 And neighbouring countries.

928 Explanatory Memorandum to the draft Act related to the approval of the Benelux Treaty on Transboundary and Inter-Territorial Cooperation (22 June 2016) Parl. St. 2015-2016, Doc 54, 1918/001.

929 Protocol supplementing the Benelux Treaty on Transboundary Cooperation between Territorial Communities or Authorities signed in Brussels on 12 September 1986 (adopted 22 September 1998) ETC n 261.

too restrictive, the 2014 version of the Benelux Treaty extends the possibility to participate in the transboundary and inter-territorial cooperation on the basis of the Treaty to (a) the States that are party to the Treaty, (b) all public authorities linked to the parties to the Treaty, (c) all public institutions, in the broadest sense of the word that have their head offices on the territory of one of the parties, including public enterprises, legal persons that are mainly financed or controlled by governments and legal persons that fulfil public tasks on the basis of a concession or legal assignment, and (d) cooperation structures between the above. Moreover, the scope is now no longer limited to the three Benelux countries, but may include a non-Benelux country, as long as one of the parties is a member state of the Benelux Union.[930] This is particularly relevant for the IRBD Scheldt, as the cooperation mechanisms provided by the Benelux Treaty are now available to regional authorities in the whole Scheldt District. Indeed, France is the only non-Benelux Union Member State of the Scheldt district, therefore the geographic criterion set in Article 1 is always met. Several cooperation mechanisms are provided in the Benelux Treaty, i.e. the administrative agreement without the creation of a joint body, the introduction of a joint body without legal personality and the creation of a public entity with legal personality, which is referred to as a "Benelux Grouping for Territorial Cooperation" (BGTC).[931] Only this last instrument is described extensively by the Treaty. Indeed, this public entity within the meaning of the Treaty implies that the entity is granted with the broadest legal capacity as granted by the legislation applicable on the territory of the party, and at least the capacity to buy and sell mobile and immobile goods, hire personnel[932] dispose of and manage its proper budget and bank account, and take legal action. This means that the authorities creating the entity may grant it governance competences.[933] The joint entity therefore has the authority to issue decisions that function as general binding rules that apply in an equal manner throughout the territory of the cross-border arrangement.[934]

---

930 Article 2 of the 2014 Benelux Treaty.

931 On the basis of the 1986 Benelux Treaty, this entity is referred to as a "joint body" ("gemeenschappelijk orgaan"), which is inspired on the Dutch legal person "joint body" (same name – "gemeenschappelijk orgaan") pursuant to the Joint Regulations Act, which will be discussed in the following chapter.

932 Either its own personnel, or the participating members may make available personnel of their organisations, but the personnel should always be under the authority of the BGTC. This in stark contrast e.g. with the Flemish sub-basin boards where personnel is still very much linked to the existing administration VMM.

933 This in contrast, for example, with the Anholt Agreement, a Treaty similar to the Benelux Treaty, but signed between the Netherlands and Germany. The Anholt Agreement excludes the possibility for the joint entity to adopt binding governance decisions that the citizens are subject to.

934 Marleen Hertoghs, 'De Grens in Je Achtertuin' in Michael Faure and Marjan Peeters (eds), *Grensoverschrijdend recht* (Intersentia 2006) 112.

The Benelux Treaty, however, does not affect internal national law in the sense that it does not create additional competences for the regional authorities, and the regional authorities, of course, may not transfer competences that go beyond their package of competences. The legal relationship between the legal subjects and the joint public entity is governed by the law of the country that would apply as if the national territorial authorities themselves had executed the competences granted to the public entity.[935] The Treaty explicitly states that the States and governments to the Treaty may assign governance powers to the entity in question.[936] The BGTC is subject to administrative and financial supervision stemming from internal law of the States involved. The parties that have created the BGTC are ultimately financially liable in case of insufficient capital of the BGTC[937] and are responsible for the obligations of the BGTC following its dissolution.[938] In addition to the BGTC, the Benelux Treaty includes softer cooperation mechanisms, i.e. the joint body without legal personality and the administrative agreement. The latter entails that parties, including existing EGTCs or BGTCs, can conclude a legally binding agreement to tackle issues of a transboundary nature, e.g. related to drainage, the creation of a cross-border road, the process of public participation for certain projects, and so forth.[939] In terms of jurisdiction, the law applicable to the administrative agreement is the law of the country on whose territory the respective aspects of the agreements are to be carried out. This means that different aspects of the agreement can be subject to different applicable jurisdictions. Through the administrative agreement, one of the participants can execute tasks of another participant, but in the name and through the applicable laws of the latter. A full transfer of competences, i.e. whereby the party executing the tasks in questions does so in its own name, is therefore not possible through the administrative agreement.

- *The Brussels Agreement*

275. Not only between the Benelux countries, at EU level, but also between France and Belgium[940], a treaty for promoting transboundary cooperation between regional authorities was signed on 16 September 2002: the Brussels

935 Article 4 of the 2014 Benelux Treaty.

936 Article 5 of the 2014 Benelux Treaty.

937 Pro rata depending on their participation, as stipulated in the statutes.

938 Article 12 of the 2014 Benelux Treaty.

939 Explanatory Memorandum to the Flemish Draft Decree related to the approval of the Benelux Treaty related to transboundary and inter-territorial cooperation. Ontwerp van decreet houdende instemming met het Benelux-Verdrag inzake grensoverschrijdende en interterritoriale samenwerking, ondertekend te 's-Gravenhage op 20 februari 2014, Parl. St. 305 (2014-2015).

940 Again, this is a mixed Treaty for Belgium, which the Flemish, Walloon and French presidents have signed along with the Belgian and French prime ministers.

Agreement.[941] The Brussels Agreement entered into force on 1 July 2005, i.e. on the first day of the second month following the date on which the last party notified the other parties that the internal requirements for entry into force had been met.[942] The goal of the Brussels Agreement is to allow territorial authorities to coordinate their decisions and jointly introduce or manage services of public interest with authorities across the border.[943]

Unlike the Benelux Treaty, the Brussels Agreement has not yet updated its legislative framework to reflect the changing realities of territorial cooperation. In terms of territorial scope, the Brussels Agreement differs from the Benelux Treaty, as the Brussels Agreement does not provide for the possibility to include a third, neighbouring country in the cooperation initiatives between territorial entities. Indeed, the Agreement only covers the territories of Belgium and France. The Brussels Agreement includes an exhaustive list of the entities that may conclude a cooperation agreement. For the Flemish and Walloon Regions, these are the provinces, the municipalities, the inter-municipal cooperation groupings, the autonomous municipal corporations, public centres for societal wellbeing (in Dutch referred to as "OCMWs") and organisations created by these public centres and, for the Flemish Region, polders and wateringues. For France, the regions Champagne-Ardenne, Lorraine, Nord-Pas de Calais and Picardie, the municipalities, the departments, their cooperation groupings on the territories of the above-mentioned regions, and their public institutions insofar as territorial communities participate in this transboundary cooperation.[944] The Brussels Agreement does not explicitly refer to the Parties themselves, as possible participants in the transboundary agreements, but Article 17 states "the provisions of this Agreement are applicable on the agreements mentioned in article 3 to which one or more Parties sign on to"[945], thereby implicitly allowing these to participate into the agreements as well. This is mainly due to the Belgian federal state structure, and more specifically to allow the French

941 Agreement between the Government of the Kingdom of Belgium, the Flemish Government, the Government of the French Community and the Government of the Walloon Region on the one hand, and the Government of the French Republic on the other hand, related to transboundary cooperation between territorial communities and local public entities, signed in Brussels on 16 September 2002. The Brussels Agreement entered into force on 1 July 2005. Ratified in France through Act no. 2005-250 of 18 March 2005, French Official Journal 19 March 2005, the Walloon ratification Decree of 6 June 2003, Belgian Official Journal of 23 June 2003 and the Flemish ratification Decree of 30 April 2004, Belgian Official Journal 9 June 2004.

942 All ratifications by Belgium took place between 2003 and 2004, the French ratified the Agreement in 2005. Article 18 Brussels Agreement. Décret n° 2005-745 du 28 juin 2005 portant publication de l'accord entre le Gouvernement de la République française, d'une part, et le Gouvernement du Royaume de Belgique, le Gouvernement de la Communauté française, le Gouvernement de la Région wallonne et le Gouvernement flamand, d'autre part, sur la coopération transfrontalière entre les collectivités territoriales et organismes publics locaux, fait à Bruxelles le 16 septembre 2002, French Official Journal 5 July 2005.

943 Article 3 Brussels Agreement.

944 Article 2 Brussels Agreement.

945 Free translation.

regions to enter into cooperation agreements with the communities and regions in Belgium.[946]

It should be noted that transboundary cooperation between the parties themselves, for example exclusively between France and Belgium as nation States, is not warranted on the basis of the Brussels Agreement. This would, and logically so, no longer be classified as regional, but as international cooperation. The Parties may expand the scope of the entities that are eligible for participating in cooperation agreements through the exchange of diplomatic notes, meaning that such initiative equals a simplified implementation agreement and does not require parliamentary approval. The Flemish Ratification Decree provides that the existence of such implementation agreement should be notified to the Parliament.[947]

The cooperation agreement may not delegate any regulatory or police competences, and the competences that the authorities in question would execute in their capacity of representatives from the Federal State, Region or Community may not be the subject of the agreement.[948] With the exception of such Article 17 cooperation, the cooperation agreements are only binding for the territorial authorities in question, and do not trigger liability of the Parties to the Brussels Agreement. The authorities may determine that a territorial authority fulfils tasks under the responsibility of another territorial authority.[949] Article 1 of the Brussels Agreement states that the transboundary cooperation in question must respect the internal laws of the Parties to the Agreement. This mainly relates to the phase prior to the conclusion of the cooperation agreement, i.e. with regard to respecting prior approval procedures stipulated by internal legislation.[950]

---

946 See Explanatory Memorandum to the Decree ratifying the Brussels Agreement, Parl. St. 2116 (2003-2004). See Jan Wouters and Maarten Vidal, 'Grensoverschrijdende Samenwerking Tussen West-Vlaanderen En Noord-Frankrijk Na de Inwerkingtreding van Het Frans-Belgisch Akkoord van 16 September 2002' (2005) Provincie West-Vlaanderen – wvi – Leiedal <https://lirias.kuleuven.be/bitstream/123456789/154985/1/GOSA+West-Vlaanderen.pdf> accessed 17 July 2017.

947 See Article 3 Flemish ratification Decree. The Council of State, in its advice issued in the run-up to the adoption of the ratification Decree and referring to the Court of Cassation, iterated that such implementation agreements do not require approval by the Parliament, if they are provided for by the basic Treaty and insofar as the implementation agreement does not exceed the boundaries set forth by the basic Treaty. See Advice of the Council of State of 13 June 2003, n 35.590/3 on the Decree ratifying the Agreement between the Government of the Kingdom of Belgium, the Flemish Government, the Government of the French Community and the Government of the Walloon Region on the one hand, and the Government of the French Republic on the other hand, related to transboundary cooperation between territorial communities and local public entities, signed in Brussels on 16 September 2002, Parl. St. 2116 (2003-2004). The Flemish ratification Decree dates from 30 April 2004 and was published in the Belgian Official Journal 9 June 2004.

948 Article 4 Brussels Agreement.

949 Article 5 Brussels Agreement.

950 For example, Article L. 1115-1 in conjunction with L. 1115-4 French General Code of Territorial Collectivities.

For example, cooperation agreements concluded by territorial authorities in France need to be submitted to a representative of the State.[951] The participating authorities should choose the law that applies to the implementation of the agreement, and in case of dispute, the courts of said jurisdiction are competent to settle the dispute.[952] If parties create a legal entity, the applicable law is that of the country where this entity has its headquarters.

Similarly to the Benelux Treaty, the Brussels Agreement provides for the possibility to create a joint entity without legal personality, i.e. without the capacity to adopt decisions that are binding on the participating members, for example.[953] For the Flemish Region, this mechanism mirrors the cooperation mechanism created in the 2001 Decree Inter-municipal Cooperation, on the basis of which two or more municipalities in Flanders may create such entity without transfer of governance powers to realise a certain project of joint municipal interest.[954] The entity does not have legal personality nor budgetary autonomy and it cannot impose binding decisions on the participants.[955] This type of cooperation mechanism is meant to facilitate e.g. the organisation of conferences or discussion groups. The cooperation agreement should state the applicable law which the entity is subject to. The regional authorities or public entities may also participate in entities with legal personality of the respective bordering country or create such entities, thus mirroring the Benelux Treaty, and the possibilities included in the protocols to the Madrid Convention.[956] This should be allowed by internal law.[957] The Flemish, Walloon and French legislative frameworks all provide for such participation. However, an important requirement is that French territorial authorities must not participate beyond 50% of costs in the capital or costs of one foreign legal entity. Such restriction does not exist in the Walloon and French legal frameworks.[958] In this scenario, depending on the geographical scale of the joint mechanism, parties would choose to establish the headquarters of the

---

951 The Prefect. Article L. 1115-1 French General Code of Territorial Collectivities.

952 Article 4(6) Brussels Agreement.

953 Article 9 Brussels Agreement.

954 See Article 6 and following of the Decree of 6 July 2001 related to the inter-municipal cooperation, Belgian Official Journal 31 October 2001.

955 Article 9 Brussels Agreement.

956 Article 10 Brussels Agreement. This needs to be allowed by the internal law of the territory in which these entities have their headquarters.

957 The legal frameworks have provided for cooperation of their proper authorities in foreign legal persons. For the Flemish Region, the participation in foreign territorial entities is allowed through the Decree Inter-municipal Cooperation, in France through Article L. 1115-4 of the French General Code of Territorial Collectivities, in the Walloon Region through Article L.1524-1 of the Walloon Code related to local democracy and decentralisation of 22 April 2004, Belgian Official Journal 12 August 2004. Vice versa, the possibility of participation of foreign authorities in legal entities of the country, only the French code has provided for, see Article L. 1115-2 of the French General Code of Territorial Collectivities.

958 Article L. 1115-4 French General Code of Territorial Collectivities.

entity in France to avoid the limit of Article L. 1115-4 of the French General Code of Territorial Collectivities.

Thirdly, the relevant entities may create a so-called "local cooperation mechanism for transboundary cooperation".[959] This type of cooperation mechanism, in comparison to the other forms of cooperation, is covered more extensively in the Brussels Agreement, e.g. in terms of what should be included in its statutes, the composition of the mechanism, and so forth. Considering the fact that the parties jointly agreed on more details with respect to this mechanism, some of the barriers linked to the internal law in question, which applies in a subsidiary manner to the joint entities, are covered. The Brussels Agreement determines the type of internal legislation that applies: the internal law related to public institutions for inter-municipal cooperation of the country in which the cooperation mechanism has its headquarters. The mechanism then has legal personality pursuant to said national law.[960] The Brussels Agreement stipulates that the mechanism needs to have a general meeting, a president and one or more vice-presidents and each of the territorial authorities or local entities needs to have a minimum of one seat in the general meeting, where none of them can have more than half of the seats.[961] The general meeting is the deliberative body of the mechanism, and the president the executive arm, ensuring the implementation of the decisions of the general meeting and acting as the legal representative. It is important that the cooperation mechanism has budgetary autonomy, with finances stemming from the participants and from its proper revenues.[962] The mechanism may also apply for loans, for which the participating authorities may ultimately be liable.[963] Finally, the mechanism is automatically dissolved when (a) it reaches its goals or (b) the period for which the mechanism was created comes to an end.[964] An example of the local cooperation mechanism for transboundary cooperation is the "GLCT Europa 1", created on the basis of French law by the departmental fire department Bas-Rhin and Ortenaukreis in Germany for the management of a transboundary fireboat.[965] The responsibilities of the entity relate to extinguishing fires and evacuating passengers, for example.

---

959 Article 11 Brussels Agreement. This type of cooperation mechanism is quite similar to the "district europeén"- a type of the French public-public cooperation mechanism of the "Syndicat mixte", within the meaning of Article L. 1115-4 of the French General Code of Territorial Collectivities.

960 The Brussels Agreement determines what should be included in the statutes of the mechanism.

961 Article 13 Brussels Agreement.

962 Article 14 Brussels Agreement. The mechanism also needs to draw up an annual budget, and so forth.

963 Article 14.3 Brussels Agreement.

964 Article 15 Brussels Agreement. The parties may also choose to dissolve the mechanism through unanimity voting.

965 This on the basis of the Karlsruhe Agreement, which is similar to the Brussels Agreement and provides for this type of instrument with the same modalities as the Brussels Agreement.

– *Euregional Co-operation Groupings*

276. The third Protocol to the Madrid Convention created an instrument that is similar to the EGTC, BGTC and the LCTC: the Euregional Co-operation Groupings (ECGs).[966] The objective of the Third Protocol is mainly to remove some of the barriers related to the discrepancies in administrative systems between the parties in establishing transboundary territorial cooperation, as it outlines the ins and outs of the joint entity to be created.[967] This Protocol was drafted in the period after the 2006 EGTC Regulation was promulgated, and provides for a legal cooperation instrument similar to the mechanism created by the Regulation. In contrast to the Benelux and Brussels Agreements and the EGTC Regulation, the Protocol is not yet in force in the countries of the International Scheldt District. Indeed, to date, only the Flemish Region has ratified the Protocol.[968] The explanatory memorandum of the ratifying Decree specifies that the Protocol is ratified, despite there already being many territorial cooperation instruments, e.g. the Benelux Treaty and the EGTC Regulation, to promote the international entry into force of the Protocol and because the Protocol creates an additional legal instrument for inter-territorial cooperation that might be useful for cooperation with entities outside the EU as well.[969] The Protocol has broadly defined the competences of the ECG, i.e. in terms of legal personality. The ECG is to have the most extensive legal capacity that can be accorded to legal persons under the national law of the State in question. The ECG has full budgetary autonomy and may conclude contracts, acquire goods[970] and initiate legal proceedings.[971] The ECG has the authority to adopt binding decisions and implement them, and the States must provide mechanisms to ensure these decisions are effectively implemented. There are limits, of course, to the powers of the ECG. The ECG may not be granted regulatory powers or establish its taxation system and like the

---

966 As the Third Protocol does not yet apply to the countries sharing the Scheldt district, it is not included as an instrument in the table. The functioning of the ECG is explained, as it is relevant for comparative reasons.

967 E.g. in terms of membership, creation, and so forth. The Third Protocol is relevant to, among other things, the existing Euregions.

968 Decree of 4 May 2012 related to the ratification of the Third Protocol to the European Outline Convention on Transboundary Co-operation between Territorial Communities or Authorities concerning Euroregional Co-operation Groupings (ECGs), Belgian Official Journal 29 May 2012.

969 See Explanatory Memorandum of the Decree of 4 May 2012 related to the ratification of the Third Protocol to the European Outline Convention on Transboundary Co-operation between Territorial Communities or Authorities concerning Euroregional Co-operation Groupings (ECGs), Belgian Official Journal 29 May 2012.

970 Immobile and mobile.

971 The ECG is also liable toward third parties for its debts and so forth. Article 2 Third Protocol to the Madrid Outline Convention. Protocol No. 3 to the European Outline Convention on Transboundary Co-operation between Territorial Communities or Authorities concerning Euroregional Co-operation Groupings (ECGs) (adopted 16 November 2009, entered into force 1 March 2013) ETS 206.

Brussels Agreement for example, the ECG must not be entrusted with tasks that the territorial authorities execute as agents of the States in question, unless duly authorised.[972] Membership of the ECG is open to both territorial communities or authorities of a signatory party and the party itself, e.g. France. Furthermore, all legal persons that operate to fulfil a certain goal of a general interest[973] are eligible to join the ECG if one of the following three conditions have been met: (i) their activity is financed by the State or territorial authority, or (ii) their management is subject to control of said entities, or (iii) half the members of their administrative, managerial or supervisory body are appointed by these entities.[974] The competent courts and the supervision procedures are linked to the headquarters of the ECG.[975]

4.2.2. Domestic legal provisions on decentralised cooperation in the Scheldt District: Theory

277. Regional authorities and public actors active in the water management realm in the IRBD Scheldt have a substantial toolkit available for transboundary water management governed by public law. As mentioned above, the Third Protocol to the Madrid Convention has not been ratified by all parties in the Scheldt, and therefore remains unavailable. However, considering the similarities between the European Co-operation Grouping and the European Grouping for Territorial Cooperation, this does not seem problematic as the toolkit available to the parties is already elaborate. From a territorial point of view, the parties can make use of all three of the instruments for agreements between regional authorities in France and Belgium and of the Benelux Treaty and the EGTC Regulation for agreements covering all three countries. Indeed, the EGTC Regulation applies directly to France, Belgium and the Netherlands. The Benelux Treaty, although strictly speaking only relevant to Belgium and the Netherlands, is available to territorial authorities in France as well, considering the widening of the scope of the Convention through the 2014 amendment.[976] The Brussels Agreement is restricted to the territories of France and Belgium. None of the instruments creates a harmonised legal arrangement for territorial cooperation, but they instead leave some room for the application of internal law in a subsidiary manner. The barriers related to the array of institutional and legal differences

---

972 Article 7.4 Third Protocol.

973 They must not have a commercial or industrial character.

974 Article 3 Third Protocol. Relevant is that territorial authorities may participate in an ECG even if its State has not yet ratified the Protocol, if said State shares borders with a State Party in whose territory the ECG has its headquarters.

975 The competent authorities may require the territorial authorities to withdraw from the ECG if the latter is acting in violation with e.g. public morality in the State in question, but such decision may be challenged in the courtroom.

976 November 2016: Benelux Treaty in its amended form has not yet been ratified in all countries.

need to be taken into account, as well as the limitations in internal law. A quick scan of the national legal frameworks in this regard is relevant.

278. The countries sharing the Scheldt District have handled regional transboundary cooperation to a varying extent. For example, whereas France provides an explicit legal framework for this issue, the Flemish Region has gone about the matter in a more minimal manner. In France, the French General Code for Territorial Collectivities deals with the topic in Articles L.1115-1 – L.1115-4.[977] French territorial authorities and their associations are authorised to conclude agreements with foreign local entities. They may also participate in public bodies governed by foreign law.[978] The main limitation stemming from French internal law relates to the cap on cost participation of these entities: their participation in costs of the foreign entity may not exceed 50% of their capital or costs.[979] The French framework provides for the creation of a European District, which is similar to the local cooperation mechanism for transboundary cooperation instituted by the Brussels Agreement.[980] The European District within the meaning of the Code is subject to the regime applicable to the "mixed syndicat", which is a French public cooperation mechanism.[981] France has also created a National Commission for Decentralised Cooperation, which was installed to improve the practice of decentralised cooperation and to create and maintain an inventory of such cooperation.[982] The Commission is meant to create a permanent dialogue between the territorial authorities and the state level. It combines 16 representatives of territorial authorities and associations thereof and Ministries involved in decentralised cooperation.[983]

---

977 The Walloon Region implemented the 2006 Regulation through its Decree of 26 April 2012. Decree related to the implementation of Regulation (EC) no. 1082/2006 of the European Parliament and the Council of 5 July 2006 related to a European Grouping for Territorial Cooperation (EGTC), Belgian Official Journal 22 May 2012. Due to the Belgian state structure, the EGTC Regulation needs to be implemented at the different levels of governance. The practical implication of this fact is that there can be a significant difference in time between the implementation in the respective territories, e.g. the Flemish Region implemented the EGTC Regulation in 2008 whereas the Walloon Region did so in 2012. Criticism regarding this mixed-treaty reality is wholly beyond the scope of this book, but it is relevant to point out its implication. The Flemish Government approved the 2006 EGTC Regulation through its Order of 18 January 2008.

978 The French Code on Territorial Collectivities establishes the approval procedure for such agreements.

979 Article L. 1115-4, Para 2 of the French Code on Territorial Collectivities.

980 See *supra*.

981 Article L5721-1 and following of the French Code on Territorial Collectivities.

982 Originally Act No. 94-337 of 24 October 1994 related to the National Commission for Decentralised Cooperation instituted by Article 134 of orientation Act No. 92-125 of 6 February related to the territorial administration of the Republic. Consolidated in Articles R1115-8 and following of the French Code on Territorial Collectivities.

983 Article R1115-9 of the French Code on Territorial Collectivities.

279. In the Flemish Region, the Decree Inter-Municipal Cooperation explicitly creates the possibility for municipalities to formally cooperate with other municipalities, and to participate in cooperation mechanisms of public law in other countries, regardless of the legal system these mechanisms are subject to.[984] Other than this provision, the Flemish Region does not explicitly regulate the participation of territorial entities in cooperation mechanisms governed by foreign law.[985] Implicitly, the possibility for municipalities and provinces to formally cooperate in a transboundary manner can be derived from Article 162 of the Constitution.[986] The Decree Inter-Municipal Cooperation also provides that supervision modalities should be agreed on between the Flemish government and the other authorities involved, and that these should be included in the statutes of the entity in question, to be submitted to the Flemish government.[987] Considering the restricted scope of Article 4 in conjunction with Article 77 of the Decree Inter-Municipal Cooperation – as it is limited to municipalities – implementation of, for example, the EGTC Regulation is necessary in order to provide an adequate legal basis for cooperation.[988] The implementing Decree of the Flemish Region and Order of the Flemish Government have designated the competent authority for approving participation of authorities in the Flemish Region to participate in an EGTC.[989]

280. In the Walloon Region, the Code for Local Democracy and Decentralisation provides that municipalities may conclude agreements with each other with regard to issues of municipal interest.[990] The municipalities may introduce a cooperation structure with legal personality to pursue a project of municipal interest.[991] With regard to transboundary cooperation, the Code

---

984 See Articles 4 and 77 Decree Inter-Municipal Cooperation. The Decree also allowed foreign legal persons to participate in cooperation mechanisms created on Flemish territory pursuant to the Decree, but the Constitutional Court annulled this provision in its judgment of 14 May 2003, no. 35/2003.

985 Ibid.

986 Case C-324/07 *Coditel Brabant SA v Commune d'Uccle and Région de Bruxelles-Capitale* [2008] ECLI:EU:C:2008:317, Opinion of AG Trstenjak.

987 In accordance with Article 30 of the Decree.

988 Explanatory Memorandum to the Decree implementing Regulation (EC) 1082/2006 of the European Parliament and the Council of 5 July 2006 related to a European Grouping for Territorial Cooperation (EGTC), Parl. St. 2007-2008, 1391, 1.

989 Order of the Flemish Government of 18 January 2008 implementing Regulation (EC) No. 1082/2006 of the European Parliament and the Council of 5 July 2006 related to a European Grouping for Territorial Cooperation (EGTC), Belgian Official Journal 5 February 2008. Decree of 21 December 2007 implementing Regulation (EC) No. 1082/2006 of the European Parliament and the Council of 5 July 2006 related to a European Grouping for Territorial Cooperation (EGTC), Belgian Official Journal 7 February 2008.

990 Article L1512-2 of the Code of Local Democracy and Decentralisation of 22 April 2004, Belgian Official Journal 12 August 2004.

991 Article L1512-2 of the Code of Local Democracy and Decentralisation of 22 April 2004.

creates the possibility for municipalities, project associations and associations of municipalities to participate in transboundary legal persons. Vice versa, foreign legal persons may participate in Walloon associations of municipalities.[992] Regional transboundary cooperation is thus provided for in the legal frameworks of the countries sharing the Scheldt, although the degree of legal coverage varies widely.

281. The Dutch Joint Regulations Act provides the framework for different types of cooperation between the relevant regional authorities; in the context of water management these are the provinces, municipalities and regional water authorities. These may conclude cooperation agreements to pursue joint interests among each other and with the other authorities referred to.[993] Bodies of an EGTC are identified as administrative authorities within the meaning of the Dutch General Administrative Law Act (GALA), i.e. "a body of a juristic person governed by public law or any other person or body vested with public authority".[994] The Act implementing the 2006 EGTC Regulation further specifies which authorities have the competence to approve the participation of Dutch entities in EGTCs. Generally speaking, i.e. for regional authorities such as provinces and municipalities, the Minister of the Interior and Kingdom Relations is the competent authority. For regional water authorities, the approval of the Minister for Infrastructure and the Environment is needed.[995]

282. The personal scope of the legal frameworks is quintessential. Indeed, the question arises which actors may make use of the instruments provided in these frameworks and on what conditions. In the Flemish Region, the competent actors in water management are the Region and its various departments, the provinces, the municipalities and the sub-basin boards. It is clear that provinces and municipalities may be formal partners of an EGTC, BGTC and LCTC. However, the participation of other relevant actors in water management, such as sub-basin boards and agencies acting as water managers is not that straightforward. The EGTC Regulation refers to bodies governed by public law within the meaning of Article 1(9) § 2 of the Public Procurement Directive.[996] The Directive 2004/18/EC has been repealed and replaced by Directive 2014/24/EU, but the latter has maintained the scope *ratione personae* of the former Directive, and therefore still includes the notion of "bodies governed by public

992 Article L1524-1 of the Code of Local Democracy and Decentralisation.

993 Joint Regulations Act of 20 December 1984, Dutch Official Journal 27 December 1984.

994 Article 1(1) Para 1 of the Act of 26 November 2009 implementing Regulation (EC) No 1082/2006 on a European grouping of territorial cooperation (EGTC) Dutch Official Journal 8 December 2009.

995 At the time of the implementing Act, this was still the Ministry of Transport, Public Works and Water Management.

996 Article 1(c) EGTC Regulation.

law".[997] In order to be qualified as such, three cumulative conditions have to be met: (i) the entity has been established to meet needs in the general interest, (ii) it has legal personality and (iii) is financed for the most part by the State, regional or local authorities or other bodies governed by public law.[998] Directive 2004/18/EC included a list that contained a catalogue of bodies governed by public law per country. In the realm of water management, several of these entities were included. More specifically, the list referred to the Flemish Environment Agency, the manager of the 1st category non-navigable watercourses, which under Flemish law is categorised as an internally independent agency with powers of jurisdiction.[999] The French Water Agencies were included as well. Polders and wateringues were viewed as central government authorities within the meaning of Annex I of the Directive, but are no longer included in Directive 2014/24/EU. For the Netherlands, in addition to the State, provinces and municipalities, also regional water authorities may participate in EGTC structures.[1000] For France, for example, the mixed syndicates and the public establishments for inter-communal cooperation are able to participate in the EGTCs. Entities without legal personality, or for example private enterprises, natural persons or associations thereof or legal persons carrying out an industrial or commercial activity cannot formally participate in the joint authorities created through the EGTC Regulation, the Benelux Treaty and the Brussels Agreement. From the perspective of governance at the hydrological scale, this mainly influences Flemish sub-basin boards, as they do not have the necessary legal personality. They may indirectly participate through the level of the Flemish Region, or, for example, whenever stakeholder meetings are organised by the joint entity in question. In the Walloon Region, river contracts have been bestowed with legal personality and can participate.

283. The added advantage in creating an administrative authority is that the principles of proper governance apply, which entail, among other things, that public participation is provided for.[1001]

---

997 Recital 10 of Directive (EU) 2014/24 of the European Parliament and of the Council of 26 February 2014 on public procurement and repealing Directive 2004/18/EC [2004] OJ L 94/65.

998 "For the most part" has been qualified as meaning "more than half" in case law of the Court of Justice.

999 See Article 10.2.1 of the Flemish Decree of 5 April 1995 related to General Provisions on Environmental Policy as amended by the Decree of 7 May 2004, Belgian Official Journal 3 June 1995.

1000 Explanatory Memorandum to the Act implementing the 2006 EGTC Regulation, Kamerstukken II 2008-2009, 31 926 n 3.

1001 A discussion of the principles of good governance in the respective countries from a comparative perspective is beyond the scope of this book.

4.2.3. *Cooperation in practice*

- *Sub-basins and sub-sub-basins: From the viewpoint of the sub-basin-specific parts of the management plans*

284. Now that the legal toolbox for decentralised cooperation and the main restrictions stemming from the national frameworks have been described, it is appropriate to zoom in on the manner in which the stakeholders at the different layers in the Scheldt District actually cooperate. Regional cooperation in the Scheldt District takes place in many different forms and mechanisms, ranging from ad hoc and informal to more institutionalised and structural. One of the main issues in cooperation between the competent authorities of these smaller hydrological units relates to the so-called "multi-level mismatch".[1002] This issue is also present in regional cooperation in the Scheldt basin. As explained in the section on national flood risk management above, competences in the context of water management are generally rather fragmented. Indeed, they are divided between the state level, provinces, municipalities, and river basin authorities. For example, from the perspective of a small municipality in the Walloon Region, i.e. Lessines, six water managers are involved.[1003] In comparison, for the municipality across the Dender located on the Flemish Territory, i.e. Geraardsbergen, three water managers are involved.[1004] As per the transfer of third-category watercourses from the level of the municipalities to the provinces, Geraardsbergen transferred its competences to the province of East-Flanders. In contrast, in the Walloon Region, the municipalities are still the managers of the third-category watercourses. This is one of the examples of the asymmetry in governance within the same sub-basin. The territorial scope of these water managers does not correspond with the hydrological or sub-basin scale. Despite the important potential coordinative role for the two entities active at the respective hydrological scales, the Flemish

---

1002 Tobias Chilla and others, 'On the Territoriality of Cross-Border Cooperation: " Institutional Mapping " in a Multi-Level Context' (2016) 20 European Planning Studies 961, 966. Marleen Van Rijswick, Herman Kasper Gilissen and Jasper van Kempen, 'The Need for International and Regional Transboundary Cooperation in European River Basin Management as a Result of New Approaches in EC Water Law' (2010) 11 ERA Forum 129. Adviescommissie Water, 'Bruggen Bouwen Nederlands Waterbeheer in Europees En Grensoverschrijdend Perspectief' (2007). Commissie van Advies inzake de Waterstaatswetgeving, 'Samenwerking over Grenzen. Relatie tussen Europese Kaderrichtlijn Water en Internationale Verdragen' (2004) <http://cawsw.nl/documenten/Samenwerking over grenzen implementatie EKW_CAW advies_22112004.pdf> accessed 13 July 2017.

1003 Namely, the 2nd category watercourse manager: the Province Hainaut, the 1st category non-navigable watercourse manager of the Walloon Region DG03, the navigable watercourse manager of the Walloon Region DG02, the municipality itself as a manager of the 3rd category watercourses and the wateringues of Tordois and Trimpont.

1004 Namely, the 2nd category watercourse managers: the Province East-Flanders, the 1st category navigable watercourse manager of the Flemish Region: Waterways and Sea Canal and the 1st category non-navigable watercourse manager of the Flemish Region: Vlaamse Milieumaatschappij.

sub-basin board Dender and the Dender River Contract do not actively weigh in on the policy making for the sub-basin with regard to flood risk management.[1005]

Table 9. Flemish Sub-basins and Types of Transboundary Cooperation

	EGTC	Structural platform or working group	Temporary forum	Informal & Ad hoc	Actions related to transboundary deliberation
Gentse kanalen FL + NL[1006]		X		X	0
Dijle & Zenne FL-W-BCR		X			2
Brugse polders FL-NL		X		X	0
Benedenschelde-Bekken FL-NL		X		X	0
Leiebekken FL-W-FR	X			X	2
Bovenschelde-Bekken FL-W-FR			X	X	5
IJzer FL-FR	X				2
Denderbekken FL-W		X			0
Demerbekken FL-W				X	1
Netebekken FL-NL				X	1

285. There is also a mismatch in the designation of water bodies: whereas the Flemish Region designated 11 sub-basins for the Scheldt, the Netherlands designated a smaller number of sub-basins, each of which logically covers a larger area. This therefore sometimes results in them having to cooperate with the competent authorities for several different sub-basins in the Flemish Region. This also makes it tricky to compare e.g. in terms of flood risks and risk calculations. In France, the Scheldt forms part of a larger basin, Artois-Picardie.[1007]

286. From the sub-basin-specific parts of the RBMP for the IRBD Scheldt for the period 2016-2021, as adopted by the Flemish Region, roughly four types of regional

1005 Mees, Suykens and others (n 525) 109.

1006 FL = Flemish Region; W = Walloon Region; FR = France; BCR = Brussels Capital Region; NL = the Netherlands.

1007 With the Somme and Manche North Sea.

cooperation can be discerned in practice, ranging from ad hoc and informal to more structural to institutionalised. It is remarkable that the stakeholders hardly make use of the cooperation instruments of the legal toolbox referred to above. Table 10 provides an overview of the sub-basin water bodies of the IRBD Scheldt, as designated by the Flemish Region, and the type of regional transboundary cooperation that they are subject to. The table also shows the action points related to transboundary deliberation or information exchange that have been included in the respective plans for the period 2016-2021.

287. Two of the sub-basins use the instrument developed in the context of EU law to encourage regional transboundary cooperation: the EGTC. The EGTC allows decentralised competent authorities of different EU Member States to work together in an entity with legal personality. The mechanism of the EGTC in general has been referred to as a *supra*-regional cooperation scheme, this as opposed to inter-regional cooperation. Indeed, the word "*supra*" refers to the degree of institutionalisation and the fact that the EGTC is has legal personality and personnel, and has its proper decision-making procedure.[1008] For the sub-basins "Leiebekken" and "IJzer", transboundary deliberations and information exchange take place through EGTCs. The EGTC for the "Leiebekken", referred to as the EGTC "Eurometropole Lille-Kortrijk-Tournai", combines fourteen entities from France and Belgium, mainly from the state level, regions, provinces and intercommunal organisations.[1009] The secretariat meetings take place twice a year and annual conferences of the 147 municipalities participating in the EGTC are also held. The material scope of the EGTC is broad: transport, tourism, the environment, etc. Land use planning, highly relevant for flood risk management, is one of the key focus areas of the EGTC.

Realising the importance of water to society, the EGTC developed in 2016 a joint vision related to "Eurometropolitan blue space", in which water is the vantage point for creating a sustainable and cohesive region with innovative actions.[1010] Although the sub-basin board for the sub-basin "Leiebekken" refers to the EGTC as the instrument through which cooperation related to transboundary water management with the Walloon Region and France will take place, the sub-basin board is not officially a partner in the EGTC and has, to date, not been involved in the functioning of the mechanism. Incidentally, formal participation of the sub-

1008 Estelle Evrard, 'The European Grouping of Territorial Cooperation (EGTC): Towards a Supraregional Scale of Governance in the Greater Region SaarLorLux?' (2016) 21 Geopolitics 513.

1009 The relevant water managers are Waterways & Sea Canal, Société Publique de Gestion de l'Eau and the Direction Régional and Direction Régionale de l'Environnement, de l'Aménagement et du Logement Service Préservation des Milieux et Prévention des Pollutions.

1010 Agency of the Eurometropole, Eurometropolitane Blauwe Ruimte: Sleutels tot Inzicht en Dialoog (2016). The actions to be taken are centred around three axes, i.e. protection, which relates e.g. to the protection of groundwater resources and biodiversity, risk management, which relates to the regulation of water levels, irrigation and flood risk protection, and valuation, which relates to cultural history and relaxation.

basin board in the EGTC is currently not possible due to its lack of legal personality. Participation of the sub-basin board with a view to transboundary coordination with its counterparts in the basin would only be possible through the EGTC's civil society forum, which is the advisory assembly of the EGTC, or through one of the official partners, e.g. a province. Internally, in the Flemish Region, the coordinator is the governor of West-Flanders.[1011] French, Walloon and Flemish River Committees are mentioned as relevant actors in the preliminary stakeholder analysis that is to be carried out in the context of the "blue space"-vision[1012], with a view to adopting an operations charter in 2018, covering priorities, challenges and actions related to the water system in the broadest sense of the word. Although the EGTC Regulation provides for the possibility to transfer governance competences from the authorities to the EGTC, this has not been done in this case. The EGTC does not have any binding decision-making powers, but is tasked with facilitating and promoting cooperation within the perimeter of the Eurometropole, between all the different authorities and stakeholders. The EGTC does not have the power, nor the ambition, to decide on specific projects – e.g. with regard to the designation of flood risk areas – but rather to create an inventory of challenges, opportunities and actions, which should then be implemented by the authorities and their subordinate entities.[1013] Although the EGTC could make a mark with regard to integrated water management in the Scheldt District, it should pay close attention to the work that is carried out by other actors, such as the river contract "Escaut-Lys" in order to avoid overlap in efforts.[1014]

For the "IJzer", cooperation takes place through the EGTC "West-Vlaanderen / Flandre – Dunkerque – Côte d'Opale". This EGTC was established in 2009. The EGTC created a specific working group for water. The sub-basin secretariat of the IJzer in Flanders is co-president of this working group, together with the French Institution "L'AGUR", the Agency for Urbanism and Development of the region Flandre-Dunkerque. For example, the working group identified the need to capture rain water as much as possible upstream in the French part of the IJzer sub-basin with a view to increasing the water supply to the downstream part of the IJzer sub-basin. On the basis of the recommendations of the working group, a study was commissioned to designate flood-control areas upstream. The FRMP for the French District "Artois-Picardie" also refers to the EGTC instrument to facilitate coordinated action at the transboundary level.[1015]

1011 Flemish Parliament, 'Commission meeting foreign policy of 1 April 2004' (2004) C172-BU17, Parl. St. 2013-2014 <http://docs.vlaamsparlement.be/docs/handelingen_commissies/2013-2014/c0m172bui7-01042014.pdf> accessed 13 July 2017.

1012 Interview with representative from the EGTC Eurometropole Lille-Kortrijk-Tournai involved in the "blue space" project, 21 October 2016.

1013 Interview with representative from the EGTC Eurometropole Lille-Kortrijk-Tournai involved in the "blue space" project, 21 October 2016.

1014 Interview with coordinator river contract Escaut-Lys, 7 November 2016.

1015 Préfet Coordonnateur de Bassin Artois-Picardie 115.

288. Several other sub-basins have some sort of structural platform for information exchange with regional authorities across the border. For the "Brugse Polders", "Benedenscheldebekken" and "Gentse Kanalen", the so-called "Kreken&Polders" is the platform in question. This platform was created in 1993, when four transboundary river basin committees were introduced by the Belgian-Dutch Commission for the Transboundary Non-Navigable Watercourses and the Benelux-Working Group Groundwater. However, for the Flemish Region, these were abolished by the Coordination Committee Integrated Water Policy in 2011 and it was decided to hold local transboundary discussions under the banner of the existing structures. These then developed into flexible transboundary working groups that work in a theme-specific and area-specific manner.[1016] For these three sub-basins, information exchange and deliberations therefore take place through the Kreken&Polders and the flexible transboundary working groups.[1017] These do not only focus on water quantity management. The river basin committees have been evaluated positively, but even with their abolishment, the contacts at the operational level across the Flemish-Dutch border remain.[1018] Financial stimulation from the EU level has played an important role in propelling cooperation forward, but has been dwindling over the past years, which has had an impact on the cooperation mechanisms.[1019]

The sub-basin Dender is shared between the Flemish Region and the Walloon Region. The only formal coordination platform for the Dender is the Interregional Consultation Watercourses, which convenes once a year.[1020] Since the floods of 2010, which affected the Dender basin, twice-yearly deliberations take place between water managers in the Walloon and Flemish Regions for the sub-sub-

---

[1016] These resort under the sub-basin structures in Flanders or the Regional Political Platforms. Secretariaat Brugse Polders, 'Stroomgebiedbeheerplan voor de Schelde 2016-2021: Bekkenspecifiek Deel Brugse Polders (2016), 22 <www.integraalwaterbeleid.be/nl/stroomgebiedbeheerplannen/stroomgebiedbeheerplannen-2016-2021/documenten/Bekken_Brugse_Polders.pdf/at_download/file> accessed 13 July 2017. Secretariaat Demerbekken, 'Stroomgebiedbeheerplan Voor de Schelde 2016-2021: Bekkenspecifiek Deel Demerbekken' (2016) <www.integraalwaterbeleid.be/nl/bekkens/demerbekken/planning-2016-2021/bekken specifiek-deel-ijzerbekken> accessed 13 July 2017.

[1017] Secretariaat Bekken van de Gentse Kanalen, 'Stroomgebiedbeheerplan Voor de Schelde 2016-2021: Bekkenspecifiek Deel Gentse Kanalen' (2016) <www.integraalwaterbeleid.be/nl/stroomgebiedbeheerplannen/stroomgebiedbeheerplannen-2016-2021/documenten/Bekken_Gentse_Kanalen.pdf/view> accessed 13 July 2017. Secretariaat Benedenscheldebekken, 'Stroomgebiedbeheerplan Voor de Schelde 2016-2021: Bekkenspecifiek Deel Benedenscheldebekken' (2016) <www.integraalwaterbeleid.be/nl/stroomgebiedbeheerplannen/stroom gebiedbeheerplannen-2016-2021/documenten/Benedenscheldebekken.pdf/view> accessed 13 July 2017.

[1018] Telephone interview with staff member of the Dutch regional water authority "Scheldt Streams", 21 April 2017.

[1019] Telephone interview with staff member of the Dutch regional water authority "Scheldt Streams", 21 April 2017.

[1020] Secretariaat Denderbekken, 'Stroomgebiedbeheerplan Voor de Schelde 2016-2021: Bekkenspecifiek Deel Dender' (2016) <www.integraalwaterbeleid.be/nl/stroomgebiedbeheer plannen/overwegingsdocumenten/od_den/at_download/file> accessed 13 July 2017.

basin "Marke". Other than that, there is no systemic platform for cooperation between the two regions with regard to the Dender sub-basin.

289. From an analysis of the various sub-basin-specific parts of the plans for the Flemish parts of the Scheldt, it is clear that informal and ad-hoc cooperation, in comparison to other more formalised methods of cooperation, is very present in the Scheldt sub-basins with respect to water quantity management.[1021] This can take the form of reciprocal invitations addressed to water managers or representatives from the sub-basin structures to the respective meetings across the border. These invitations are issued when drawing up the agendas, not systematically.[1022] The sub-basin-specific parts with respect to the "Bovenschelde" mention that structural deliberations should take place for the Maarkebeek catchment (shared between Flanders and Wallonia) to cover water quantity and water quality management.

– *The Scheldt Estuary: Importance of the Flemish-Dutch Scheldt Commission*

290. In addition to the regional cooperation at the sub-basin and sub-sub-basin levels, bilateral interaction also exists, the most important mechanism being the Flemish-Dutch Scheldt Commission (FDSC), which oversees the Scheldt Estuary. The governance of the Scheldt Estuary and the governance of the IRBD Scheldt as a whole should be seen as two parallel processes, as there is little interaction between these two governance mechanisms.[1023]

Considering the independent character of the governance mechanism of the Estuary, and the consideration that some elements thereof are interesting to river basin management in general[1024], this will be explained in this section. The Scheldt Estuary is the part of the Scheldt District where freshwater and seawater mix and which is composed of the Seascheldt and the Western Scheldt. The Flemish-Dutch Scheldt Commission also oversees the channel of Gent-Terneuzen and the Scheldt-Rhine Channel. The Commission stems from the Technical Scheldt Commission, which was created in 1948 to oversee scientific studies with regard to the Scheldt, and was composed of Dutch and Flemish civil servants. At the end of the 1990s, a third deepening of the Scheldt had been requested by the Flemish government.[1025] The Dutch government, in response, requested that such discussions were to take place in a more integrated framework, also taking into account nature and flood-safety concerns.[1026]

1021 The same conclusion can be drawn for water quality management, but this is not the focus of this research.

1022 See sub-basin specific parts of the RBMPs.

1023 Elli Louka, *Water Law and Policy Governance without Frontiers* (Oxford University Press 2008) 329.

1024 E.g. Scheldt Council, e.g. five-year re-evaluation of Treaty, see *infra*.

1025 Kathleen De Wit, Youri Meersschaut and Marc Sas, 'Third Enlargement Program of the River Scheldt' (2000) <www.vliz.be/imisdocs/publications/224252.pdf> accessed 13 July 2017.

1026 Interview with staff member Secretariat-General Benelux, 25 April 2017.

291. The Treaty with regard to cooperation for the policy and management of the Scheldt Estuary was concluded in 2005 and entered into force in 2008.[1027] The Treaty requires the Flemish-Dutch Scheldt Commission to evaluate the extent to which the goals set in the Treaty have been accomplished and to make suggestions to the parties in terms of measures that can be taken to increase the likelihood of reaching said goals and the modification of the Treaty.[1028]

The Long Term Vision 2030 reflects an integrated approach to safeguard the accessibility, safety and nature in the Estuary.[1029] On the basis of the Long Term Vision, the "development outline 2010" has been introduced as well as a joint monitoring programme.[1030] The Development Outline is based on the target scenarios included in the LTV. Through the Development Outline, Flanders and the Netherlands have agreed on safety issues in the Scheldt basin, and concerns and opportunities related to sustainability in the Scheldt Estuary.[1031] The Development Outline provides for specific projects for the short and medium term, as first steps towards realising the desired scenario for 2030. In the Flemish Region, the Updated Sigma Plan of 2005 executes two of the Long Term Vision themes, i.e. safety and nature.[1032] This means there is a clear link between the governance regime for the Estuary, i.e. Estuary Treaty and the FDSC, and the Sigma Plan in Flanders. Another important aspect in the governance of the Estuary relates to the "agenda for the future", which entails the development of a research programme addressing specific challenges the Estuary will be facing in coming years, e.g. increased tides, higher concentrations of sludge, and so forth.[1033]

In 2014, the Flemish-Dutch Scheldt Commission has established the Scheldt Council, which combines the expertise from regional and local governments, water authorities, harbours, environmental organisations, and agriculture, and which is presided over by the Secretariat-General of the Benelux Union.[1034] This key role played by the Benelux Union facilitates the independence of the Council vis-à-vis the FDSC. The Scheldt Council functions as an advisory body on the

---

1027 Treaty related to cooperation with regard to the policy and management in the Scheldt Estuary.

1028 Article 9 Estuary Treaty, 21 December 2005 in Middelburg.

1029 Technische Scheldecommissie, 'Long-Term Vision Scheldt-Estuary' (2001) <www.vnsc.eu/uploads/2012/01/ltv-schelde-estuarium-18-januari-2001.pdf> accessed 14 May 2014.

1030 In the Scheldt Estuary, the Long Term Vision has been developed under the auspices of the Technical Scheldt Commission, with three supporting pillars, i.e. safety, accessibility and naturalness. <www.vnsc.eu/uploads/2015/06/evaluatiemethodiek-schelde-estuarium-fase-2-1-v26032015-1.pdf> accessed 25 April 2017.

1031 Wil de Jong, Transborder Governance of Forests, Rivers and Seas (Earthscan, 2010) 154.

1032 See <www.wenz.be/nl/projecten/Sigmaplan> for an overview of the different projects.

1033 Flemish-Dutch Scheldt Commission, 'Agenda for the Future' (2015) <www.vnsc.eu/uploads/2015/11/vnsc-agendavoordetoekomst-hr.pdf> accessed 25 April 2017.

1034 Decision of the Political College of the Flemish-Dutch Scheldt Commission related to the Creation and Activation of the "Scheldt Council", 15 April 2014 <www.benelux.int/files/4013/9754/6191/intellingsbesluitSchelderaad.pdf> accessed 15 April 2017. Interview with staff member Secretariat-General Benelux, 25 April 2017.

basis of "joint fact finding" in cooperation with the FDSC. The recommendations provided by the Council are non-binding, but the FDSC needs to supply substantiation if its policies deviate from said recommendations. There is no direct participation from the public in the Scheldt Council, but citizens are indirectly involved through participation of the stakeholders such as municipalities.

In terms of the link between the ISC and the FDSC, the secretary-general of the TSC is a member of the Flemish delegation in the ISC, and the ISC has an observer status in the FDSC. The FDSC Treaty also provides that the Commission should maintain relations with the ISC.[1035] As mentioned, there is not much interaction between the FDSC and the ISC, and in fact, the need for more contact with the International Scheldt Commission was identified as an action point in a 2013 report on the management of the Scheldt Estuary as issued by the Flemish-Dutch Scheldt Commission.[1036]

4.2.4. Cooperation in theory and cooperation in practice: A good match?

292. There is a large array of available legal instruments to be used to structure cooperation in a systematic way. There is room for improvement in terms of matching cooperation in practice with cooperation in theory at the sub-basin and sub-sub-basin levels in the Scheldt District.[1037] A review of the sub-basin-specific parts of the Flemish RBMPs has shown that the importance of cross-border cooperation is widely recognised by the relevant actors. Until now, informal, ad-hoc, and project-based cooperation has been dominant, with flexible transboundary working groups reflecting the more structural route. There is room for improvement in the establishment of links between the different water bodies in the international Scheldt River Basin District. The structural, institutionalised regional cooperation platforms do not include a feedback mechanism regarding the level of the International Scheldt Commission, nor does the latter play a coordinating role in this regard. The functioning of the International Scheldt Commission will be discussed in the following section.

4.3 INTERNATIONAL COOPERATION THROUGH THE ISC

4.3.1. Legal basis for cooperation

293. The international cooperation with regard to the Scheldt was initiated in 1994. This cooperation was formalised through the Agreements on the Protection

[1035] Article 8 Estuary Treaty.

[1036] Vlaams-Nederlandse Scheldecommissie, 'Evaluatie Verdrag Beleid en Beheer Schelde-Estuarium' (2013) <www.vnsc.eu/uploads/2015/06/evaluatie-beleid-en-beheer-schelde-estuarium-1.pdf> accessed 26 April 2017.

[1037] This section only covers a limited review of this connection; the reader is referred to the evaluation section at the end of this Chapter.

of the Rivers Meuse and Scheldt.[1038] The 1994 Scheldt Agreement[1039] was limited to the qualitative aspects of water management. Indeed, Article 3(6) of the Agreement stated that the contracting parties protect and, as far as possible, improve the quality of the aquatic ecosystem of the Scheldt, among other things by management measures and steering the use of the river.[1040] As is the case with most flood-related measures, flood events encouraged specific action in this regard. Following the floods of 1993, the Working Group on Floods was created in the same year. As mentioned above, the Scheldt Agreement had been adopted prior to the entry into force of the WFD and FD. The Agreement therefore does not constitute the implementation of these Directives, but is based on international agreements, specifically the UNECE convention. The triangular relationship between the international level, EU level and national level discussed in Chapter II of this book presents itself here.

The adoption of the WFD did provide an impetus for the revamping of the cooperation structures for the Scheldt River. In 2001, the Ministerial Declaration of Liège was signed, on the basis of which the countries (and in the case of Belgium, the regions) agreed to work together in the context of the implementation of the WFD.[1041] The Declaration further stipulates that IRBDs for the Scheldt and Meuse are established, in accordance with maps that were attached as annexes to the Declaration. The Declaration expressly states that all EU Member States are individually responsible for the implementation of the Water Framework Directive, but that a coordinated approach with regard to the Scheldt river basin is necessary. Following the Ministerial Declaration of Liège, the Treaty of Ghent was concluded in 2002. The Treaty further provides for the multilateral cooperation within the meaning of the WFD.

294. The Treaty establishes the principles on the basis of which the parties should act in the context of water management. These are well-known principles in European environmental law, i.e. the precautionary and prevention principles, the principle which entails that environmental damages should be remedied at the source and the polluter pays principle.[1042] The Scheldt Treaty provides that "States work together" to coordinate the implementation of the requirements of the WFD for the International River Basin District, adopt a single management plan for the WFD, to discuss and coordinate the prevention and protection related

---

1038 Agreements on the Protection of the Rivers Meuse and Scheldt (Belgium (Brussels Capital Region, Flemish Region and Walloon Region), France and the Netherlands) (adopted 26 April 1994) 851 ILM 1995. 851.

1039 The Meuse and Scheldt Agreements refer to two international treaties: the UNECE Water Convention and the OSPAR Convention. Convention for the Prevention of Marine Environment of the North-East Atlantic (1993) 32 ILM 1069.

1040 Free translation of the authentic Dutch text. Only the Scheldt will be mentioned hereafter.

1041 Ministerial declaration of Liège (2001) <www.cipm-icbm.be/files/files/DECLARATION_MINISTERIELLE_DE_LIEGE_n.pdf> accessed 28 March 2014.

1042 Article 2 Scheldt Treaty.

strategies in the context of flood risk management, taking into account factors such as ecology and spatial planning, and to coordinate measures with regard to pollution control. The parties "shall" take the necessary measures to implement the Treaty and the advice and recommendations of the Commission and inform each other in this regard, improve the quality of aquatic ecosystems, strengthen the exchange of information and commit to quickly informing each other in case of calamitous pollution or high water.[1043] This means that most prominent in terms of cooperation is the information exchange and notification requirement pursuant to Article 3 of the Treaty. The obligation to adopt a single River Basin Management Plan is merely an obligation of efforts as opposed to an obligation of results, which mirrors the EU provision in question.[1044]

4.3.2. Organisational structure

295. The International Scheldt Commission has a rotating biennial presidency, with an annual plenary meeting, meetings of delegation leaders and working groups on several key issues, such as floods, water quality, public participation. The activities are supported by the ISC secretariat, which is staffed by approximately four people.

4.3.3. The ISC: Advisory body without binding powers

296. The International Scheldt Commission (ISC) plays a central role in the governance of the River on the basis of the Scheldt Agreement.[1045] It should be noted that the ISC has not been designated as the competent authority in the implementation of the Directives. Instead, the competent authorities have been designated at the level of the respective national River Basin Districts: in Belgium alone, there are three competent authorities for the Scheldt District.[1046] In the ISC, federal and regional governments are represented, namely France, Belgium, the Walloon Region, the Flemish Region, the Brussels Capital Region and the Netherlands.

297. The above-mentioned 2002 Treaty of Ghent broadened the scope of competences of the ISC.[1047] Even though the Commission's tasks were increased by the 2002 Ghent Treaty in comparison to the 1994 Agreement, the competence of this Commission is mainly limited to advisory tasks.[1048] The International

---

1043 Article 3 Scheldt Treaty.

1044 See Chapter II of this book.

1045 The same goes for the International Meuse Commission.

1046 Namely, the Coordination Commission Integrated Water Policy of the Flemish Region, the Government of the Walloon Region, and the Government of the Brussels Capital Region.

1047 International Scheldt Treaty (n 594). Moreover, the Commission was renamed the International Scheldt Commission.

1048 Jacko Van Ast, *Interactief watermanagement in grensoverschrijdende riviersystemen* (Ph.D. Thesis, Erasmus University Rotterdam 2000) 460. Article 4 of the Scheldt Treaty.

Scheldt Commission does not function as a supranational body with binding powers, as its role is more restricted to providing non-binding recommendations and offering a platform for coordination.[1049] When reading Article 3(2)(a) of the Scheldt Treaty, which stipulates that the Parties should take the necessary measures to implement the advice, recommendations and decisions on their territory, one would think that the Commission does have binding authority, as stated by Peeters.[1050] However, this provision should be read in conjunction with Article 4, which deals with the tasks of the Commission and which clearly states that the Commission may only publish decisions with respect to its internal organisation, and must use the tool of advice and recommendations when it concerns the parties with respect to the implementation of the goals of the Treaty.[1051] Moreover, the Explanatory Memorandum of the Dutch Act for the ratification of the 2002 Scheldt Treaty explicitly rejects any legally binding decision-making power of the Commission. The advice and recommendations should be adopted through unanimity voting.[1052] On the basis of the Scheldt Agreement, the International Commission for the Protection of the Scheldt River does not have competences beyond the quality of the water.[1053]

298. The ISC does not have binding legal powers, except with respect to its own operations and functioning, such as personnel matters.[1054] Moreover, the Commission only has legal personality with regard to the execution of its tasks as assigned by the Agreement.[1055] Furthermore, the Commission does not have a lot of room to manoeuvre in this regard, as their competences do not go beyond what is strictly necessary to accomplish their tasks.[1056] Implementation throughout the basin entirely depends on the willingness of States.[1057] Moreover, in contrast to the Scheldt Estuary Treaty discussed above, the Scheldt Treaty does not include a follow up-mechanism for the International Scheldt Commission to evaluate if and how its recommendations have been implemented, nor do the parties have any sort

---

1049 In contrast to, for example, the Commission for the Protection of the Rhine.

1050 With respect to the Meuse Treaty, which has a similar scope and wording as the Scheldt Treaty. Marjan Peeters, 'The Joint Governance of Transboundary River Basins: Some Observations on the Role of Law' in Michael Faure and Song Ying (eds), *China and International Environmental Liability: Legal Remedies for Transboundary Pollution* (Edward Elgar 2008) 204.

1051 Article 4(2) of the Scheldt Treaty.

1052 Article 5(4) of the Scheldt Treaty.

1053 Article 5 of the Scheldt Treaty.

1054 Andre Nollkaemper and Carel de Villeneuve, 'Recht van Internationale Waterlopen' in Nathalie Horbach, René Lefeber and Olivier Ribbelink (eds), *Handboek Internationaal Recht* (T.M.C. Asser Press 2007) 769-799.

1055 Gilissen (n 548) 42.

1056 Ibid. Article 5(4) of the 2002 International Scheldt Agreement. The Treaty of Ghent also leaves room for regional cooperation at a smaller scale. Indeed, Article 4 Para 5 of the Treaty stipulates that the coordination work for transboundary sub-basins within the IRBD Scheldt may take place within a suitable regional framework.

1057 Leo Santbergen, *Ambiguous Ambitions in the Meuse Theatre* (Eburon 2013) 196.

of obligation to report back on the implementation process and its output. Finally, the ISC does not have financial autonomy, as it depends on financial contributions from the participating States. As opposed to the Delaware governance scheme, which will be addressed in Chapter IV, the ISC does not manage projects in the basin or issue permits to safeguard its financial independence.

299. Regional cooperation at the sub-basin levels is not coordinated by the ISC. The 2002 Scheldt Treaty refers to cooperation at sub-basin level, but stipulates that such cooperation may take place within the appropriate regional framework without reserving a role for the ISC.[1058] The sub-basin authorities do not have particular affinity with the ISC, as the latter functions at a more abstract level.

4.3.4. Lack of focus on water quantity management from a legal perspective

300. Although the material scope of the Scheldt Treaty has broadened in comparison to the 1994 Scheldt Agreement, water quantity is not an important element in the scope.

301. The Treaty aims to implement the Water Framework Directive, which means that water quantity is of limited importance in the scope. The Treaty does not refer to the implementation of the Floods Directive, as it precedes the entry into force of the Directive, but does aim to offer a platform for information exchange on the prevention of flood risks and protection against floods.[1059]

The Scheldt Treaty refers to the principle of sustainable development, by stipulating that the contracting parties are to consult each other to safeguard the conditions for sustainable development of the Scheldt and the drainage area.[1060] It is somewhat confusing what the scope is of the concept of "sustainable development" within the meaning of the Scheldt Agreement.[1061] The Agreement, in its preamble, refers to the UNECE Water Convention, which adopts the definition of sustainable development given by the Brundtland Report: "water resources shall be managed so that the needs of the present generation are met without compromising the ability of future generations to meet their own needs".[1062] However, one cannot claim that the principle of sustainable development as adopted by the Brundtland report and the UNECE Water Convention applies unequivocally to the Scheldt Agreement. Indeed, the scope of this principle applies not only to the quality-related aspects of water resources,

[1058] Article 4(5) Scheldt Treaty.
[1059] Article 2 Scheldt Treaty. The Scheldt Treaty does not refer to the implementation of the Floods Directive.
[1060] Free translation of the Dutch version. See Article 3(6) of the Scheldt Treaty.
[1061] Frank Maes, 'The Content of the Agreements on the Protection of the Rivers Scheldt and Meuse' (1997) Revue Belge de Droit International 2, 678.
[1062] Article 2(5)(c) UNECE Water Convention.

but also to the use of water.[1063] However, the field of application of the Scheldt Treaty does not include water quantity management, i.e. there are no provisions on flow management or use, and the Treaty only briefly refers to floods. The mere reference to sustainable development does not alter this limited scope. As has been emphasised in Chapters I and II of this book, and time and time again in international literature, the interdependence between water quality and water quantity is incontestable. Moreover, floods can cause and exacerbate pollution of the watercourses. In this regard, the International Scheldt Commission has also stated that adequate quantitative water management can contribute to the restoration of ecosystems. This is not reflected, however, in the legal framework governing the International Scheldt River Basin District as embodied by the International Scheldt Commission.

4.3.5. "Significant increase of flood risks"

302. An important part of the activities taking place in 2013 and 2014 at the level of the ISC related to the drawing up of the transboundary part (also referred to as the "common part") of the second-generation River Basin Management Plans issued following the first-generation Flood Risk Management Plans. The Parties are to report on the results at the level of the River Basin District, and should come up with a joint interpretation of the concept of "significant flood risks upstream of downstream in other countries, in the same river basin or sub-basin area, caused by measures created by a Member State in the Flood Risk Management Plan". Indeed, on the basis of Article 7(4) of the Floods Directive, Flood Risk Management Plans in one Member State in the river basin must not significantly increase flood risks upstream or downstream in the same river basin or sub-basin, unless these measures have been coordinated and the States in question have come to an agreed solution in this regard.[1064] The 2013-2015 work plan for the IRBD Scheldt was adopted in a plenary session in the ISC. This joint interpretation of the "significant increase in transboundary flood risks" is one of the actions listed. However, the plan states that parties have not been able to come up with this joint interpretation, so that the different interpretations are listed in the common part of the Flood Risk Management Plan.[1065] These different interpretations were to be evaluated in the course of 2014 / 2015. At the time of writing, this information was not yet available. The final umbrella plan for the IRBD does not refer to the interpretation of the notion "significant increase".

1063 Frank Maes (n 1061) 679.
1064 Article 7(4) Floods Directive.
1065 See work plan of the ISC for 2013-2015: International Scheldt Commission, 'Work plan for the period 2013-2015' <www.isc-cie.org/images/Documents/PLEN_1301_WERKPLAN2013-2015_REV1.pdf> accessed 13 May 2014.

5. CONCLUSION: EVALUATION OF THE SCHELDT REGIME

303. This concluding section aims to bring all the pieces of the puzzle together and evaluate the river basin mechanism of the Scheldt, and what this tells us in terms of the facilitating EU legal framework. Where relevant, examples are given from other IRBDs (partly) located in the EU, to provide a point of reference for the Scheldt. The benchmarks used have been explained in Chapter I.

5.1. EQUITABLE AND REASONABLE UTILISATION AND THE NO-HARM RULE

– *ERU and No-Harm have not explicitly been provided for*

304. The principle of equitable and reasonable utilisation and the no-harm principle are the cardinal principles of international (customary) water law, deeply embedded in the UNECE Water Convention and the UN Watercourses Convention. This has been explained in Chapter I. All three States sharing the Scheldt as well as the European Union have ratified the UNECE Water Convention and are subject to international customary water law. As a result, the principles are *a priori* applicable in the District. The question then arises how these principles have been translated in the EU legal framework and operationalised in the Scheldt District.

305. As was determined by Van Kempen in the context of water quality management, the Water Directives are mute with respect to the distribution of usage in shared waters.[1066] The same conclusion holds true in terms of water quantity management. This can be explained by the consideration that EU water law is focussing on all water bodies within the EU, not just the regulation of the transboundary aspects thereof. This, as mentioned, is in contrast with the UNECE Water Convention and the UN Watercourses Convention, which explicitly provide for flow management and the use of shared waters. This difference can be explained by the consideration that the WFD and FD aim to address water management and flood risk management directly, whereas the Conventions target transboundary issues associated with water such as conflicts of use, competing uses, and protection of shared waters.[1067] Whereas the WFD is very elaborate

---

[1066] Jasper Van Kempen, *Europees Waterbeheer: Eerlijk Zullen We Alles Delen?* (Boom Juridische Uitgevers 2012) 361.

[1067] Alistair S Rieu-Clarke and Flavia Loures, 'The Role and Relevance of the UN Convention on the Law of the Non-Navigational Uses of International Watercourses to the EU and Its Member States' (IHP-HELP Centre for Water Law, Policy and Science, 2010) 400.

with respect to environmental objectives associated with achieving "good status", the transboundary aspect is less developed.

The lack of explicit reference to the principle of equitable and reasonable utilisation in the WFD and the FD is mirrored in the Scheldt regime. Indeed, the Scheldt Treaty does not regulate the principles and approaches related to the use of the Scheldt District. In contrast to e.g. the Rhine Convention, the Scheldt Treaty does not set specific targets that should be reached throughout the river in terms of flood risk reduction or water retention. The Scheldt Treaty does refer to the principle of sustainable development, which lies at the basis of the equitable and reasonable utilisation principle, by stipulating that the parties should consult each other to warrant sustainable development of the Scheldt.[1068] No specific agreements have been made in terms of flow management or principles underpinning the use of waters across the borders. With respect to the Meuse, parties have addressed water quantity issues in a separate Treaty. In the Scheldt River, the broadening and deepening of the Western Scheldt have equally been dealt with in a separate Treaty.[1069]

306. The no-harm principle entails that States should take all appropriate measures to prevent, control and reduce any transboundary impact.[1070] In the Water Directives, the no-harm principle is mainly illustrated with regard to the solidarity principle in flood risk management, where Member States are not allowed to include measures in their FRMPs that may cause significant harm to another State sharing the same basin.[1071] The meaning of the word "significant" has not been elaborated at EU level, but, as mentioned in Chapters I and II, international water law provides helpful guidance for interpretation. The harm must be qualified as going beyond a minor inconvenience to be considered significant.[1072] An exception to this prohibition to cause a significant increase of risk is made for the scenario where such measures have been coordinated. The FD neither attaches legal consequences to the inclusion or lack of inclusion of such measures in the FRMPs, nor does it determine the platform or provide for the instruments through which this coordination may occur. States sharing an IRBD are not obligated to conclude agreements on the meaning and application of "significant increase in flood risks" with respect to the district in question.

The solidarity requirement has been translated into the legal regimes of the Scheldt parties, although in a rather vague fashion. In the Netherlands, the discourse that it should be avoided to pass on flood risks to adjacent areas

---

1068 Free translation of the Dutch version. See Article 3(6) of the Scheldt Treaty.

1069 I.e. with respect to the Estuary, see the section on the Flemish-Dutch Scheldt Commission.

1070 Article 2(1) of the UNECE Water Convention.

1071 Article 7(4) FD.

1072 Commentary to the Draft Articles on the Law of the Non-Navigational Uses of International Watercourses, in ILC, 'Report of the International Law Commission on the work of its 46th session' (2 May-22 July 1994) UN Doc A/49/10, 94, para 15.

constitutes one of the elements in choosing measures to be included in the FRMP. The Flemish Decree Integrated Water Policy has literally translated the solidarity requirement of the FD, by stipulating that measures leading to a significant increase in other areas of the River Basin District are prohibited. However, the question remains what constitutes a "significant increase of risk", as this has not been explained, neither in the national frameworks or the management plans, nor in the roof report at the level of the IRBD. The Scheldt parties could not reach agreement on the meaning of this notion, and it was stated in a plenary meeting report that it should suffice to determine the threshold "significant increase" on a case-by-case basis.[1073] Other than this "significant increase of risk" notion, the WFD and FD do not expressly set a requirement for States sharing IRBDs to prevent, control and reduce transboundary impacts.[1074]

– *Presence of procedural requirements underpinning ERU and No-Harm*

307. However, the lack of explicit provisions on the principle of equitable and reasonable utilisation and the partial lack with respect to the no-harm principle at EU level do not necessarily imply that the principle is not endorsed by the EU legal framework through other provisions. Indeed, there are procedural aspects to both the principle of equitable and reasonable utilisation and the no-harm principle, with the duty to cooperate constituting a catalyst for the realisation of the principle.[1075] The specific procedural duties underpinning the two equitable and reasonable utilisation and the no-harm notions and operationalising the duty to cooperate are the duty to exchange information, to notify other States where relevant and to perform consultation and negotiation exercises.[1076]

308. The duty to cooperate forms part of the WFD and FD. Member States should achieve coordinated implementation of the environmental objectives set in Article 4 of the WFD and the associated programmes of measures. As such "Article 4" objectives do not form part of the FD, a similar duty to coordinate does not exist with respect to flood risk management. In addition to this requirement, the duty to cooperate is mirrored in the specific procedural obligations, such

---

1073 International Scheldt Commission, 'Report of the 21st plenary meeting' (2013) < www.isc-cie.org/images/Documents/PLEN_1301_PV.pdf> accessed 13 July 2017.

1074 Gabor Baranyai, 'The Water Convention and the European Union: The Benefits of the Convention for EU Member States' in Attila Tanzi and others (eds), *The UNECE Convention on the Protection and Use of Transboundary Watercourses and International Lakes* (Brill Nijhoff 2015) 88.

1075 Owen McIntyre, 'The Principle of Equitable and Reasonable Utilisation' in Attila Tanzi and others (eds), *The UNECE Convention on the Protection and Use of Transboundary Watercourses and International Lakes* (Brill Nijhoff 2015) 156.

1076 Owen McIntyre, 'A Comparative Analysis of the Legal Frameworks That Govern Europe's Transboundary Waters' in Werner Scholtz and Jonathan Verschuuren (eds), *Regional Environmental Law Transregional Comparative Lessons in Pursuit of Sustainable Development* (Edward Elgar 2015) 387.

as the adoption of international RBMPs and FRMPs, which mainly entail due diligence obligations with respect to cooperation.

However, the national and regional legal frameworks of the Scheldt countries have insufficiently incorporated cooperation and coordination into their legal and policy system. Although references are made, the laws implementing the FD and WFD and/or the RBMPs and FRMPs, do not do so systematically as the relevant provisions are scattered throughout various laws and plans. The Flemish DIWP has transposed the provision of the FD urging Member States to cooperate with the aim of producing one single FRMP.[1077] However, the DIWP also states that an FRMP exclusively for the Flemish part of the Scheldt basin should be adopted, should this coordination fail. The Walloon Region includes a similar provision. The obligation to cooperate with other countries in the International Scheldt District in the Flemish and Walloon Regions can therefore be considered as an obligation of best efforts. In France and the Netherlands, explicit references to international cooperation in the context of the FRMPs are not included in the legal framework, but the management plans do include references.

309. In addition to the publication of management plans, relevant procedural requirements are in place at EU level. These different procedural elements have been described in Chapter II.[1078] Notification requirements have been established by the EU legal framework, and follow particularly from the EIA and SEA Directives. RBMPs and FRMPs fall under the scope of the SEA Directive, and projects related to water quantity management often fall under either Annex I or Annex II of the EIA Directive. The legal frameworks of the Scheldt countries also provide for the notification of cross-border competent authorities. Information exchange is required with regard to e.g. flood risk cartography. This is effectuated through the various working groups of the ISC. The Scheldt Treaty also includes specific information exchange requirements, e.g. with respect to calamitous water pollution.[1079] Information exchange has also taken place in the context of the EU-funded Scaldit and ScaldWIN projects, illustrating the importance of financial incentives. In comparison to the UNECE Water Convention, the provisions of the Water Directives on consultation between States sharing IRBDs are rather limited. For example, whereas the Water Convention includes a mechanism where States should obligatorily enter into consultations upon the request of one of the States, this is not provided for in the EU Directives.[1080]

1077 The same goes for the RBMP as the implementation of the WFD.

1078 See the discussion of the third pillar, procedural provisions.

1079 Article 2 (c) Scheldt Treaty.

1080 The Scheldt countries can therefore make use of the "consultation-upon-request" procedure as established by the UNECE Water Convention. Gabor Baranyai, 'The Water Convention and the European Union: The Benefits of the Convention for EU Member States' in Attila Tanzi and others (eds), *The UNECE Convention on the Protection and Use of Transboundary Watercourses and International Lakes* (Brill Nijhoff 2015) 98.

310. Tanzi calls for a contextual hermeneutical approach to the application of the no-harm principle, stating that its operationalisation should be led by the specific conditions of the river basin in question.[1081] This is indeed the right approach, as the EU level is not equipped to exhaustively interpret the meaning of the "no-harm" principle for all IRBDs shared by EU (and non EU) Member States. The same goes for the equitable and reasonable utilisation principle, which should form part of a mutual agreement between the States sharing the basin in question.[1082] However, neither the equitable and reasonable utilisation principle nor the no-harm principle have been explicitly operationalised in the context of the legal regime governing the IRBD Scheldt. In this regard, it may be concluded that the WFD and FD fall short of providing the right tools and incentives for States sharing IRBDs.

5.2. INTEGRATED RIVER BASIN MANAGEMENT AT THE HYDROLOGICAL SCALE AND NESTED GOVERNANCE

311. From legal analysis, supported by empirical research, it can be concluded that there is substantial room for improvement with regard to integrated hydrological scale management in the IRBD Scheldt. Horizontal, i.e. between the various competent authorities, and vertical, from the sub-sub-basin to the international district level, links and coherence should be strengthened.

– *Multi-level governance*

312. Within countries, competent authorities for the watercourses or parts thereof, are often not those operating on the hydrological scale. Competences related to water management are fragmented and divided between the state level, provinces, municipalities, and functionally decentralised authorities such as river basin authorities.[1083] Especially in the Flemish and Walloon Regions, there is often a lack of actual governance on the hydrological scales. These rather function as coordination and advisory platforms.

Comparing countries, there is a mismatch in scales and administrations, due to differences in how water bodies have been designated and, of course, institutional differences.[1084] This is a common issue in transboundary water

---

1081 Attila Tanzi and Alexandros Kolliopoulos, 'The No-Harm Rule' in Attila Tanzi and others (eds), *The UNECE Convention on the Protection and Use of Transboundary Watercourses and International Lakes* (Brill Nijhoff 2015) 145.

1082 Owen McIntyre, 'The Principle of Equitable and Reasonable Utilisation' in Attila Tanzi and others (eds), *The UNECE Convention on the Protection and Use of Transboundary Watercourses and International Lakes* (Brill Nijhoff 2015) 153.

1083 As will be discussed in Chapter IV, the same holds true for the United States.

1084 Also reported e.g. by Gilissen in Gilissen, Internationale En Regionaal Grensoverschrijdende Samenwerking in Het Waterbeheer.

management and has been referred to as the "multi-level mismatch".[1085] For example, Dutch regional water authorities operating on the hydrological scale have legal personality, have binding decision-making power and have autonomy in terms of financial resources. Flemish sub-basin boards do not have legal personality and depend on the Flemish administration for human and financial resources and project initiation. An important part of the responsibilities of the latter relate to the fulfilment of procedural requirements.[1086] This is similar to the Walloon Region, where the water management tasks are with the department of navigable and non-navigable watercourses (DG02 and DG03), the provinces and municipalities. The river contracts are in place to bring together the relevant actors and stakeholders. They are different from the sub-basin boards, in the sense that their creation follows a bottom-up process with top-down support, legal personality and a limited degree of financial autonomy. In France, hydrological scale governance is important with the Water Agencies, River Basin Committees, Local Water Committees and the possibilities for local authorities to delegate actions to more basin-oriented entities such as the EPTB. In France, river contracts exist as well, but have a different role to play in comparison to the Walloon river contracts.

– *Gap between theory and practice in terms of territorial cooperation instruments*

313. Beyond water management, there is a broad range of mechanisms for regional cooperation that territorial authorities can use in the Scheldt District, with the aim of formalising cooperation at the more local levels. These mechanisms have been explained in the sections above and mainly relate to the EGTC Regulation, the Brussels Agreement and the Benelux Treaty. Together, they present a comprehensive toolbox of available instruments, such as the EGTC, the BGTC, the LCTC, entities without legal personality, and so forth. Except for the EGTC, which was not created specifically for water management purposes but has an impact on parts of the geographical scope of the sub-basin authorities, these instruments are not used by the competent authorities.

Regional cooperation in the Scheldt District takes place in many different forms and mechanisms, ranging from ad hoc and informal to more institutionalised and structural. An analysis of the different sub-basin-specific parts of the plans for the Flemish parts of the Scheldt shows that informal and ad-hoc cooperation,

---

1085 Tobias Chilla and others, 'On the Territoriality of Cross-Border Cooperation: "Institutional Mapping" in a Multi-Level Context' (2016) 20 European Planning Studies 961, 966. Marleen Van Rijswick, Herman Kasper Gilissen and Jasper van Kempen, 'The Need for International and Regional Transboundary Cooperation in European River Basin Management as a Result of New Approaches in EC Water Law' (2010) 11 ERA Forum 129. Adviescommissie Water, 'Bruggen Bouwen Nederlands Waterbeheer in Europees En Grensoverschrijdend Perspectief' (2007).

1086 Namely providing the sub-basin specific parts of the RBMPs and FRMPs.

in comparison to other more formalised methods of cooperation, is rather dominant in the Scheldt sub-basins. This may take the form of reciprocal invitations addressed to water managers or representatives from the sub-basin structures to the respective meetings across the border. These invitations are issued when drawing up the agendas, not systematically.[1087] These mechanisms can be effective at the more local scale, and there surely are examples of successful projects of local cross-border cooperation in the Scheldt District. However, the disadvantage of these methods is that these processes strongly depend on the dynamics and level of engagement of the persons involved in the deliberations. The sustainability of this type of processes in the context of transboundary water management, taking into account the increasing pressure on water resources and the projections with respect to flood risks in the future, can therefore at least be questioned in this regard.

As mentioned, there are examples of successful cooperation initiatives throughout the Scheldt District, e.g. with regard to the nature area "Zwin", the polders of Kruibeke, and so forth.[1088] These cooperation practices are not necessarily based on formalised structures, but are often project based and/or take place at the operational level. EU funding such as INTERREG, for example for the Scaldit and ScaldWIN projects, plays an important role in encouraging cooperation. When such funding dwindles, territorial cooperation efforts tend to do so as well.[1089]

314. The need for increasing transboundary cooperation is quite consistently referred to in the various basin and sub-basin management plans.[1090] The roof report of the River Basin Management Plan for the IRBD Scheldt refers to the necessity to reinforce interregional and international cooperation, to exchange information and to streamline methods and measures.[1091] However, clear actions and mechanisms to achieve this cooperation are disparate and inconsistent. For example, reference can be made to the transboundary influx of pollution loads carrying dangerous substances in the "Kleine Nete" and the "Grote Nete" from the Walloon Region to the Flemish Region, and the lack of data in this regard.[1092] Another example is that the sub-basin-specific part of the sub-basin "Leiebekken" refers to the EGTC Eurometropole, in the context of which transboundary cooperation is to take place for the period 2016-2021. However, the sub-basin secretariat is not a formal partner in the EGTC and can therefore

1087 See sub-basin specific parts of the RBMPs.

1088 Telephone interview staff member of the Dutch regional water authority "Scheldt Streams", 21 April 2017.

1089 Ibid.

1090 Of the Flemish sub-basin authorities.

1091 International Scheldt Commission, 'Roof Report 2016-2021' (2015) <www.isc-cie.org/images/Documents/ODB2-PFPG2_RAPPORT_NL-FR_VDEF.pdf> accessed 17 July 2017.

1092 Sub-basin specific parts for the sub-basin Demer.

only be involved through the formal partners of the Flemish Region or in the context of the interest-group meetings that are planned in the framework of the "blue space" project. The sub-basin board cannot become a formal partner due to, among other things, its lack of legal personality.

315. This means that although there are several examples and good practices in terms of cross-border cooperation related to water issues in the broad sense[1093], there is ample room for improvement in terms of approaching this more systematically.

– *Lack of thorough basin-wide overview and integrated management*

316. Several elements conducive to resilient river basin management are in place in the IRBD Scheldt: an internationally binding agreement, an institutional mechanism in the form of the ISC, joint monitoring programmes with regard to pollution and warning systems, and a common roof report that is updated every six years.[1094] However, several pieces of the puzzle are missing, which means that the Scheldt mechanism cannot be considered as a truly integrated river basin management regime. When push comes to shove, the governance of the IRBD Scheldt is the sum of the legal and governance output of the respective national River Basin Districts. The competent authorities within the meaning of the WFD and FD are the national competent authorities. The 2002 Scheldt Treaty is drafted in such a way as to maintain a maximum of sovereignty and a minimum of state engagement. The ISC only has limited room to manoeuvre and lacks decision-making power. The ISC is a vehicle that mainly serves information exchange and discussions. There is a lack of follow-up mechanisms to evaluate whether and how parties have implemented the advice.[1095]

317. In terms of integrated water management, from a horizontal integration perspective, there is fragmentation because of the large number of government agencies responsible for different elements of water management. In the EU, there is a fundamental and constitutional discrepancy in the governance of water management, because of the different decision-making procedures. Water quantity management is subject to the unanimity procedure in the Council, whereas water quality management is subject to qualified majority voting. From the vertical integration perspective, Member States' sovereignty claims are more prominent with regard to water quantity management than water quality management. Water quantity management at EU level is only dealt with in the

1093 See the section on "cooperation in practice" above.

1094 The procedural requirements of notification and consultation also apply in the Scheldt, as has been reported in the context of the first benchmark "equitable and reasonable utilisation and the no-harm principle".

1095 This relates to the final benchmark, explained below.

Water Framework Directive to the extent that it is necessary to achieve the water quality related goal. Notwithstanding the fact that the scope of the 2002 Scheldt Treaty is broader than its 1994 predecessor, this focus on the quality aspect has been translated into the Scheldt regime. Yearly reports on water quality in the District were issued by the ISC, as well as a report on the development of water quality in the period 1997-2007. This is not the case for scarcity and droughts. There are, however, programmes for joint flood risk management in place, especially in terms of cartography and the roof FRMP.

- *Nested governance*

318. In line with the subsidiarity principle, local issues at the sub-basin and sub-sub-basin scale should be dealt with at these levels. Moreover, inter-scalar links are paramount to achieve a resilient legal framework for governance of the IRBD.[1096] However, no link is provided, in the relevant legal frameworks in any of the three dimensions, between what happens at the regional level and what happens at the international level within the hydrological unit of the Scheldt River. Various actors have stated that this link is also missing in practice.[1097] A mechanism should therefore be developed to strengthen this synergy between regional and international cooperation.[1098] The entity operating at the level of the IRBD, here the International Scheldt Commission, should have a clear vision regarding the cooperation mechanisms existing at the lower-level hydrological units in the district. In this regard, a feedback mechanism to transfer information from the lower scales to the scale of the IRBD would be relevant.

5.3. MEANINGFUL PUBLIC PARTICIPATION

319. Public participation is considered to be an important part of the WFD and FD, and has been given a prominent position in the Directives. The three-tier approach to public participation embodied by the WFD and FD provides a good basis for citizens' involvement.

320. In terms of the transboundary aspect, the WFD and FD do not set requirements for Member States to organise transboundary public participation or to coordinate the results of internal participation processes with the basin countries.[1099] However, through the existence of EIA and SEA related

1096 As will be discussed in the evaluation section.

1097 This also relates to the link between the Flemish Dutch Scheldt Commission and the International Scheldt Commission. This has mainly been deduced from talks with sub-basin boards and (regional) water authorities.

1098 See Chapter V of this book.

1099 See Chapter II of this book and e.g. Article 14 WFD.

requirements, transboundary public participation is to some extent applicable. The legal frameworks of the Netherlands, Belgium and France do provide for transboundary consultation of the competent authorities in the Scheldt District, but not for participation. Participation is dealt with exclusively by the individual parties to the Scheldt Treaty, and although processes and transparency mechanisms are enshrined in the applicable legal frameworks, the common denominator is that actual participation is rather limited mainly because of a lack of flood risk awareness. The instrument of "citizens' juries" as proposed by the working group in the context of the Common Implementation Strategy, has not been tested in the Scheldt District. Basin-wide "participation" is limited to the availability of the umbrella part of the RBMP and FRMP and their annexes for the District on the website of the International Scheldt Commission. Non-governmental observers are allowed to attend the technical meetings taking place under the auspices of the ISC as well as the plenary meeting.[1100] Cross-border public participation is (i) not explicitly provided for in the relevant Article in the WFD and (ii) not translated into the national legal regimes.[1101] Other Directives, in particular the EIA Directive, provide for cross-border participation as explained in Chapter II.

321. Other IRBD mechanisms in the EU do provide for more basin-wide public participation. For example, the Rhine Convention provides that the ICPR must inform the public with respect to the state of the Rhine and the output produced by the Commission, and must publish reports to this end.[1102] The ICPR has allowed stakeholders to provide input on the draft management plans, e.g. the draft international Flood Risk Management Plan in 2014, following which the Commission published its responses to the comments received.[1103] Another example is the transboundary River Eems, for which a document has been drawn up explaining to the citizens of Germany and the Netherlands which routes they can follow to actively participate in the adoption of plans for the IRBD.

5.4. MONITORING AND OPPORTUNITIES FOR LEARNING

322. Adaptability is a key component of the WFD and FD, through the six-year planning cycle and the requirement for authorities to update the plans based

---

1100 International Scheldt Commission, 'Roof Report 2016-2021' (2015) <www.isc-cie.org/images/Documents/ODB2-PFPG2_RAPPORT_NL-FR_VDEF.pdf> accessed 17 July 2017, 60.

1101 It is relevant to note that Article 2.5 of the Directive 2003/53 excludes from its scope the plans and programmes of the Water Framework Directive.

1102 Article 8(4) of the Rhine Convention.

1103 Internationale Commissie ter Bescherming van de Rijn, 'Reactie van de ICBR / Het Coördineringscomité Op de Zienswijzen T.a.v. Het Concept-ORBP 2014' (2015) <www.iksr.org/fileadmin/user_upload/Dokumente_nl/Rijn_actueel/R%C3%A9actie_ORBP_2014.pdf> accessed 27 April 2017.

on monitoring results.[1104] Monitoring is a particularly important element of the WFD, both for surface water and for groundwater. The monitoring programmes for surface water should relate to the ecological and chemical status and ecological potential and, to the extent relevant to this status, the volume and rate of flow and for groundwater, the monitoring programme should relate to the chemical and quantitative status.[1105] The Floods Directive provides that a description of the manner in which the progress in implementing the Flood Risk Management Plans should be included in the plans, but does not include an obligation to monitor (transboundary) flood risks as such.[1106] Monitoring mechanisms have also been put in place in the Scheldt District, both in the Scheldt Estuary through the Flemish-Dutch Scheldt Commission, e.g. the MONEOS motoring tool, and in the Scheldt basin district through the International Scheldt Commission, e.g. the WASS, homogenous monitoring net related to water quality. For the IRBD Scheldt, the monitoring tools mainly relate to water quality.

323. The other aspect of this benchmark relates to the question how the information ensuing from the monitoring programmes is processed, e.g. cyclical evaluations, adaptation of targets, and so forth. The Scheldt Treaty does not include adaptive principles or mechanisms. However, considering the fact that the Treaty does not include specific substantive provisions such as targets or allocation formula in terms of water quantity, the lack of specific re-evaluation mechanisms is not earthshattering. Individually and taken together, Scheldt parties are subject to the cyclical adaptation processes of the relevant plans and programmes stemming from the WFD and FD. Individually, as each of the Scheldt parties adopt their separate plans; and together, as the roof report also needs to be updated every six years. The Treaty governing the Scheldt Estuary does refer to an evaluation mechanism with respect to the extent to which the goals of the Treaty are reached, i.e. on a five-year basis.[1107] The WFD itself is adaptable through instruments such as the Common Implementation Strategy and the comitology procedure, where the regulatory committee assists in adapting elements of the Directive to scientific and technical progress.[1108]

5.5. EFFECTIVENESS AND ENFORCEMENT

324. This final benchmark is twofold as it entails both effectiveness and enforcement and dispute resolution. A precondition for achieving resilience

1104 Barbara Beijen, Marleen van Rijswick and Helle Tegner Anker, 'The Importance of Monitoring for the Effectiveness of Environmental Directives: A Comparison of Monitoring Obligations in European Environmental Directives' (2014) 10 Utrecht Law Review 126, 131.

1105 Article 8 WFD.

1106 Annex, II.1 of the Floods Directive.

1107 Article 9 Estuary Treaty.

1108 Articles 20 and 21 WFD.

in a water management regime is that it should be effective.[1109] Compliance of agreements in shared waters needs to be followed up on and enforced, if need be.[1110] As a last resort, dispute settlement routes should be available to the parties sharing the waters.[1111]

325. Enforcement of provisions applicable in an EU transboundary river basin, here the International Scheldt District, is mainly a question of Member States' discretion. Specifically, in terms of cooperation requirements, the sections above have shown that there is a limited number of solid enforceable provisions. Reference can be made to the obligation to coordinate with the goal of producing a single management system, be it river basin or flood risk: this, as explained in Chapter II, is clearly an obligation of best efforts. If Member States "fail to produce" one single River Basin Management Plan or Flood Risk Management Plan, it is sufficient for them to adopt plans for the part of the IRBD located on their territories. Needless to say, enforcement of cooperation in this regard will inevitably be equally weak. One of the cooperation provisions that can be considered as an obligation of result is Article 3(4) WFD, which entails the duty for Member States sharing an IRBD to achieve coordinated implementation of the environmental objectives included in Article 4, and in particular of all programmes of measures established to comply with the Directive.[1112] However, the FD does not include a similar duty to coordinate as stipulated by Article 3(4) WFD, mainly because the FD does not set specific objectives similar to those included in Article 4 WFD.

326. The Water Framework Directive provides for the possibility for the EU Commission to intervene with regard to issues pertaining to the international district level. However, on the basis of the Directive, the EU Commission does not have binding decision-making power. The 2002 Scheldt Treaty does not enable the International Scheldt Commission to (a) render decisions that are binding on the parties and (b) follow up on and enforce any recommendations. In the Rhine River, the parties are obligated to report back to the Rhine Commission with respect to the implementation of its decisions, i.e. the regulatory measures the parties have taken in this regard, the results and the associated problems.[1113] The

1109 Peter Driessen and others, 'Toward More Resilient Flood Risk Governance' (2016) 21 Ecology and Society. Marleen van Rijswick and others, 'Ten Building Blocks for Sustainable Water Governance: An Integrated Method to Assess the Governance of Water' (2014) 39 Water International 725, 736.

1110 Marjan Peeters (n 1050) 217.

1111 Patricia Wouters, 'International Law – Facilitating Transboundary Water Cooperation' (2012) TEC Background Papers n 17 <www.gwp.org/globalassets/global/toolbox/publications/background-papers/17-international-law---facilitating-transboundary-water-cooperation-2013-english.pdf> accessed 13 July 2017.

1112 Hey and Van Rijswick (435) 244.

1113 Article 11(3) of the Rhine Convention.

parties need to motivate the reasons why the Commission's decisions have not been or have only partly been implemented, and there is a consultation-at-request procedure.[1114]

327. Until now, because of the lack of transboundary mechanisms in the WFD and the FD, the Directives refer to and rely on the international river mechanisms for almost all aspects of cooperation except for dispute resolution. Indeed, because of the exclusive jurisdiction of the CJEU, as discussed in Chapter II, it is difficult for EU Member States to use the dispute resolution mechanisms included in the UNECE Water Convention. When an issue arises in an IRBD, States have two formal options at the EU level, based on the WFD and FD: the open-ended mediation procedure centred around the EU Commission as defined in Article 12 WFD, and the EU dispute resolution route, i.e. non-compliance proceedings before the CJEU, either through the Commission or directly.[1115] The exclusive jurisdiction of the CJEU on the basis of Article 344 TFEU should be taken into account, possibly restricting the use of the range of settlement instruments under international water law. Just as the Rhine Convention does, the Scheldt Treaty stipulates that the parties should negotiate or find another method of dispute resolution if a dispute arises between them with regard to the interpretation or the implementation of the Treaty.[1116] The Scheldt Treaty does not go beyond that provision, whereas the Rhine Convention provides for an arbitration procedure which can be initiated at the request of one of the parties.[1117]

6. LOOKING BACK AND LOOKING FORWARD

328. This Chapter has built further upon Chapter II by focussing on the international and inter-regional cooperation mechanisms in a specific river basin, the Scheldt. Through the analysis of the legal frameworks in place at the national, regional and international, regional levels, it has been possible to conduct a thorough evaluation of the river basin mechanism in place. The Chapter has clarified the compatibilities of one aspect of water quantity management, i.e. flood risk management, between the Scheldt States, and has analysed the institutional interplay between existing administrative authorities and hydrological scale authorities. Furthermore, this Chapter has contributed to legal knowledge by setting forth the territorial cooperation routes, stemming from different levels of governance, that are available to States in a particular transboundary water body. From the perspective of sub-basin management plans, the coherence

---

1114 Article 11(4) of the Rhine Convention.
1115 Although Member States are reluctant to take each other to the CJEU.
1116 Article 8 Scheldt Treaty and Article 16 Rhine Convention.
1117 Article 16 Rhine Convention.

between theory and practice in this regard has been illustrated. Finally, the international mechanism governing the IRBD Scheldt, i.e. the Scheldt Treaty and the International Scheldt Commission, has been scrutinised. This combination of the national, regional and international analysis has enabled an encompassing view of the legal landscape of the River.

329. Although many good practices for flood risk management and territorial cooperation projects can be identified throughout the different jurisdictions in the Scheldt District, there are options for improvement in terms of, especially, integrated river basin management. Taking into account climate-change projections, increased urbanisation and land sealing, a growing population and even issues such as climate refugees, the current governance approach of International River Basin Districts in the EU is likely to be insufficient from a legal point of view. The mechanisms to prevent conflict in the first place, and, should conflict arise, to deal with it in the most efficient manner possible, should be improved. Now that top-down (Chapter II) and bottom-up (Chapter III) bottlenecks have been identified and issues have been identified, this book will look beyond the EU and into the management of a river basin in the United States. The Delaware Compact and Commission is seen as a best practice in water management. This will supply further data to feed the analyses and enable normative recommendations with the goal of improving cooperation in International River Basin Districts, especially from the perspective of the EU legal framework.

CHAPTER IV

CASE STUDY: TRANSBOUNDARY WATER GOVERNANCE IN THE UNITED STATES WITH A FOCUS ON THE DELAWARE BASIN

1. INTRODUCTION

330. The main question with regard to river basin management relates to the identification of the mechanism or process used to address the issues inherent to transboundary water bodies, and once these are identified, the appropriateness of the level of authority at which this should be done. Inter-jurisdictional water quantity management issues have existed in the United States since the country was formed. Over time, the states and the federal level have been seeking solutions to the question how to appropriately govern water resources that cross the state boundaries, i.e. in terms of finding the appropriate scale, the material scope, and so forth. The manner in which these transboundary water resources have been managed, has undergone great changes over the past century.

331. In Chapters II and III of this book, fragmentation in the context of water resources management in the EU and its Member States, here with respect to the Scheldt basin, has been identified as an important issue. Within one International River Basin District in the EU, a vast number of competent authorities operate, both at the hydrological scale and beyond, with different characteristics, e.g. in terms of legal personality, decision-making powers, levels of government, and so forth. The US has struggled with similar issues: (i) functions of the federal agencies often overlap, (ii) different aspects of water management are regulated by the states and the federal level, and where these are regulated by the federal level, they are implemented differently in the states, (iii) water quantity and water quality are governed by different institutions and there is often a lack of integration, and (iv) different water doctrines exist in the East and West of the country.[1118] In the US

[1118] Jeffrey P Featherstone, 'Water Resources Coordination and Planning at the Federal Level: The Need for Integration' (2011) 104 Journal of Contemporary Water Research and Education 12; RW Adler and M Straube, 'Watersheds and the Integration of US Water Law and Policy:

especially, different aspects of water resources management belong to competence spheres of different federal and state agencies[1119], e.g. the EPA Office of Water is involved in broad aspects of water management, but the Drinking Water Office is involved in drinking water quality, Wastewater Office, Science/Tech Office and Watershed Management in surface water quality, and so forth.[1120]

332. The primary responsibility and competence with regard to water quantity management lies with the states. The federal level has several tools to intervene, based on the Constitution and case law.[1121] Over time, the awareness grew that the Supreme Court adjudication mechanism for transboundary water disputes was wholly inappropriate, as the Court itself declared.[1122] As described by Frankfurter and Landis in 1925, complex multiple-state water-related controversies should not be settled by the courts, but through interstate agreements, i.e. legislation in lieu of adjudication.[1123]

333. In order to understand the types of interstate cooperation governance models in general, and the governance mechanism for the Delaware River specifically, the legal system for water law in the US will described out below. A brief historical overview will be given of the federal – state dynamics with respect to water quantity management, as this is necessary to understand the multi-level governance challenges in question.

Bridging the Great Divides' (2000) 25 William & Mary Environmental Law and Policy Review 1 <http://scholarship.law.wm.edu/wmelpr/vol25/iss1/3> accessed 13 July 2017.

1119 Gerald Kauffman, 'Governance, Policy, and Economics of Clean Water in the Delaware River Basin' (University of Delaware 2014) 56.

1120 John Hoornbeek, Policy-Making Institutions and Water Policy Outputs in the European Union and the United States: A Comparative Analysis, (2004) 11 Journal of European Public Policy 461.

1121 Lynn A Mandarano, Jeffrey P Featherstone and Kurt Paulsen, 'Institutions for Interstate Water Resources Management' (2008) 44 Journal of the American Water Resources Association 136 <http://doi.wiley.com/10.1111/j.1752-1688.2007.00143.x> accessed 13 July 2017.

1122 *New York v New Jersey*, 256 US 296 (1921). The Court stated: "We cannot withhold the suggestion, inspired by the consideration of this case, that the grave problem of sewage disposal presented by the large and growing populations living on the shores of New York Bay is one more likely to be wisely solved by co-operative study and by conference and mutual concession on the part of representatives of the states so vitally interested in it than by proceedings in any court however constituted".

1123 Felix Frankfurter and James M. and Landis, 'The Compact Clause of the Constitution. A Study in Interstate Adjustments' (1936) 45 Yale Law Journal 701.

2. UNITED STATES LEGAL FRAMEWORK FOR WATER MANAGEMENT: BACKGROUND

2.1. CLIMATOLOGICAL BACKGROUND TO RIVER BASIN MANAGEMENT

334. Water scarcity, droughts and floods are common phenomena in the United States, which will further intensify caused by climate change and other factors such as an increasing population. The figure below illustrates the projected changes to the water cycle in the East and West of the country for the coming years.

Figure 13. Projected Changes to the Water Cycle in the US

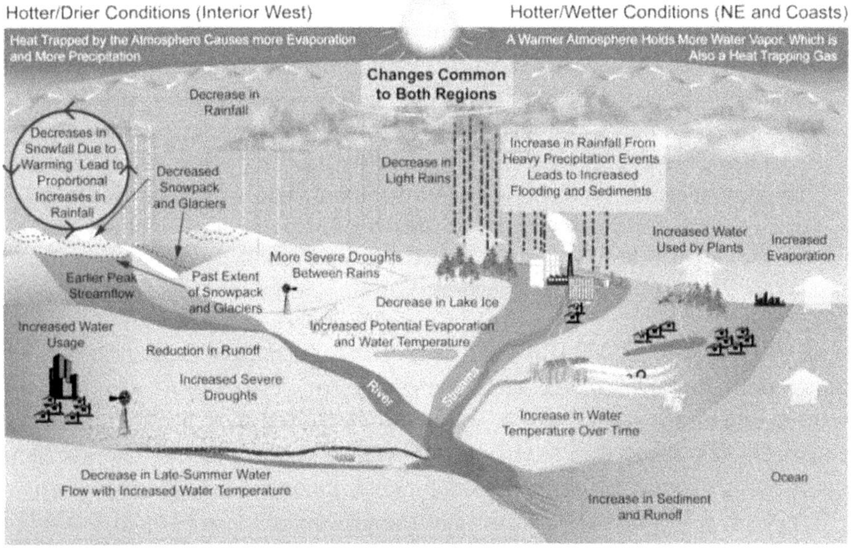

Source: USGCRP 2009

Water supply is expected to decrease in most areas of the country, which will be compounded by a rise in water demand, due to factors such as population growth.[1124] Indeed, taking into account climate-change projections, the total water demand in the United States is said to increase by 26% from 2005 to 2060. The key message in the latest national climate assessment with regard to the water cycle is that both wet and dry extremes are projected to intensify. The number and intensity of heavy precipitation events are expected to increase throughout

1124 Noah Hall, 'Interstate Water Compacts and Climate Change Adaptation' (2010) 5 Environmental & Energy Law & Policy Journal 237, 243.

the United States.[1125] In turn, dry spells are projected to cover longer periods of time, especially in the southern and north-western regions. Certain parts of the country depend on groundwater for 100% of their water supply. A decrease in soil moisture during the winter and spring seasons might negatively influence aquifer recharge.[1126] A key message throughout this book relates to the intricate link between water quantity and water quality. The risks posed to water quality due to projected changes in water temperature, the intensification of precipitation and flooding and at the same time the increase of dry spells are significant.

2.2. RIVER BASIN MANAGEMENT

335. Whilst the EU is making a tremendous effort to introduce an integrated river basin management approach across the EU Member States, this concept goes back a long way in the history of the US. River basin planning initiatives were taken in the US in an early stage, primarily driven by the ideas of John Wesley Powell, who planned to introduce the "United Watershed States of America". In 1878, Powell, then head of the US Geological Survey, proposed to designate the political boundaries in the western United States on the basis of hydrological ones.

> "I want to present to you what I believe to be ultimately the political system which you have got to adopt in this country, and which the United States will be compelled sooner or later ultimately to recognize. I think each drainage basin in the arid land must ultimately become the practical unit of organization, and it would be wise if you could immediately adopt a county system which would be convenient with drainage basins".[1127]

This idea was rejected, but is at the root of river basin management in the United States today. Roosevelt stated "each river system from its headwaters in the forest to its mouth on the coast is a unit, and should be treated as such".[1128] In 1999, the Environmental Protection Agency (EPA) referred to the watershed approach as constituting a coordinating framework for environmental management, in which efforts both from the public sector and the private sector are aimed at solving issues within hydrological boundaries, and taking into account surface and groundwater flow.[1129]

1125 US National Climate Assessment, 'Climate Change Impacts in the United States' (2014) 71 <http://s3.amazonaws.com/nca2014/high/NCA3_Climate_Change_Impacts_in_the_United%20States_HighRes.pdf> accessed 13 July 2017.

1126 Ibid 76. On a side note, aquifers are often not properly monitored in the US, which adds to the complexity of projecting changes in light of climate change.

1127 'Report on the Lands of the Arid Region of the United States, with a More Detailed Account of the Lands of Utah. With a More Detailed Account of the Lands of Utah. With Maps.' (1978) H.R. Exec. Doc. 73, 45th Cong., 2d Sess. Daniel Kemmis, *This Sovereign Land: A New Vision for Governing the West* (Island Press 2001).

1128 Charles McKinley, *Uncle Sam in the Pacific Northwest* (University of California Press 1952) 66.

1129 EPA Office of Water, 'Watershed Approach Framework' (1996) EPA 840-S-96-001 <https://www.epa.gov/sites/production/files/2015-06/documents/watershed-approach-framework.pdf> accessed 13 July 2017.

2.3. COMPETENT AUTHORITIES IN US WATER MANAGEMENT

Table 10. Competent Authorities in US River Basin Management

<table>
<tr><td rowspan="12">River basin</td><td rowspan="6">Federal level: US President and competent departments / agencies</td><td colspan="3">Environmental Protection Agency</td></tr>
<tr><td colspan="3">Department of Defense (Corps of Engineers)</td></tr>
<tr><td colspan="3">Department of Commerce (National Oceanic and Atmospheric Administration)</td></tr>
<tr><td colspan="3">Department of Agriculture (Natural Resources Conservation Service & US Forest Service)</td></tr>
<tr><td colspan="3">Homeland Security (Federal Emergency Management Administration)</td></tr>
<tr><td colspan="3">Department of the Interior (US Geological Survey, National Park Service, Fish & Wildlife Service)</td></tr>
<tr><td rowspan="5">States</td><td colspan="3">Example: Application to the Delaware River Basin</td></tr>
<tr><td>New York</td><td>Counties and SWCDs</td><td>NYC & towns</td></tr>
<tr><td>New Jersey</td><td>Counties and SWCDs</td><td>Cities and towns</td></tr>
<tr><td>Delaware</td><td>Counties and SWCDs</td><td>City and towns</td></tr>
<tr><td>Pennsylvania</td><td>Counties and SWCDs</td><td>Cities and towns</td></tr>
<tr><td>Sub-basin initiatives and regional agencies</td><td colspan="3">e.g. Partnership for the Delaware Estuary, Delaware River Basin Authority, Delaware Valley Regional Planning Commission, the Upper Delaware Council Inc.</td></tr>
</table>

336. Table 10 shows the competent authorities with regard to water management in the United States. The responsibility for water management lies with the federal level, states, counties, cities and local governmental agencies. As will be explained in the following sections, the question whether the focus with regard to water management lies with the federal level or the state levels, strongly differs for water quantity and water quality. Although the US Constitution only refers to two types of government, the federal and state governments, states also have more local units of governance to exercise their powers under the Constitution.[1130]

337. The federal government is divided into three branches: the legislative, the executive, and the judicial. The legislative branch, i.e. the US Congress consisting of the Senate and the House of Representatives, creates the laws, which should then be applied by the executive branch, consisting of the President, Vice-President, Cabinet, departments, agencies and other committees. The judicial branch, i.e. Supreme

[1130] Rossana Rosado and Andrew Cuomo, 'Local Government Handbook' (2011) New York State, Department of State 2011 <www.dos.ny.gov/lg/publications/Local_Government_Handbook.pdf> accessed 13 July 2017.

Court and other federal courts, in turn, examines laws brought before its court. On the basis of the 10th amendment to the US Constitution, the powers not delegated to the federal level are reserved to the states and the people.[1131] The states also consist of the three branches mentioned above, where the executive branches of the states are presided by a governor, the legislative branch is responsible for law-making and the approval of state budget for example, and the judicial branch consists of state courts and state Supreme Courts. At the more local level, a distinction is generally made between counties and municipalities, which is an umbrella term and can refer e.g. to cities, towns and boroughs.[1132] Local governments must be bestowed with competencies from the State. For example, the local governments operating in the Delaware River Basin have been granted such water resources planning powers.[1133]

338. As illustrated in the table above, at the federal level, water resources management is fragmentation-ridden. Water quantity and water quality are governed by different agencies, and within these two categories, many sub-categories exist, e.g. water allocation and flood risk management. The actors most important with regard to water resources management at the federal level are the following. The US Army Corps of Engineers (USACE), as part of the Department of Defense, plays an important role in water management through the public engineering activities and planning, forecasting and preparation in the context of water resources management and flood risk management. Within the realm of the USACE, an Institute for Water Resources has been created, which aims to develop planning methods and tools in water resources planning and policy. The Army Corps of Engineers plays a huge role in water resources management in river basins such as the Delaware. For example it developed a survey between 1956 and 1961 of the water resources of the Delaware, which formed the basis of the comprehensive plan developed by the Delaware River Basin Commission, as will be discussed in the following sections.[1134] On the basis of the Water Resources Development Act of 1986, the Corps of Engineers was given the power to carry out a study with regard to flood risk management in the Delaware River Basin. The Federal Emergency Management Agency operates under the auspices of Homeland Security and is the central actor in emergency management, e.g. it has developed the National Flood Insurance Program. The Environmental Protection Agency is the federal agency whose aim is to protect human health and the environment. There are ten EPA regions, where the EPA regional offices aim to ensure the implementation

---

1131 "The powers not delegated to the United States by the Constitution, nor prohibited by it to the States, are reserved to the States respectively, or to the people."

1132 The meaning of the term is therefore different from the term used in the countries of the Scheldt basin.

1133 Kauffman (n 1119) 69.

1134 Vincent Terenzio, 'Development and Effect of the Delaware River Basin Compact' (1962) 54 American Water Works Association 1445 <www.mnawwa.org/about/councils/Letterhead20102011.pdf> accessed 13 July 2017.

of the policies defined by the Agency. The EPA administers, among other things, the Clean Water Act. At the Department of Commerce, the National Oceanic and Atmospheric Administration operates, in which the National Weather Service is housed, which provides weather forecasts for the country. The US Geological Survey is an important player in water resources management in the US, as it monitors and provides scientific data on, among other things, water-related issues. For example, it has developed a National Stream Flow Information, which provides detailed information on water levels and temperature and so forth. Finally, the Department of Agriculture is important, as the Natural Resource Conservation Service aims to promote nature conservation on private lands.

2.4. FEDERAL-STATE DYNAMICS

339. Both in the EU and in the US, environmental protection laws came to life starting from the 1970s.[1135] Whereas in the EU, the competences of the EU institutions to legislate with regard to the water quality and water quantity management are based on specific constitutional provisions of primary law[1136], this is not the case in the US. Federal competence to issue rules and regulations with regard to the environment and water resources mainly stems from its power to regulate interstate commerce and through Judgments of the Supreme Court. Federal water-related legislation in the United States therefore has a more reactive character, in the sense that federal intervention constitutes a response to specific problems justifying or necessitating national level action.[1137] In this regard, water management constitutes an exception to how environmental law is shaped in general, i.e. through cooperative federalism in which the federal level establishes norms which should in turn be implemented at the level of the states.[1138]

This state dominance is especially apparent with regard to issues related to water rights and allocation mechanisms, which are therefore governed differently in the fifty states. For the category of water quality, federal-level intervention is more significant due to the Clean Water Act (CWA).[1139] However, at the implementation level there is fragmentation because of varying approaches adopted by the respective states.[1140] The allocation of quantities of water is a competence of the State, where

1135 John a Hoornbeek (n 1120) 465.

1136 Namely, Article 191 and 192 TFEU. See *supra*, Chapter II.

1137 John a Hoornbeek (n 1120) 461. Although the need to address environmental issues at the EU level also relates to the consideration that national level action would be less beneficial because of transboundary externalities.

1138 Noah D Hall, 'Toward a New Horizontal Federalism: Interstate Water Management in the Great Lakes Region' (2006) 77 U. Colo. L. Rev 405.

1139 33 U.S.C. §§ 1251-1387 (1994).

1140 The CWA is the primary federal law to regulate water pollution, which includes an elaborate permitting system on the basis of cooperative federalism. Water quality as such is beyond the scope of this book, therefore, the CWA will not be discussed in detail. Adler and Straube (n 1118).

the CWA expressly stipulates that the Act does not supersede this jurisdiction. In turn, the CWA provides that federal agencies must cooperate with state and local agencies with the aim of preventing pollution in a manner that is coherent with the programmes for managing water resources.[1141] Equally, a State may submit an application to the EPA for the delegation of authority to administer its own pollution discharge permit system.[1142] Section 401 of the CWA stipulates that applications for federal permits for discharging into waters of a State must be accompanied by said State's water quality certification process.[1143] Through this certification, the applicant should comply with appropriate state-law requirements in this regard. When the Supreme Court had to review the question whether the creation of a minimum stream-flow requirement by a certain State in this certification exceeded said State's scope of competence provided for by Section 401, the Court answered in the negative.[1144] The bottom line of this judgment is that not only "discharges" should be in line with relevant state law, but "activities" in general.[1145] The Court also emphasised, and rightly so, that the argument that the CWA only concerns water quality cannot be supported because quantity and quality are inextricably linked.

This state-oriented water policy development can also be explained by the fundamental differences in how the East and the West have approached the matter at hand, i.e. the different doctrines applicable in the respective areas. In the West, the applicable doctrine is that of prior appropriation, which implies a "first come-first served" mechanism, in terms of using or diverting quantities of water, based on so-called "beneficial use" considerations. The beneficial-use concept should identify whether the legislative framework for water resources applicable in the State in question would actually protect the use of water against appropriations of a later date. Beneficial use can, among other things, relate to appropriation for agricultural purposes. The prior appropriation doctrine has been described as creating negative incentives with respect to the conservation of water as, if a holder does not use the full allocation, said holder may (partially) lose his rights to the use of water, also referred to as "use it or lose it".[1146] In the East, the riparian rights doctrine prevails. This doctrine implies an intricate relationship between the land and the water, in the sense that the landowner is allowed to use the waters adjacent to the land, as long as it constitutes "reasonable utilisation". The water

1141 33 U.S.C. § 1251 (g).

1142 33 U.S.C. § 1342 (b). See Sandra Zellmer, 'United States: The Emergence of Environmental Considerations' in Joseph W Dellapenna and J Gupta (eds), *The Evolution of the Law and Politics of Water* (Springer Science 2009) 216.

1143 33 U.S.C. § 1341.

1144 *Jefferson County v Washington Dep't of Ecology*, 511 US 700 (1994).

1145 For a full discussion of the pre-emption questions with respect to the Clean Water Act, see Robert V Percival, 'The Clean Water Act and the Demise of the Federal Common Law of Interstate Nuisance' (2004) 55 Alabama Law Review 717.

1146 Richard Margerum, 'Integrated Water Resources Management in the United States: The Rogue and Willamette River Cases' in Jeroen Frank Warner, Arwin van Buuren and Jurian Edelenbos (eds), *Making Space for the River* (IWA Publishing 2012) 79.

doctrine in the East revolves around the right to use the water, not the right to own the water. One indicator shedding light on this difference between allocation mechanism in the East versus the West of the country relates to geographical circumstances, i.e. the East has historically had more water abundance than the West.[1147] It should be noted that, within the East and within the West, the interpretation of concepts such as "beneficial use" and "reasonable utilisation" also differs from State to State. As this book will mainly focus on the governing regime for the Delaware River, which is situated in the East of the country, the riparian rights doctrine is the most relevant.

In addition to regulation of water resources per se, the federal level influences water development through the regulation, for example, of large-scale dams and energy projects funded by the federal government and managed by federal agencies such as the US Army Corps of Engineers.[1148]

2.5. HISTORY OF WATER GOVERNANCE IN THE UNITED STATES

340. This section gives a concise overview of the timeline of water management in the United States. This overview should be seen as providing a context to enable the further comparison between river basin management in the EU and US.

341. The legal constellation regarding water policy has developed in the United States from a state-based federalism to cooperative federalism.[1149]

2.5.1. State-based federalism

342. State-based federalism was dominant in the 18th century until the beginning of the 20th century. In the beginning of the 19th century, efforts in the context of the management of water resources were very much led by the states.[1150] Several Acts passed in Congress encouraged states to address issues related to water rights.[1151] The Federal Government sought to improve waterway transportation and issued the Gallatin Report in 1808, proposing measures to

1147 Jonathan Deason, Theodore Schad and George William Sherk, 'Water Policy in the United States: A Perspective' (2001) 3 Water Policy 175 <www.sciencedirect.com/science/article/pii/S1366701701000113>.

1148 Adler and Straube (n 1118) 12.

1149 Andrea Gerlak, 'Federalism and US Water Policy', The Use of Interstate Compacts to Resolve Transboundary Water Allocation Issues (Edward Elgar 2014) 42.

1150 I.e. not by the Federal Government. See Beatrice Holt Holmes, 'A History of Federal Water Resources Programs, 1800-1960' (1972) US Department of Agriculture, Economic Research Service 3 <https://archive.org/details/historyoffederal1233holm> accessed 13 July 2017.

1151 For example, the Carey Act 1894, 43 U.S.C. § 641. See Amy K. Kelley, 'Staging a Comeback – Section 8 of the Reclamation Act' (1984) 18 U.C. Davis L. Rev 97.

improve canals and rivers as transportation routes across states.[1152] Gallatin argued that these types of fundamental and structural measures should be taken at the federal level, stating, "The national legislature alone, embracing every local interest, and superior to every local consideration, is competent to the selection of such national objects".[1153] However, the proposed measures were blocked due to constitutional limitations in terms of federal involvement in internal improvements.[1154] Fourteen years after the Gallatin Plan, the Supreme Court stated, in *Gibbons v Ogden*, that Congress did have authority over navigation on the basis of its power to regulate interstate commerce.[1155] In the same year, 1824, the General Survey Act was passed in Congress, giving the US President the competence to designate civil engineers and officers of the Corps of Engineers to draw up survey plans of roads and canals he deemed important.[1156] The Rivers and Harbours Act further enabled the US Army Corps of Engineers to improve waterways for navigational purposes. At the end of the 19th century, the Army Corps of Engineers had a broad regulatory responsibility, including water diversions.[1157]

2.5.2. More prominence for the federal level

343. The second era of water policy in the US had its starting point in 1901 with the election of Roosevelt and was marked by a more prominent role of the federal level.[1158] Roosevelt was a proponent of the focus on ecological units of governance in the context of water resources, and argued for the acknowledgment of the river basin as the unit for water-resources administration.[1159] The Reclamation Act of 1902 freed up funding for structural measures in water-distribution facilities in 16 states.[1160] Legislation with regard to flood risk management was passed following floods in 1915 and 1916, namely the Flood Control Act of 1917, on the basis of which the Corps of Engineers could carry

---

1152 Albert Gallatin, Report of the Secretary of the Treasury; on the Subject of Public Roads and Canals; made in pursuance of a Resolution of the Senate, of March 2, 1807 (Washington: R.C. Weightman, 1808) (The Gallatin Plan) <http://oll.libertyfund.org/titles/2046> accessed 13 July 2017 724-779.

1153 The Gallatin Plan, 75.

1154 Carter Goodrich, the Gallatin Plan after One Hundred and Fifty Years (1958) 102 Proceedings of the American Philosophical Society 436, 440.

1155 This judgment is considered as a landmark case for the use of the "commerce clause". Holmes (n 1158) 3.

1156 Ibid. In this period, these surveys were authorised by specific congressional enactments, as opposed to a comprehensive Act. This changed through the enactment of the first Rivers and Harbors Act of 1826, providing a basis for specific improvements and surveys.

1157 Holmes (n 1158) 4.

1158 Andrea Gerlak (n 1149) 43.

1159 Jeffrey Featherstone, 'Existing Interstate Compacts: the Law and the Lessons' (2001) 4 Tol. J. Great Lakes L. SCI & Pol'Y. 271, 275.

1160 Amy K. Kelley, 'Staging a Comeback – Section 8 of the Reclamation Act' (1984) 18 U.C. Davis L. Rev 97.

out flood-control works on the Mississippi and Sacramento rivers.[1161] The Flood Control Act broadened the scope of the activities on the rivers from navigation improvements to navigation and flood-related improvements. The activities of the US Army Corps of Engineers also constituted a factor in the federal prominence. In 1920, the Federal Water Power Act was adopted in Congress. As mentioned before, the concept of river basin management has a long-standing history in the United States. Around 1927, Congress gave the go-ahead for a nation-wide comprehensive river basin scrutiny, following which it adopted the Boulder Dam Act.[1162]

In 1936, the Flood Control Act was adopted, which stipulated that flood control was an activity pertaining to the federal level.[1163] This Act represented the first nation-wide flood-control programme.[1164] Several floods triggered the adoption of this Act, especially the nation-wide floods of 1935. Until 1926, surveys carried out following floods were carried out in light of navigation concerns. In 1933, the Tennessee Valley Authority (TVA) was created through a Congressional Charter signed by President Roosevelt.[1165] The TVA was shaped as a basin-wide authority with direct reporting channels to the President, and was given a great deal of independence and flexibility to plan and construct projects in the context, among other things, of flood control and economic development.[1166] In 1939, attempts at federal coordination of planning of water resources were realised through the agreement between the Army Corps of Engineers, the Bureau of Reclamation and the Department of Agriculture, on the basis of which river basin surveys needed to be subject to consultation of these bodies.[1167] This agreement led to the creation of the Federal Interagency River Basin Committee, which is popularly referred to as "Firebrick".[1168]

During this period of time, Congress also approved several interstate Compacts with regard to river basin management.[1169] These types of river basin approaches were part of the "New Deal" administration, where the focus was comprehensive planning, i.e. planning both natural resources and human

---

1161 Holmes (n 1158) 7.

1162 Samuel Hays, Conservation and the Gospel of Efficiency: The Progressive Conservation Movement 1890-1920 (University of Pittsburgh Press 1999) 199.

1163 22 June 1936, c. 688, 49 Stat. 1570.

1164 Although legislation in the context of flood control was already passed in 1917. See Joseph L. Arnold, 'The Evolution of the 1936 Flood Control Act', (Office of History, United States Army Corps of Engineers 1988) <www.publications.usace.army.mil/Portals/76/Publications/EngineerPamphlets/EP_870-1-29.pdf> last accessed 22 July 2014.

1165 Tennessee Valley Authority Act 1933, Pub. L. No. 73-17, 48 Stat. 58.

1166 Ludwik Teclaff, 'Evolution of the River Basin Concept in National and International Law' (1996) 36 Natural Resources Journal 359, 366.

1167 Featherstone (n 1159) 276.

1168 Peter Black, *Readings in Soil & Water Conservation* (Irvington Publishers 1974) 46.

1169 For example the Colorado River Compact: Boulder Canyon Project Act, 45 Stat. 1057 (1928). See Kevin Heron, 'The Interstate Compact in Transition: Cooperative State Action to Congressionally Coerced Agreements' (1985) 60 St. John's Law Review 1, 7.

and institutional resources, and on multi-purpose projects. In 1961, the federal government together with the riparian states created the Delaware River Basin Commission, which was the first time for the federal and state administrations to partner up and empower a single agency to govern interstate waters.

2.5.3. Major federal acts as milestones and more equal footing for states

344. Between 1960 and 1980, the third era, states again gained more equal footing in the water context, as more initiatives were taken to increasingly involve states in river basin management. For example, the Water Resources Planning Act (WRPA) of 1965[1170] attributed equal standing to states and the federal government in the context of the river basin commissions.[1171] On the basis of the WRPA, the President could authorise the creation of regional river basin commissions, either following a request of a riparian State on whose territory the river basin area is located, or at the request of the Water Resources Council.[1172] The WRPA established the Water Resources Council, created with the aim of providing the administration with advice related to water.[1173] The so-called "Title II river basin commissions" were to serve as the principal bodies for the coordination of the plans at the different layers, namely federal, state, interstate, local and non-governmental for the development of water and land related resources in a specific area, river basin or group thereof. Moreover, the WRPA includes a "Water Framework Directive – like" provision, ordering that these commissions should "prepare and keep up to date, to the extent practicable, a comprehensive, coordinated, joint plan for Federal, State, interstate, local and nongovernmental development of water and related resources".[1174] The commissions should also suggest long-term priorities for data analysis and for establishing projects. The powers of these Title II-Commissions were restricted to coordination and planning, to the exclusion of activities such as regulation, development and management.[1175] Pursuant to the WRPA, the Commissions should consist of a Chairman, also serving as coordinating officer of the federal members of the Commissions and acting as representative of the federal government in negotiations with the states, and one member for each Federal department and from each State lying wholly or partially within the area, river basin or group thereof.[1176] Several of these river basin commissions were established on the

1170 Water Resources Planning Act of 1965, Pub. L. 89-90.
1171 Ibid 45.
1172 Featherstone (n 1159) 277.
1173 Title I of the Water Resources Planning Act of 1965. See Gerlak (n 1149) 44.
1174 Sec. 201 (b) Water Resources Planning Act of 1965. The plan should also evaluate possible alternative means to achieve optimal development of water and land resources.
1175 Featherstone (n 1159) 277.
1176 Sec. 202 (a)(b)(c) Water Resources Planning Act of 1965. Also, one member appointed by an interstate agency created by an interstate Compact should be on the Commission, and, if

basis of the WRPA, but the new administration did not continue funding for the commissions pursuant to the Act and the commissions were dissolved through a presidential executive order in 1981.

2.5.4. Decrease of federal funding

345. In the fourth era, between 1980 and 1990, which overlapped with Reagan's presidency, new emphasis was on state responsibilities. This era saw a decrease in federal funding for water resources projects or federal participation in regional river basin planning.[1177]

2.5.5. Push in river basin management

346. Several policy initiatives furthered river basin management at the turn of the 21st century. Reference can be made to the publications by the EPA, such as the "Watershed Protection Approach" in 1995[1178] and the "Renewed Commitment to Watershed Management" in 2002. The Army Corps of Engineers also emphasised the watershed approach during that period.[1179] In December 2016, the Water Infrastructure Improvements Act for the Nation (WIIN) was passed in Congress.[1180] The WIIN includes the 2016 Water Resources Development Act. This includes the requirement for the Army Corps of Engineers to make publicly availably all data related to the planning, design, construction, operation and maintenance of water resources development projects, taking into account confidential business information.[1181] The Act sets up and authorises the funding for restoration and conservation programmes for specific river basins across the country. One of the programmes set up by this Act relates to the Delaware River Basin, which will be explained in the section on the Delaware River Basin Compact below.

347. The body of water governance has gone through tremendous changes and is still a dynamic, quickly changing constellation. The main role that the federal government has played in water resources management in the different eras revolves around funding and technical expertise for water supplies.[1182]

---

applicable, of any international commission created by a treaty with jurisdiction extending to the relevant waters.

1177 Gerlak (n 1149) 47.

1178 Carol Browner, 'Watershed Approach Framework' (1996) EPA <https://www.epa.gov/sites/production/files/2015-06/documents/watershed-approach-framework.pdf> accessed 14 July 2017.

1179 Kauffman (n 1119) 16.

1180 S.612 – 114th Congress (2015-2016).

1181 Section 1135 modifying Section 2017 of the Water Resources Development Act, 33 U.S.C. 2342.

1182 James Huffman, The Federal Role in Water Resources Management (2008) 17 N.Y.U. Environmental Law Journal 696.

2.6 PREVENTING AND SOLVING INTERSTATE WATER DISPUTES

348. In the US, water quantity management in interstate waters can be approached in three ways.[1183] First, federal legislation could stipulate provisions related to the use and apportionment of these interstate waters. Second, the states may bring a case before the Supreme Court who, on the basis of equitable apportionment, can address the questions with regard to the allocation of water. Third, states may enter into a contract with regard to their shared waters and the manner in which the water resources in these water bodies are to be managed.[1184] The first option has not materialised, because the federal government has focused on water quality management, and the focus with regard to water quantity management is with the states. The second option has been discouraged by the Supreme Court itself, stating that a cooperative approach to water management is far more beneficial and efficient than court proceedings.

349. States sharing river basins in the US have often applied to the original jurisdiction of the Supreme Court to obtain solutions relating to water apportionments.[1185] Looking at how the body of water law in the United States came about, differentiation should be made between the East and the West of the country. Indeed, the bulk of litigation related to water allocation took place in the West, as this part of the country is traditionally faced more often with issues related to the scarcity of water resources in comparison to the East, where precipitation is abundant and water is a-plenty. Water-related litigation in the East has focused more strongly on water quality issues. The Supreme Court's original jurisdiction to apportion interstate streams dates from the beginning of the 20th century, and was announced by the Court in *Kansas v Colorado*. The Court aims to safeguard equitable distribution of water resources between states.[1186] In order for the Court to accept jurisdiction with regard to the dispute, it must be proven that one State has clearly caused harm to the other. However, very regularly, the Court has urged the states to solve their dispute through other channels than expensive and lengthy litigation. In *Texas v Florida*, Justice Frankfurter referred to the limits of the litigation-based solving of problems, including the episodic character of litigation, the fact that it only has a limited scope, and that it does not fully take into account policy considerations.[1187] In *New York v New Jersey*, the Court argued that the problem in question would be more wisely solved

1183 N Hall, 'Interstate Water Compacts and Climate Change Adaptation' (2010) 5 Environmental & Energy Law & Policy Journal 237, 256.

1184 As will be discussed below, this contract, in most cases, is subject to congressional approval.

1185 The Court has jurisdiction over interstate water disputes on the basis of Article III, § 2 Constitution.

1186 John Johnson, *United States Water Law: An Introduction* (CRC Press 2009) 452, 11.

1187 *Texas v Florida*, 306 US 398, 428 (1939) (Frankfurter, J., dissenting).

through cooperative efforts and mutual concessions by the respective states, in comparison with proceedings in any court.[1188]

Despite these repeated attempts of the Court to point the states in the direction of Compacts rather than litigation, states have repeatedly submitted their water-related issues to the scrutiny of the Court. The Court carried out its equitable apportionment exercise for the first time in 1907, with regard to diversions on the Arkansas River resulting in a dispute between Kansas and Colorado.[1189] Through this judgment, the Court asserted its original jurisdiction over water-related disputes between states, resulting in equitable apportionment.[1190] In *Colorado v New Mexico*, or the *Vermejo* case, the Supreme Court held that equitable apportionment is the governing doctrine over disputes with regard to interstate waters.[1191] As uttered by Justice Holmes in his opinion to the Supreme Court in *New Jersey v New York* "the different traditions and practices in different parts of the country may lead to varying results but the effort always is to secure an equitable apportionment without quibbling over formulas".[1192]

350. There are several mechanisms embedded in the US Constitution for the federal government to oversee and influence actions and measures taken by states in relation to water resources. Congress has extensive powers in the context of interstate commerce to intervene in the relationships between states.[1193] One of the basic claims from the federal government on the waters of the US follows from the Commerce Clause, on the basis of which the Unites States can regulate commerce "with foreign nations and among the several states, and with the Indian tribes".[1194] Furthermore, power is granted to the federal level through the Supremacy Clause on the basis of which federal legislation takes precedence over state constitutions and state laws.[1195] Combined with the Supremacy Clause, is the doctrine on federal pre-emption, which implies that federal law may prevail over state and local law in certain cases. It can both be express, i.e. when Congress has made its legislative intent clear here, and

1188 *New York v New Jersey*, 256 US 296, 313 (1921).

1189 Dan Tarlock, 'The Law of Equitable Apportionment Revisited, Undated and Restated' (1985) 56 U. Colo. L. Rev 381, 385.

1190 The Court has original jurisdiction, as opposed to appellate jurisdiction, over these types of disputes on the basis of Article III of the Constitution, which states: "In all cases affecting ambassadors, other public ministers and consuls, and those in which a state shall be party, the Supreme Court shall have original jurisdiction". Through the Judiciary Act of 1789, the original jurisdiction of the Supreme Court was granted an exclusive character. NB: Appellate jurisdiction depends on the creation of federal courts of a lower level than the Supreme Court.

1191 Albert E Utton, 'In Search of an Integrating Principle For Interstate Water Law: Regulation versus the Market Place' (1985) 183 Natural Resources Journal.

1192 *New Jersey v New York*, 283 US 336 (1931) (Holmes, 343).

1193 Caroline Broun and others, The Evolving Use and the Changing Role of Interstate Compacts: A Practitioner's Guide (American Bar Association, 2006) 2.

1194 US Const. Article 1, § 8, cl. 3.

1195 US Const. Article 6, cl. 2.

implied, which is further sub-divided into field and conflict pre-emption.[1196] Federal pre-emption has been defined as a "simultaneous expansion in power of a higher level of government and reduction in power of a lower level of government".[1197] Reference can be made to a case involving pollution of waters caused by a paper mill across state borders. More specifically, property owners living on the Vermont side of Lake Champlain, affected by the pollution caused by the mill located on the New York side of the lake, initiated proceedings before a Vermont state court on the basis of the common Vermont state law principle of "private nuisance".[1198] In the end, when the case was brought before the Supreme Court, it found that the Clean Water Act pre-empted the state law of Vermont, i.e. the affected State.[1199] In another case, the Supreme Court also considered that the nature of the CWA was comprehensive, so that it pre-empted the federal common law on interstate nuisance with respect to pollution as well.[1200] The federal pre-emption not only extends to congressional statutes, but also to regulations issued by federal agencies, although courts will apply the pre-emptive intent test more stringently in the latter case.[1201] In additions to the instances of pre-emption pursuant to the CWA, in the realm of water law, federal pre-emption would be quite rare as the focus of law and policy making is with the states rather than at the federal level.[1202] Even in the context of the CWA, a cooperative federalism approach has been adopted, which leaves the states with substantive powers in this regard.

In addition to the commerce clause, the federal government can influence water policy pursuant to its spending power under the Constitution, e.g. with respect to the construction of reservoirs or dams. Furthermore, pursuant to the property clause of the Constitution, "Congress shall have Power to dispose of and make all needful Rules and Regulations respecting the Territory or other Property belonging to the United States".[1203] This is relevant, as federal lands constitute approximately 28% of all land in the US[1204] The property clause provides Congress with the authority to protect public domain, giving it the competence to regulate activities taking place on these lands, keeping the public interest in mind.[1205]

1196 Paul S Weiland, 'Federal and State Preemption of Environmental Law: A Critical Analysis' (2000) 24 Harvard Environmental Law Review 237. In case of implied pre-emption, the courts refer to the language of the Act in question or legislative history.

1197 Ibid. In turn, pre-emption may also exist in the relationship state-local level.

1198 *Int'l Paper Co. v Ouellette*, 479 US 481 (1987).

1199 The Court stated that only the law of the State where the pollution originated applied.

1200 Percival (n 1145) 717. *City of Milwaukee v Illinois*, 451 US 304 (1981).

1201 Amy K Kelley, 'Federal Preemption and State Water Law' (1996) 105 Journal of Contemporary Water Research and Education 5.

1202 Ibid 5.

1203 US Const. Article 4, § 3, cl. 2.

1204 Daniel Farber, *Environmental Law in a Nutshell* (9th edn, WestAcademic Publishing 2014) 71.

1205 Ibid.

351. The most prominent instrument in solving disputes related to water issues in the US is the Compact. This instrument is embedded in the US Constitution, which provides for the possibility for states to enter into interstate Compacts, provided they have obtained consent from Congress. Indeed, the latter is a *conditio sine qua non* for entry into force of any Compact: "No state shall, without the consent of the Congress, [...] enter into any agreement or Compact with another state".[1206] This Compact Clause and the ensuing requirement to obtain consent of Congress aims to safeguard the balance of the federal-state power divide.[1207] The Compact Clause is quintessential in the analysis of US transboundary water law.

352. To elaborate extensively on the history of the interstate Compact in the US would take us too far beyond the scope of this chapter. Suffice it to say that it originates from before the American Revolution, as it was used by British North American colonies to settle disputes related to boundary issues during colonial times.[1208] It is important that the Compact is a contract that states enter into through state legislation.[1209] Once these Compacts have received federal approval, the Compact becomes federal law, which can be enforced in the federal court. Indeed, in *Culyer v Adams*, the Supreme Court stated "Because congressional consent transforms an interstate Compact within the Compact Clause into a law of the United States, the construction of an interstate agreement sanctioned by Congress under the Compact Clause presents a federal question".[1210] In earlier judgments, however, the Court had held that congressional consent does not automatically imply transformation of these Compacts into federal law. In this line of reasoning, Compacts would then be similar to contracts rather than statutes or treaties of the United States.[1211] As Compacts form part of the statutory law of the states involved, they take precedence over subsequent state law.[1212]

353. In general, not all Compacts require congressional consent. The rule of thumb in this regard is that all Compacts that have an impact on the federal-state balance should be subject to consent of the congress.[1213] If the Compact results in

1206 US Const. Article 1, § 10, cl. 3.

1207 The origins of the Compact clause, i.e. in the sense that it is an instrument allowing for federal - interstate agreements, are the Articles of the Confederations and Perpetual Union (Article VI) of 1781: "A state may not enter into any treaties or alliances with another state without the approval of Congress".

1208 Patricia Florestano, 'Past and Present Utilisation of Interstate Compacts in the United States' (1994) 24 Publius 13, 14.

1209 Noah Hall, 'Toward a New Horizontal Federalism: Interstate Water Management in the Great Lakes Region' (2006) 77 U. Colo. L. Rev 405, 411.

1210 *Culyer v Adams*, 449 US 433, 438 (1981).

1211 *Hinderlider v La Plata River and Cherry Creek Ditch Co.*, 304 US 92 (1938).

1212 Kevin Heron, 'The Interstate Compact in Transition: From Cooperative State Action to Congressionally Coerced Agreements' (1985) 60 St. John's Law Review 1, 6.

1213 Patricia Florestano, 'Past and Present Utilisation of Interstate Compacts in the United States' (1994) 24 Publius 13, 14.

the increase of political power of the states to the detriment of the supremacy of the federal level, congressional consent is required.[1214] One of the indicators to assess whether consent is required, is whether or not the Compact would touch on areas where Congress can legislate. In terms of interstate resources, i.e. transboundary waters, this question is often answered affirmatively. As mentioned above, several clauses bring these resources under the realm of federal legislative power.[1215] Indeed, there is consensus on the fact that agreements settling jurisdictional issues over water, allocation questions with respect to trans-state waters, et cetera are of such a nature as to trigger participation, to some degree, of the federal level.[1216] Several judgments of the Supreme Court have acknowledged the federal role in interstate water management. For example, in *Arizona v California*, the Court held that the federal government has the authority to allocate water in interstate waters.[1217] As water allocation forms an intricate part of most interstate water Compacts, it is difficult to overlook the federal level. This point is also proven by practice: all major interstate water Compacts have been subjected to congressional consent and include one or more federal representatives as members of the competent joint body.[1218]

The requirement of congressional consent also comes with presidential approval, as is required by the Constitution for any Order, Resolution, or Vote approved by the Senate and House[1219], although Congress has the possibility to override the presidential veto. When President Roosevelt issued a veto against the Republican River Compact in 1942 following approval by Congress, out of concern that the Compact exceedingly affected the federal competence in this context, the Compact was re-negotiated.[1220] The question whether a Compact is classified as a contract or as federal law has an impact on the leeway of the Court to adjudicate. The Court has more room to manoeuvre in the interpretation of contracts than it has in interpreting federal laws.

Congressional consent does not imply that the federal level should have an equal voice in the negotiations of interstate Compacts as the states. Indeed, it is in line with the spirit of the Constitution to grant states the necessary degree

---

1214 *Virginia v Tennessee*, 148 US 503 (1893).

1215 Zachary Lorne McCormick, The Use of Interstate Compacts to Resolve Transboundary Water Allocation Issues' (Oklahoma State University 1973) 40.

1216 Patricia S Florestano, 'Past and Present Utilisation of Interstate Compacts in the United States' (1994) 24 Publius: The Journal of Federalism 13, 15.

1217 "Congress has put the Secretary of the Interior in charge of a whole network of useful projects constructed by the Federal Government up and down the Colorado River, and it has entrusted him with sufficient power, principally the § 5 contract power, to direct, manage and coordinate their operation. This power must be construed to permit him to allocate and distribute the waters of the mainstream of the Colorado River [...]". *Arizona v California*, 373 US 546 (1963).

1218 See, among other things, the Delaware River Basin Compact, the Colorado River Compact, and the Susquehanna River Basin Compact.

1219 Zachary Lorne McCormick (n 1215) 39.

1220 Ibid 40.

of discretion and freedom to resolve regional issues amongst themselves.[1221] The federal government does not have the constitutional authority to draft the central elements of the Compact, as this device is meant to develop and grow as a "grassroots" process.[1222]

354. The federal-state relationship in terms of public domain and ownership of water and streams is subject to a complex amalgam of judgments and court opinions. Certain landmark rulings have specified principles, such as *Gibbons v Ogden*, which established control of the federal government over waters pursuant to its competences over commerce and navigation.

As mentioned above, the commerce clause of the Constitution, which relates to the power to regulate interstate commerce, is the main basis for federal involvement in interstate water issues. The commerce clause can limit state action, not only resulting from action from Congress, but also in case of an inactive Congress.[1223] This doctrine is referred to as the "dormant commerce clause", pursuant to which state regulations that influence interstate commerce can be invalidated, even when Congress has not issued any legislation in this regard.[1224] In *Sporhase v Nebraska ex rel. Douglas*[1225], the defendants called on the dormant commerce clause to challenge an anti-export statute issued by Nebraska. Indeed, the defendants, Sporhase and Moss, sought to transport water originating from a well in Nebraska to irrigate their farmland on Colorado territory.[1226] Nebraska had unilaterally issued a reciprocity requirement, in that a permit for water export would only be issued if the State to which the water is transported, has also granted reciprocal rights to transport groundwater from that State to the State of Nebraska.[1227] In this manner, Nebraska interfered with interstate commerce and violated the dormant commerce clause.[1228]

355. Another doctrine with an impact on interstate water Compacts is the federal reserved rights doctrine, established by *Winters v United States* in 1908, which implies that when the federal level reserves public land e.g. for the preservation of national parks, it also reserves the amount of water necessary for fulfilling the purpose for which the reservation was created. In *Cappaert v United*

1221 Heron (n 1212) 22.

1222 Frankfurter and Landis (n 1123) 702.

1223 Daniel Farber (n 1204).

1224 Ca Klein, "The Dormant Commerce Clause and Water Export: Toward a New Analytical Paradigm" (2011) 35 Harvard Environmental Law Review 131, 134.

1225 *Sporhase v Nebraska ex rel. Douglas*, 458 US 941 (1982).

1226 The land straddled the Colorado-Nebraska border but the issue was the irrigation of land on the Colorado side of the border.

1227 Neb. Rev Stat. § 46-613.01.

1228 The Court nuanced the judgment by emphasising that Nebraska had failed to indicate that its reciprocity requirement was linked to conservation concerns, e.g. that an export ban would be necessary considering structural water shortage issues, etc.

States, the federal reservation is demonstrated.[1229] The federal government had declared a cave inside the Death Valley National Monument as being part of the national monument, and had thus, according to the Court, effectively reserved the groundwater within the cave. The reason for this reservation was that a certain amount of water necessarily needed to be present in the pool in the cave, because of the presence of breeding ground for fish. This entitlement by the federal government could be carried out despite the lack of property rights of the federal level under state laws.[1230] However, in *United States v Mexico*, the Court held that the implied purpose should be interpreted restrictively.[1231]

356. As described above, water law in the US is characterised by a complex scheme of property rights. These are governed differently in the East and the West of the country with respect to surface water, i.e. by the riparian doctrine and the prior appropriation doctrine respectively, and there are five legal schemes that apply to groundwater.[1232] In this regard, rules and regulations issued with the goal to protect the environment or to safeguard water resources are at times confronted with existing property interests of individuals. In accordance with the Takings Clause of the Constitution, the government should compensate property owners for damages occurring when the government places a certain burden on said property in the public interest.[1233] The question of non-fault liability of the government and how the compensation regime should be formed vis-à-vis the infringement of property rights in the context of water management is a topical one in EU Member States as well. For example, reference can be made to the application of the principle of equality before public burdens in the Netherlands to the situation where a person suffers damage resulting from measures taken in the context of e.g. flood protection such as dams or dikes.[1234] In the US, there are several types of water-takings, which have been subject to litigation. Takings cases brought by the injured (ex-) property owner vis-à-vis the government, is referred to as the inverse condemnation doctrine. Courts need to establish whether there is in fact property,

---

1229 *Cappaert v United States*, 426 US 128 (1976).

1230 *United States v Mexico*, 438 US 696 (1978). Daniel Farber (n 1204) 305.

1231 Zachary Lorne McCormick (n 1215) 400.

1232 The rule of capture, correlative rights, reasonable utilisation doctrine, prior appropriation, and the Second Restatement of Torts' Doctrine of Reasonable utilisation. See John D Echeverria, 'Water and Takings' (32nd Annual Water Law Conference American Bar Association 2014).

1233 Article 5 of the Constitution: "[...] nor shall private property be taken for public use, without just compensation".

1234 For example, when, as a result of this measure, the property is devaluated. See Willemijn van Doorn-Hoekveld, 'Compensation in Flood Risk Management with a Focus on Shifts in Compensation Regimes Regarding Prevention, Mitigation and Disaster Management' (2014) 10 Utrecht Law Review 216 <www.utrechtlawreview.org/index.php/ulr/article/view/URN:NBN:NL:UI:10-1-115825>; Willemijn J Van Doorn-Hoekveld and others, 'Distributional Effects of Flood Risk Management – a Cross-Country Comparison of Preflood Compensation' (2016) 21.

and whether that property has been "taken".[1235] Types of water-takings include flooding, for example when private property is permanently inundated.[1236] Other types of government actions brought before courts with the goal of qualifying said action as a regulatory taking may relate to restrictions placed on surface water diversions or groundwater withdrawals stemming from environmental considerations.[1237] Several doctrines influence and limit water rights on the basis of which the existence of a taking might be claimed. The public trust doctrine is an example of such limitation, on the basis of which the states have the responsibility to protect tidal and navigable waters and the land underneath these waters in light of the public's use. The public trust doctrine therefore provides counterbalance to private use and ownership of resources.[1238] The so-called "navigation servitude" implies that there can be no government liability on the basis of the Takings Clause for measures promoting navigation. The federal government therefore does not need to provide compensation if it influences property rights while exercising its power over navigation.[1239] This implies, for example, that a structure such as a dam placed in the water of a navigable river should be removed without public compensation if Congress exercises its power over navigation to safeguard the right of navigation, as was the case in *United States v Chandler-Dunbar Co.*[1240] As this competence is based on the Commerce Clause, pursuant to the Supremacy Clause, the servitude trumps potentially conflicting property law issued at the level of the states.[1241]

357. There are various incentives for states to involve the federal level in their interstate water management. As discussed above, federal jurisdiction in these matters is one of them. Another "carrot" in this regard is the benefits that federal involvement brings with it, namely the added value of enforcement of the requirements stipulated in the Compact.

In the US, too, there is concern that the federal powers may nibble away at the sovereignty of the states. The American author Tribe formulated an often-

---

1235 Echeverria (n 1232).

1236 John D Echeverria, 'The Intersection of Water Law and Takings Doctrine' (Proceedings of the 60th Annual Rocky Mountain Mineral Law Institute, Rocky Mountain Mineral Law Foundation 2014).

1237 Ibid 8B-6. There has also been debate on the question whether court judgments could give rise to judicial takings, for example when a court alters a water allocation system adopted by the State.

1238 Erin Ryan, 'The Public Trust Doctrine, Private Water Allocation, and Mono Lake: The Historic Saga of National Audubon Society v Superior Ct.' (2015) 45 Environmental Law. Following a judgment rendered by the California Supreme Court, *National Audubon Society v Superior Court*, the scope of protection pursuant to the public trust doctrine was broadened to also embrace purposes of a recreational and ecological nature. Russell M Mcglothlin and Scott S Slater, 'No Fictions Required: Assessing the Public Trust Doctrine in Pursuit of Balanced Water Management' (2014) 17 University of Denver Water Law Review 53.

1239 Melissa Estes, 'The Effect of the Federal Endangered Species Act on State Water Rights' (1992) 22 Environmental Law 1027.

1240 *United States v Chandler-Dunbar Co*, 229 US 53 (1913).

1241 Since *Gibbons v Ogden*, 22 US 9 Wheat. 1 1 (1824).

cited reflection of this concern in 1978: "Of course, no one expects Congress to obliterate the states, at least in one fell swoop. If there is any danger, it lies in the tyranny of small decisions – in the prospect that Congress will nibble away at state sovereignty, bit by bit, until someday essentially nothing is left but a gutted shell".[1242] Indeed, the federal-state sovereignty balance or tension is at the heart of the difficulties of interstate water management.[1243]

2.7. RIVER BASIN MANAGEMENT: TYPES OF GOVERNING ENTITIES

358. In terms of organisations with responsibilities for interstate water management, the federal and state authorities have experimented with various types of governance.[1244] Roughly five types of competent authorities have been created over time. The first type (i) leaves all the power with the federal level, with a single federal authority setting the rules for the interstate watershed. An example of this model is the Colorado River Compact, where the Secretary of Interior acts as the single leading authority promulgating top-down type allocation rules.[1245] At the other end of the spectrum, (ii) watershed councils are primarily governed by states with little interference of the federal level. These have been promoted by federal agencies such as the Environmental Protection Agency, and are informal, consensus-oriented, grassroots organisations focused on the watersheds, i.e. at a more local scale than the river basin level.[1246] These councils are similar to the Walloon river contracts, "contrats de rivières". The functioning of the watershed councils mirrors Ostrom's polycentric governance model, as it reflects the consideration that there should be different management units ranging from the local sub-sub-basins to the sub-basin to the river basin, as such constituting nested watersheds.[1247] (iii) The Regional Authority operates quite similarly to the Single Federal Administrator, but is focused on a centralised regional agency. The Tennessee Valley Authority is an example of the Regional Authority, which is centralised and has a broad package of competences. The Water Resources Planning Act (WRPA) established the so-called (iv) "Title II-Commissions", which were established at the initiative of the Water Resources Council and following approval of the President. Several Title II-Commissions were established on the basis of the WRPA, such as the Missouri and the Upper Mississippi. The commissions in question had both state and federal

---

1242 Laurence Tribe, *American Constitutional Law* (The Foundation Press, Inc., 1978) 302.
1243 Gerald Kauffman (n 1119) 26.
1244 Ibid 27.
1245 Kauffman (n 1119) 26.
1246 Jerome Delli Priscoli, 'Case Study of River Basin Organizations' Oregon State University <www.transboundarywaters.orst.edu/research/case_studies/River_Basin_Organization_New.htm> accessed 14 July 2017.
1247 Thomas Dietz, Elinor Ostrom and Paul C Stern, 'Struggle to Govern the Commons' (2003) 302 Science 1907.

representatives, but were mistrusted by the states as the focus remained with the federal level.[1248] These Commissions were dissolved through presidential executive order in 1981. (v) The fifth model of interstate basin organisation is the Federal-State Interstate Water Commission, which will be studied in detail in the following sections. Indeed, the Delaware River Basin Compact, the subject of analysis in this chapter, is an example of this type of organisation.

Of these five types of governance mechanism mentioned above, the interstate Compact commissions are considered to have the farthest-reaching degree of power, as they are based on an agreement that is transposed into both federal and state law, which grants them binding powers, and they have a permanent office, staff and regular meetings.[1249]

3. THE DELAWARE RIVER

Figure 14. The Delaware River Basin

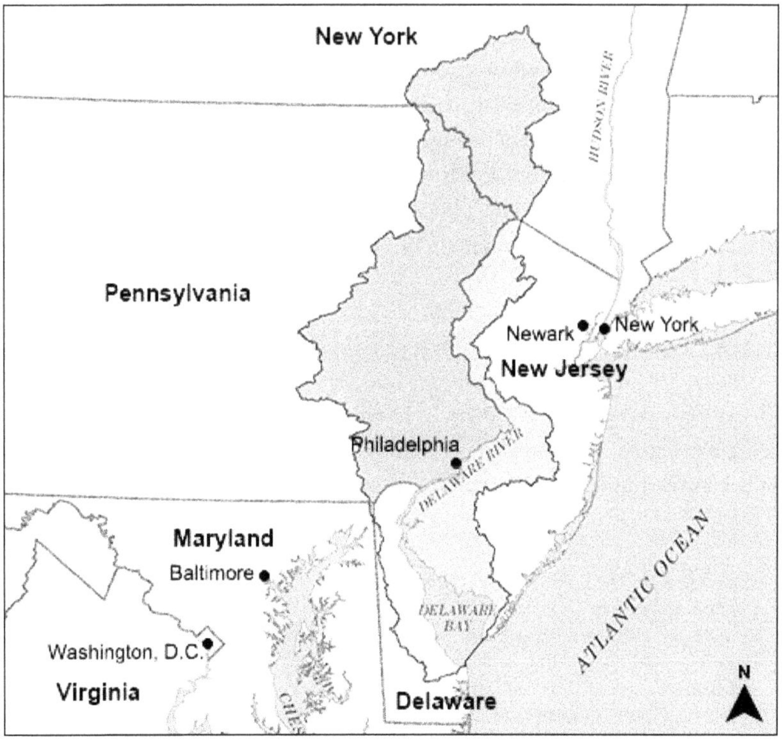

Source: DRBC

[1248] Mandarano, Featherstone and Paulsen (n 1121) 140.
[1249] Gerald Kauffman (n 1119) 346.

3.1. JUSTIFICATION OF CHOICE

359. The Delaware River Basin and its water resources have been subject to conflicts, landmark litigation, and various attempts at forming formal joint-governance structures since the emergence of the United States as an independent country. The Delaware River Basin is shared between four states and is faced with significant multi-level governance challenges. In 1783, efforts of a bi-state Committee created by New Jersey and Pennsylvania resulted in an agreement that was passed in both states' legislation, prohibiting the installation of dams on the main stem of the river.[1250] Over time, the governance structures related to the Delaware River Basin have developed toward what is perceived to be an effective centralised management scheme. Once referred to as one of the most polluted areas of the US, it is now seen as a best practice for integrated river basin management in the country.[1251] The Delaware River Basin Compact (the Delaware Compact) represents the first Compact in the United States in which the federal government participates as an equal member with the signatory states in an agreement for river basin management.[1252] This means that when it was concluded, it was a novel governance tool.[1253] Studies have shown that the Delaware Compact, due to its adaptive nature, which allows tuning into new and unexpected developments, is an appropriate tool to face future climatic pressure, among other things.[1254] It arose in response to the urgent needs of the states to resolve issues regarding water quality, water allocation and flooding.[1255] Whereas the scope of many interstate Compacts is restricted to water allocation, the Delaware Compact addresses all aspects of water resources management on a basin-wide basis.

3.2. NATURAL CHARACTERISTICS OF THE DELAWARE

360. As depicted in figure 15, the Delaware is shared between the states of New York, Pennsylvania, New Jersey and Delaware.[1256] The headwaters of the Delaware River are located in New York and the waters of the Delaware eventually flow into

1250 Susan Senecah, The Environmental Communication Yearbook (Routledge 2013).
1251 Gerald Kauffman (n 1119), 86.
1252 Prior to the Delaware River Basin Compact, the federal government was also involved in Compacts with States, but its role was more limited to providing technical assistance to States, or to safeguard its navigational stakes. Frank Grad, 'Federal-State Compact: A New Experiment in Co-Operative Federalism' (1963) 63 Columbia Law Review 825, 838.
1253 Gerald Kauffman (n 1119).
1254 Watermark Initiative, 'US Water Stewardship: A Critical Assessment of Interstate Watershed Agreements' (2009) <http://watermarkinitiative.com/wp-content/uploads/2013/10/WMI_Report_09-13.pdf> accessed 10 May 2017.
1255 Emily Jeffers, 'Creating Flexibility in Interstate Compacts' (2009) 36 Ecology Law Quarterly 209, 219.
1256 Mandarano, Featherstone and Paulsen (n 1121) 364.

Delaware Bay. When the Compact was created, lawmakers observed that nineteen federal agencies, fourteen interstate agencies and forty-three state agencies were involved in the management of the basin's water resources, together with 38 counties and 838 municipalities. The four states involved and their counties, cities and towns implement diverse land-use practices and regulations affecting water quality, floods, and so forth.[1257] Physiographically, the basin can be divided into four distinct regions, which may be referred to as the Upper Region, the Central Region, the Lower Region and the Bay Region. These may be further divided into sub-basins.

Figure 15. Watersheds of the Delaware River Basin

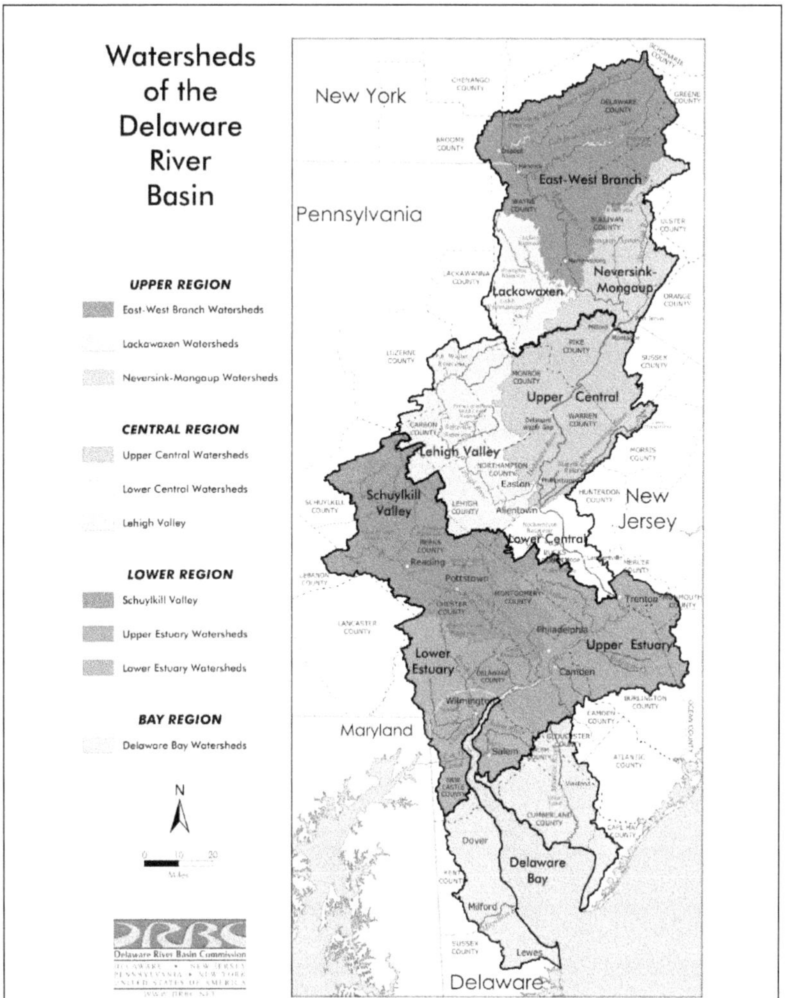

Source: The Delaware River Basin Commission

[1257] Gerald Kauffman (n 1119) 69.

361. The Delaware River Basin spans 36,570 km^2 and supplies drinking water to 16 million people.[1258] Half of New York City's drinking-water supply originates from the Delaware Basin. The basin is home to more than 8 million people. The Delaware River is the longest undammed river in the United States east of the Mississippi. The Delaware River is sensitive to flooding. In 1995, today's equivalent of USD 2.8 billion in damages were caused by floods that followed two hurricanes taking place in a short time.[1259] The consecutive floods of 2004, 2005, and 2006 cost nine lives and caused USD 500 million in property damage.[1260] The Delaware experiences approximately three periods of drought every ten years.[1261]

3.3. GOVERNANCE OF THE DELAWARE IN THE 20th AND 21st CENTURIES

362. During the 20th century, governance of the water resources of the Delaware Basin developed from reliance on judicial allocation toward various forms of basin-wide interstate cooperation. The basin was the subject of a famous United States Supreme Court decision in 1931 in *New Jersey v New York*, which has been described as "the most significant original litigation over water to arise in the eastern states".[1262] A subsequent Supreme Court decision in 1954 modified the Supreme Court's 1931 decree. One of the pillars underpinning the Delaware Compact is the Supreme Court's 1954 allocation, the terms of which have been preserved by the Compact. The litigation arose from a dispute between New York (NY) and New Jersey (NJ) resulting from a decision by NY to construct dams on major headwater streams of the Delaware River in order to provide water for NY City. The Supreme Court allocated 440 million gallons per day to the State of New York in 1931, through application of the "doctrine of equitable apportionment".[1263] New Jersey brought the case back to the court twenty years later, when New York City undertook to expand its storage in the Delaware's headwaters and to divert as much as 800 mgd out of the basin to serve its growing population. The 1954 decision (or "decree") allowed New York City to divert 800 mgd, provided that

---

1258 By Ken D Bovee and others, 'A Decision Support Framework for Water Management in the Upper Delaware River' (2007) <www.fort.usgs.gov/sites/default/files/products/publications/21938/21938_508.pdf> accessed 14 July 2017; Gerald Kauffman (n 1119) 2.

1259 United States Army Corps of Engineers.

1260 Anthony DePalma, 'A Fight over Taming the Delaware' *The New York Times* (20 April 2008) <www.nytimes.com/2008/04/20/nyregion/nyregionspecial2/20delawarenj.html?pagewanted=all> accessed 13 July 2017.

1261 Partnership for the Delaware Estuary, 'Comprehensive Conservation and Management Plan for the Delaware Estuary' (1996) <https://s3.amazonaws.com/delawareestuary/pdf/CCMP.pdf> accessed 4 May 2017.

1262 Joseph Dellapenna, 'Interstate Struggles over Rivers: The Southeastern States and the Struggle over the 'Hooch' (2005) 12 N.Y.U. Environmental Law Journal 828, 841.

1263 Times 3.7854 for the amount in litres.

from its three Delaware Basin reservoirs it released sufficient water to maintain a flow at Montague, NJ (the approximate location of the border between New York, New Jersey and Pennsylvania) of 1750 cubic feet per second (cfs). Because New York City was not expected to require its full allocation for some years, the 1954 decree also required the City to annually release downstream over a period of no more than 120 days a quantity of water known as the excess release quantity (or "ERQ") equal to 83 percent of the difference between the continuous safe yield of all the City's reservoirs and the City's estimated consumption for the year. New Jersey was given the right to divert as much as 100 mgd from the basin until it had constructed additional storage of its own. The decree also provided for the appointment of a "River Master" by the Chief Hydraulic Engineer of the US Geological Survey to administer the provisions of the decree related to yields, diversions and releases. The Supreme Court retained jurisdiction over the dispute, indicating that any of the parties could seek re-apportionment of the basin's waters if it deemed needs and circumstances to have changed. However, this adjudicated solution can be considered inherently static and inflexible, in the sense that it made no provision for the parties to modify operations in response to shifting conditions or values without resuming their litigation.

363. Back in 1934, the US Army Corps of Engineers created a report arguing for a body charged with overseeing water resources projects in the Delaware Basin. The focus of the report was on the identification of sites for the construction of multi-purpose dams for water supply, quality, hydroelectric power, flood risk management, and irrigation, but no appropriations were approved to implement the projects. Following the publication of this report, the Interstate Commission on the Delaware River Basin (INCODEL) was created, which operated on the basis of voluntary cooperation.[1264] The legislation introducing INCODEL did not follow the procedure for congressional approval. Indeed, INCODEL was not created by an interstate Compact under Article 1, section 10, clause 3 of the US Constitution, but on the basis of legislation issued in parallel at the level of the states in question.[1265] INCODEL focussed primarily on pollution-related issues, and not on allocation questions. The INCODEL's lifespan was therefore limited, and the states carried on in their efforts to reach a more sustainable agreement. However, INCODEL did demonstrate that interstate cooperation to improve water quality was feasible.

The states entered into the Delaware River Basin Compact in 1961, upon the enactment of concurring legislation by the four basin states and the US Congress.[1266] The Compact created the Delaware River Basin Commission. The Commission's five members are the governors of New York, Pennsylvania, Delaware and New Jersey, serving *ex officio,* and a federal member, who originally was to be appointed by the

1264 Grad (n 1252) 826.
1265 Ibid 826.
1266 Delaware River Basin Commission Compact, Pub. L. No. 82-573, 66 Stat. 738 (1961).

President. A congressional amendment enacted in 1997 provided that from then on, the federal member would be the commander of the North Atlantic Division of the US Army Corps of Engineers, serving *ex officio*. The Compact in effect preserved the terms of the Decree for a period of 100 years, but it also provided a mechanism for modifying the decree without resuming litigation. This meant that the Commission may modify the Supreme Court decree of 1954 with the unanimous consent of the parties to the decree (the four states and New York City).

An immediate impetus for the negotiation of the Compact was an external shock event: hurricanes Connie and Diane in 1955, affecting the people living in the Delaware basin. The Compact was approved in Congress on 16 September 1961.[1267] It aimed at addressing pressing issues related to pollution and flooding and created a body with comprehensive authority for river basin management.

3.4. THE NUTS AND BOLTS OF THE DELAWARE GOVERNANCE MECHANISM

364. As mentioned above, the Delaware Compact constitutes an important milestone in US water resources management, as it pioneered in bringing together the federal and state levels in a joint governing body competent to address pressing issues in an interstate river basin. The win-win character of joining forces between the federal government and the states was emphasised at the time of Congress' ratification of the Compact: "The establishment of a single agency to coordinate Federal interests in the Delaware River Basin is of as much importance as the joining together of the four states and the resultant coordination of their various State activities".[1268] The Delaware River Basin Compact provides for direct federal participation in the Delaware River Basin Commission (The Delaware Commission), by including as a member a representative of the federal government.[1269] The Commission also consists of the governors of the four states, which commonly act through delegates.[1270] The participation in the Commission is based on equity, i.e. every member has a vote. There are approximately 35 permanent staff members, with an annual budget of approximately $6.5 million.[1271] Pursuant to the Supreme Court Decree of 1954, under the auspices of the United States Geological Survey, a Delaware River Master has been appointed to ensure compliance with the 1954 Supreme Court Decree. The Director of the US Geological Survey then created

1267 Pub. L. No. 87-328, 75 Stat. 688 (1961). The Compact was signed into law by President John F. Kennedy on 27 September 1961.

1268 Senate Committee on the Judiciary Report n 87-854, "Delaware River Basin Compact" (August 31, 1961).

1269 Article 2, § 2(2) the Delaware Compact.

1270 Article 2, § 2(2) the Delaware Compact. Also see Dellapenna (n 1262) 843.

1271 See Kauffman (n 1119) 87.

an advisory committee consisting of one representative per State to assist the River Master by providing technical and policy-related assistance.

The Compact refers to the fact that the water resources of the basin are subject to the sovereign right and responsibility of each of the Parties, and that these parties have agreed to jointly exercise these powers of sovereignty in the common interest of the citizens of the region of the Delaware River.[1272] The parties have ceded a substantial part of their sovereignty over the interstate waters to the Delaware Commission, the entity that can be considered as the manager of the water resources of the Delaware.

On 10 December 2016, the Delaware River Basin Conservation Act was adopted in Congress, and forms part of the legislative package related to the Water Infrastructure Improvements Act for the Nation (WIIN), replacing the Water Resources Development Act.[1273] The Delaware River Basin Conservation Act aims to provide a coordinated approach to implement restoration and protection activities throughout the basin, and to provide a technical assistance project and a competitive grant to enable such activities conducted by state and local governments and by non-profit organisations.[1274] The Act explicitly refers to the Delaware River Basin Commission as the agency charged with providing a unified approach to the management of the river system throughout the basin. The Act more specifically requires the United States Fish and Wildlife Service to establish a non-regulatory restoration programme in consultation with the relevant entities operating within the basin, such as the Delaware River Basin Commission and the Partnership for the Delaware Estuary.[1275]

3.4.1. The scope of the Delaware Compact: comprehensiveness first

365. The Delaware Compact takes as its starting point the "basin", which is defined as "the area of drainage into the Delaware River and its tributaries including the Delaware Bay".[1276] The territorial scope of the Delaware Compact thus focuses on hydrological boundaries, in keeping with the river basin approach. In terms of the hydrological scope of the Compact, both surface water and groundwater are included. With regard to the material scope, the Delaware Compact incorporates

1272 Article 1(3) (b) Delaware Compact.

1273 S.612 – 114th Congress (2015-2016).

1274 Section 3501 to 3507 of the Water Infrastructure Improvements Act for the Nation Act.

1275 Specifically, within 180 days following the enactment of the Act. For financial means to become available, federal funds need to be appropriated for the Delaware River Basin Restoration Program. This restoration programme will be referred to as the "Delaware River Basin Restoration Grant Program" with the goal to provide funding to the relevant actors in the basin, "to the extent that funds are available". Section 3504 (a) of the Water Infrastructure Improvements Act for the Nation Act.

1276 Let us recall the definition of river basin of the WFD: "the area of land from which all surface run-off flows through a sequence of streams, rivers and, possibly, lakes into the sea at a single river mouth, estuary or delta." See Article 2(13) WFD.

the governance of natural resources related to the waters in the Delaware and the uses of land, and focuses on all aspects of water management. The Compact aims to provide a means through which all demands of the basin are met, in terms of diversions, the use and reuse of the water resources, allocation questions, et cetera.[1277] Indeed, the Compact includes a comprehensive set of issues related to the management of water resources, such as flood control, water pollution, land use and water use. It grants far-reaching powers to the Delaware Commission, as will be explained in the sections below. The Compact has been concluded for renewable periods of one hundred years, which implies that Parties have waived their rights to return to the Supreme Court for apportionment during the period in which the Compact applies.[1278]

3.4.2. Powers of the Delaware Commission: far-reaching

366. Much ink has flowed over the issue of the powers that are or should be granted to a joint body governing transboundary waters. Indeed, one of the intrinsic characteristics of the federal-interstate Compact such as the Delaware Compact is the package of competencies of the administrative entity responsible for implementing and enforcing the Compact, in this case the Delaware Commission.[1279]

367. Two aspects of the Delaware Commission are analysed below with a view to providing lessons for the EU framework, namely (i) the competencies of the Commission and (ii) the composition of the Commission with a particular focus on the federal-State dynamics.

3.4.3. Composition and funding of the Commission: Federal and state dynamics

368. As mentioned above, the Delaware Compact pioneers in bringing together the federal and state levels to govern an inter-jurisdictional water body in the US. There are five *ex officio* members, i.e. commissioners, which are the governors of the four states involved and the representative from the federal level, i.e. a delegate from the Army Corps of Engineers. There are approximately 35 permanent staff members, allocated full-time to the activities of the Commission.[1280] Supporting the work of the Commission are advisory committees through which information can be exchanged on various topics. The meetings organised by these advisory committees, which may consist of representatives from watershed organisations,

---

1277 Article 1(3)(d) Delaware Compact.

1278 Comptroller General of the United States, 'Federal-Interstate Compact Commissions: Useful Mechanisms for Planning and Managing River Basin Operations' (United States General Accounting Office 1981) 13.

1279 The model Compact that has been developed by the Utton Transboundary Resources Center also stresses the importance of this point. Cf. *infra*, Part V.

1280 Specialised in finance, water resources management, etc.

industry, municipalities, or government agencies, are open to the public. For example, with regard to flood risk management, a flood advisory committee has been created, which convenes three to four times a year.

369. The Commission faces a significant institutional challenge with respect to funding. Although the Compact provides that the Parties "covenant and agree to include the amounts [allocated to each to meet the Commission's expense budget for each fiscal year] for the support of the ... budget in their respective budgets next to be adopted" contributions have fallen short of approved allocations for nearly two decades. The allocation methodology to determine which State should contribute what, is based on equitable apportionment.[1281] However, there is no set formula for cost sharing. This is determined by the Commission through unanimous vote. For example, in its first cost-appropriation decision at the first meeting of the Commission in 1961, it was determined that New Jersey, New York, Pennsylvania and the United States contribute 24% and Delaware 4%.[1282] This cost allocation has been regularly revised over the years, and – in general – is based on population and area covered by the signatory states. For the fiscal year ending June 2017, the DRBC issued a resolution, following a public hearing on its expense and capital budget, apportioning the respective contributions totalling the amount of $2,964,500.[1283]

Up until the year 1997, the federal government was a regular contributor – in terms of financial and other resources – to the functioning of the Commission. However, in 1997, through the adoption of the Emergency Supplemental Appropriations Act, the federal level ceased its funding contributions.[1284] At that point in time, funding was halted both for the Compact and for the federal commissioner represented in the Delaware Commission.[1285] From that year on, the federal representative in the Commission has been the US Army Corps of Engineers.[1286] The reasoning behind the fact that the federal level pulled out from

1281 Article 13(3) Delaware Compact.

1282 Richard Gore, 'Delaware River Basin Commission Administrative Memorandum: Equitable Apportionment of the Commission Current Expense Budget (General Fund)' (2012). As mentioned above, federal funding ceased to exist in 1997.

1283 With the apportionment for Delaware constituting $447,000, New Jersey $693,000, New York $359,500, Pennsylvania $750,000 and the federal government $715,000. See the Resolution of the DRBC to apportion among the Parties the amounts required for the support of the Current Expense and Capital Budgets for the fiscal year ending 30 June 2017, No. 2016-6.

1284 The appropriation bills did not include funding for the DRBC. See Section 3001 (a) of the 1997 Emergency Supplementary Appropriations Act for recovery from natural disasters, and for overseas peacekeeping efforts, including those in Bosnia. 2 USC. 59, 61 and 65f.

1285 Mandarano, Featherstone and Paulsen (n 1121) 144.

1286 Instead of a neutral representative of the federal government. There was state opposition with regard to the cessation of federal funding, see e.g. the assembly resolution of the State of New Jersey No. 162 introduced 5 June 1997 <ftp://www.njleg.state.nj.us/19961997/AR/162_I1.htm> accessed 11 November 2016. This assembly resolution aims at urging the Congress of the United States not to adopt legislation that would cut the federal funding and at opposing

the Delaware governance mechanism is that Congress argued that interstate river management was primarily a state concern and that there was more added value to these Commissions for the states than for the federal level.[1287] Prior to the adoption of the Emergency Supplemental Appropriations Act, the federal level contributed approximately 20-25% of the share in funding. According to the DRBC, the federal shortfall in funding from the year 1996 to the year 2016 amounts to $12,854,250.[1288]

The Water Resources Reform and Development Act was signed by President Obama on 10 June 2014. The Water Resources Reform and Development Act explicitly states "The Secretary shall allocate funds to the Susquehanna River Basin Commission, the Delaware River Basin Commission, and the Interstate Commission on the Potomac River Basin to fulfil the equitable funding requirements of the respective interstate Compacts".[1289] This Act specifies that the amount of federal funding should correspond to the amount set by the Commission on the basis of the Compact as signed by Congress. When refusing to allocate funds, the Secretary of the Army must explain to the Committee on Environment and Public Works of the Senate and the Committee on Transportation and Infrastructure of the House of Representatives the reasons why it has not done so, and evaluate the impact on the various areas of jurisdiction of the Commission, in this case the Delaware River Basin Commission: water supply, water quality, flood-loss reduction, and so forth.[1290] However, as this Act only entails an authorisation, as opposed to an appropriation, federal funding has not been released.[1291] This funding reality clearly has its impact on the power and influence of the federal level over the functioning of the Delaware Commission.

Currently, more than a third of the funding of the Delaware Commission comes from state contributions, with permits and fees, and grants and contracts constituting the other two categories of financial support.[1292] However, the withholding of funding by the federal government has had a domino effect on funding by other members of the Commission. The DRBC has a water supply use charge system in place, which helps supplement funding by the Parties. The Commission early on purchased storage in federal reservoirs in the basin in order to augment flows in case of low flows in the main-stem Delaware River, simultaneously set up a fund, and started charging a fee for withdrawals of surface

the elimination of the President of the United States' ability to appoint a neutral federal representative to the Commission.

1287 Gerald Kauffman (n 1119) 73.

1288 See <http://nj.gov/drbc/about/budget.html> accessed 15 June 2017.

1289 Water Resources Reform and Development Act of 2014 (Public Law 113-121), Section 4001 (b)(1).

1290 Ibid Section 4001 (b)(3).

1291 Telephone interview with senior staff member Delaware River Basin Commission, 15 December 2016.

1292 Comparable to the Dutch regional water authorities, as these have their own financial support scheme.

water.[1293] This financial source has safeguarded a steady flow of funding in light of the volatility of members' contributions.

3.4.4. Powers of the Commission: General

370. The Commission has extensive powers to attain the goals set by the Compact. It may effectuate all aspects of project management, from the planning stage to the implementation phase, through to operation and maintenance.[1294] For example, the Delaware Commission is empowered to develop projects for the use of the water for supply reasons[1295], which means that it may construct, acquire and operate dams, reservoirs and all other facilities to this end.[1296] The actions may also be directed toward regulating the flow of the water, and storing and releasing it. The Parties are then prohibited from carrying out any activities that may influence the flow level, for example, when the Commission has ordered the waters to be released from storage with the goal of augmenting the flow. Before any project can be executed, the Commission must review all existing rights, plans and programmes with relevance to the project. The Compact also provides for a public participation requirement, by stipulating that the Commission should organise a public hearing, but does not specify the procedure to be followed.[1297] The drafting of and the modifications to the Comprehensive Plan, the water resources programme, the budget plans, and the contracts drawn up for the Commission to sell or dispose of hydroelectric energy or water resources require the organisation of a public hearing, of which the citizens should be given notice at least 10 days before it takes place.[1298] The Delaware Commission ensures vertical and horizontal communication, the former through bi-monthly meetings with the commissioners, the latter through information dissemination to stakeholders such as the citizens and NGOs.[1299] For example, after issuing draft regulations on natural gas drilling, approximately 69,000 public comments were accepted.[1300]

371. The Commission may also establish standards of planning, design and operation of all projects affecting the water resources in the basin. Other

---

1293 This is higher for consumptive users than for non-consumptive users. This fund amounts to approximately $20 million, part of which goes into the operating fund of the Commission that has ensured its financial viability. Telephone interview with senior staff member Delaware River Basin Commission, 15 December 2016.

1294 Namely planning, designing, acquiring, constructing, reconstructing, completing, owning, improving, extending, developing, operating and maintaining. See Article 3(6)(a).

1295 Of domestic, municipal, agricultural and industrial nature.

1296 Article 4(2) Delaware Compact.

1297 Article 4(4) Delaware Compact.

1298 Article 14(4) Delaware Compact.

1299 Gerald Kauffman (n 1119) 84.

1300 These are available on the website of the DRBC: <www.nj.gov/drbc/programs/natural/draftregs-dec2010_comments.html> accessed 15 June 2017.

mandates of the Commission are, for example, conducting and sponsoring research on water resources, coordinating groundwater-level forecasting data, and distributing information and reports related to water problems of the basin. Financial resources constitute one of the basic requirements for any agency or commission aiming to develop short-term to long-term visions and implement them effectively and efficiently. On the basis of the Delaware Compact, the Delaware Commission has been granted borrowing power for any of the purposes set in the Compact.[1301] It may negotiate loans, grants and services that are available from public and private sources. Bonds are payable exclusively out of the revenues of the Commission.[1302] The concept of "costs" is broadly defined: all costs connected with issuing and disposing of the bonds, the amounts required for the establishment of an operating, construction, reserve and sinking fund, and all expenses related to planning, designing and constructing of any facility in the context of the Compact.[1303] What the Commission cannot do is pledge credit of any of the signatory party or bestow payment obligations on them.[1304] Finally, the Commission may exercise all powers needed to carry out its express powers.[1305]

372. A provision of paramount importance to the functioning of the DRBC, is Article 3(8) of the Compact, which entails that, whenever a project may have a substantial effect on the water resources of the basin, it must be submitted to the Commission for review.[1306] The main parameter for evaluating whether a project should be approved or not, is whether it would substantially impair or conflict with the Comprehensive Plan.[1307] The DRBC has been sued for presumably overstepping its authority. One example relates to natural gas drillings, where the Commission considered that all facilities targeting shale formations in the basin constituted "projects" that should be reviewed under Article 3(8) of the Compact. Based on this consideration, in 2009 the Commission decided that all gas drilling activities in the basin needed to be submitted to its review. Since the Commissioners could not find agreement on regulations with regard to natural gas drilling, the Commission suspended its consideration of drilling projects in the basin until such regulations were adopted. In May 2016, a corporation wishing to drill in a certain area along the Delaware River brought suit against the Commission on the basis that it had exceeded its authority under the Compact, and Article 3(8) specifically.[1308] On 23 March 2017, the court dismissed the case with prejudice, declaring that the activities of the plaintiff did constitute a

---

1301 Article 12 Delaware Compact.
1302 Article 3(6)(c), (d) and (f) and Article 12(1) Delaware Compact.
1303 Article 12(2) Delaware Compact.
1304 Article 12(3) Delaware Compact.
1305 Article 3(6)(h) Delaware Compact.
1306 On the basis of Article 3(8) of the Compact.
1307 See below, Section c.
1308 The plaintiffs filed a notice of appeal to the US Court of Appeals for the Third Circuit in April 2017. The gas site is the Marcellus Shale formation. See the Natural Gas Drilling Index Page of the Delaware River Basin Commission: www.nj.gov/drbc/programs/natural/ and news

"project" within the meaning of the Compact, which resulted in the requirement to apply to the Commission in order to determine whether the activities would have a substantial effect on the water resources of the basin.[1309] The Administrative Manual stipulating the rules of procedure lists a series of actions which are deemed not to have a substantial effect on the water resources of the basin, and which may therefore legitimately be taken without prior consultation of the DRBC.[1310] The Commission is required to set uniform standards and procedures to evaluate the benefits and costs for all projects with an impact on the basin.[1311] This approval or disapproval should be based on a procedure determined by the Commission. Each final decision of the Commission with regard to a project is subject to judicial review in any court of competent jurisdiction.[1312] Potential plaintiffs also have the possibility to challenge actions approving projects that the plaintiffs believe are in conflict with the Comprehensive Plan. The Commission can also establish advisory committees consisting of citizens and representatives of governments – from the federal to the local levels – and other stakeholders.[1313] Pursuant to the Compact, the Commission is also required to develop cost-sharing mechanisms between the signatory states.[1314] The Parties are then required to provide capital funds required for the projects that are to be implemented.[1315]

373. The Delaware River Basin Water Code consolidates certain actions taken by resolution of the Delaware River Basin Commission.[1316] The integration of the resolutions into the Water Code transforms them into rules.[1317] The Code fills in the details on the procedure to be followed when the Commission issues a drought warning, the quantities to be diverted, and so forth. The Water Code is divided into four parts, i.e. general DRBC policies, the conservation and utilisation of water resources in the basin, water quality standards for the basin, and application of standards. The Delaware River Basin Compact grants the Commission the authority to allocate the waters of the basin to and among the states.[1318] In light

article on the law suit: <https://stateimpact.npr.org/pennsylvania/2016/05/18/wayne-county-landowners-sue-drbc-to-allow-gas-drilling-in-nepa/#more-39848> accessed 10 June 2017.

1309 Wayne Land and Mineral Group, LLC. V Delaware River Basin Commission, 3:16-CV-00897 (2017). At the time of writing, the appeal proceedings were pending.

1310 See § 401.35 of the Administrative Manual: Rules of Practice and Procedure, 18 C.F.R. 401 (2015).

1311 Article 11(4) Delaware Compact.

1312 Article 3(8) Delaware Compact.

1313 Article 3(10) Delaware Compact.

1314 Article 11(4) Delaware Compact: these cost sharing and reimbursement formulas e.g. relate to procedures for the allocation of project costs, establishing a system of accounts, etc.

1315 Obviously not the costs that go beyond the project, i.e. the Commission cannot subject the signatory States to mandatory payment obligations other than those connected to the specific project or facilities. See Article 12.20 Delaware Compact.

1316 Delaware River Basin Water Code, 18 C.F.R. Part 410 (2013).

1317 This is beneficial for the functioning of the Commission because the federal and state courts give the Commission a great degree of discretion when interpreting these rules, which is important for water supply security in the basin.

1318 Article 3(3) Delaware Compact.

of this competence, during drought emergencies the powers of the Commission are amplified. The Code provides that the Commission must give first priority to those uses that sustain human life, health and safety, and second priority to water necessary for livestock, thereafter allocating the remaining water based on the doctrine of equitable apportionment. The Code provides tailored action plans for situations of drought conditions for the relevant sub-basins, e.g. it determines the actions to be taken in case of lower-basin drought warnings, and the drought-management strategy for the Christina River Basin. During a drought or water-supply emergency declared by the Commission, following the unanimous consent of the decree parties, said action plans reduce the diversions, releases and flow targets established pursuant to the Supreme Court Decree and the Compact.

For the Christina River Basin, the Water Code determines a strategy entailing that, in the event of a drought, the Christina River Basin Drought Management Committee is to be activated, which consists of representatives of the relevant departments of the two states and relevant counties involved: Pennsylvania and Chester County, and Delaware and New Castle County.[1319] This Committee should then carry out the relevant analyses with regard to the drought event, and provide advice to the Governors of the respective states. For example, the Governors of the two states may decide that a drought warning or drought emergency is warranted on the basis of the analysis of the Committee. The Delaware River Basin Water Code, determining the drought management strategy for the Christina River Basin, then provides that a ban on non-essential use of water must be implemented in the area, which is voluntary at first and mandatory in the event of drought emergency. When the Drought Committee deems that such a ban should be activated, "the Governors of Pennsylvania and Delaware shall be urged to take simultaneous and identical actions in order to further the cause of water conservation in the Christina Basin".[1320] The Water Code thus includes, through the drought-management strategy, a protocol clarifying the steps to be taken in the event of a drought, and a division of responsibilities. These provisions gain traction by the fact that they should be included as an annex to the drought contingency plans of the two states, where the statutory penalty and enforcement provisions applicable to the parent plans should apply to the strategy determined under the auspices of the DRBC.[1321] This is to create clear synergy between the DRBC and the states involved regarding actions at sub-basin levels.

3.4.5. Legal character of the Commission

374. The Commission has legal personality and can therefore negotiate contracts, initiate proceedings and in turn be sued, borrow money, acquire, control and sell

1319 Article 2(5)(7) C Delaware River Basin Code.

1320 Article 2(5)(7) F Delaware River Basin Code.

1321 Article 2(5)(7) G Delaware River Basin Code.

property, exercise all corporate powers pursuant to the goals of the Compact, and own and operate any facility or project and regulate on the use thereof.[1322]

The Commission may issue rules and regulations to effectuate the provisions of the Compact and develop instruments to enforce these rules and regulations. There are two main conditions for these to be effective: a public hearing needs to be conducted prior to their adoption and the states need to integrate them into their respective legal frameworks.[1323] The Compact explicitly provides for sanctions in case of violation by any person, corporation or association of the provisions of the Compact or any rule or regulation stipulated by the Commission. This works as a sort of cascade mechanism. The violation should be punished on the basis of the implementing legislation at the level of the states, in which the penalty should be enshrined. However, if no such penalty is provided for, the Compact provides that the violator is liable to a penalty ranging from $50 to $1,000 per day for each offence. The Commission may sue the violator before any court of the competent jurisdiction in its own name to recover the payment.[1324] This provision implies that the Commission cannot impose fines on its own, it needs to bring the violator before a competent court to enforce the financial penalty. However, under its regulations, the Commission may enter into a settlement with any violator, and such settlement may provide for a payment.

3.4.6. Plans and programmes

375. As is the case for the competent authorities governing (inter)national river basin districts in the European Union, there are a number of procedural requirements included in the Compact which the Delaware Commission should comply with. The Commission has three main tasks in this regard: (i) to develop a Comprehensive Plan, (ii) to draft a Water Resources Program, and (iii) to issue an annual current expense and capital budget. The Commission also issues rules of practice and procedure, which govern the adoption of and amendments to these plans, programmes and other measures taken on the basis of the Compact.

376. The Comprehensive Plan developed by the Delaware Commission plays a major role in the governance of the river basin. It relates to short and long term development and usage of the basin and its water resources, and should be subject to consultation with water users and interested public bodies.[1325] In this regard, it should be noted that, in a way, the Compact promoted integrated water resources management early on, as the scope of "water resources" in the context of the Compact includes "water and related natural resources in, on, under, or above

1322 Article 14(1) Delaware Compact.
1323 Article 14(2) Delaware Compact.
1324 Article 14(17) Delaware Compact.
1325 Article 3, § 3(2)(a) Delaware Compact.

the ground, including related uses of land, which are subject to beneficial use, ownership or control".[1326] The Compact grants extensive powers to the Commission to initiate comprehensive planning throughout the basin.[1327] It has the authority to issue permits for diversion and withdrawal related actions, to regulate on hydro-electric power generation in the basin, to restrict land use in floodplains, and so on. In general, neither federal agencies nor state and local agencies may authorise the construction, acquisition or operation of a project or facility if it has not been included in the Comprehensive Plan drawn up by the Commission.[1328]

377. Secondly, the Commission should draft a Water Resources Program, which is based on the Comprehensive Plan and which details the quantity and quality of the water resources needs of the area and the existing and proposed projects to fulfil those needs as well as a separate statement on the projects that the Commission plans to carry out. The Water Resources Program is a strategic document, setting forth the policy lines of the Commission over a period of three years.

378. Thirdly, the Commission should issue an annual current expense and capital budget consistent with the water resources programme.[1329] In these documents, the Commission stipulates its estimated expenses for administration, operation, maintenance and repairs, and a separate statement including cost-allocation for each project.[1330]

3.4.7. Flood risk management

379. Also in the context of flood risk management, the Delaware Compact grants explicit authority to the Commission. Flood risk management in the Delaware River Basin reflects an interplay between the federal National Flood Insurance Program developed by the Federal Emergency Management Agency[1331], measures adopted and data and information, through, for example Geographical Information Systems, developed and consolidated by the Delaware River Basin Commission, state programmes and regulations and local programmes. These local programmes have been developed by local communities participating in the National Flood Insurance Program. In the context of this National Flood Insurance Program, communities have implemented local flood-management ordinances and flood-hazard plans.[1332]

1326 Article 1, § 1(2)(i) Delaware Compact.
1327 Mandarano, Featherstone and Paulsen (n 1121).
1328 Article 11(1) Delaware Compact.
1329 Article 3(2) Delaware Compact.
1330 Article 13(3) Delaware Compact.
1331 The three components of the National Flood Insurance Program are flood-hazard mapping, floodplain management and flood insurance.
1332 Army Corps of Engineers and NJDEP, 'Delaware River Basin Comprehensive Flood Risk Management Interim Feasibility Study and Integrated Environmental Assessment for New Jersey'.

380. In terms of powers of the Commission, the Delaware Compact enables the Commission to make a categorisation of lands along the Delaware based on their susceptibility to floods and set standards for floodplain use to safeguard public health, safety and property.[1333] The Commission may adopt recommended standards related to the nature and extent of the uses of land in flood-prone areas along the Delaware and its tributaries.[1334] The Compact further stipulates that the standards cannot restrict or impair the power of the states[1335] to adopt zoning and other land-use regulations "not inconsistent herewith".[1336] It therefore seems that states can only issue zoning and land-use rules that are consistent with the standards issued by the Commission.

381. In the 1970s, the Delaware Commission conducted a floodplain mapping exercise for 151 municipalities. In practice, the number of zoning provisions stipulated by the Commission is rather minimal, as land-use planning is seen as a local issue in the US with local control and the member states and federal government are reluctant to establish requirements by means of an interstate agency.[1337] As mentioned above, the Delaware River has flooded on several occasions in the past ten years, resulting in several casualties and economic damage. One of the factors influencing the floods was the fact that the reservoir on Delaware tributaries operated by New York City, which provides the City with half of its daily water supply, is kept too full resulting in overflows in times of heavy precipitation. The counterargument put forward by NYC is that if the reservoir were kept only 80% full, which is what has been suggested by the downstream riparian citizens, the City would face serious problems in times of drought. The Delaware Riverside Conservancy, which was created as a lobby group in this regard, directed its lobbying to the Delaware Commission, as opposed to directly to NYC, which is a confirmation of the power of the Commission.[1338] In 2007, through the Commission, the parties agreed to the adoption of a so-called "Flexible Flow Management Program", which relates to the use of supply reservoirs in the context of flood risk management. This agreement was drawn up in response to the flood events of 2004 and 2006.[1339] The Flexible Flow Management Program is updated every few years to ensure adaptation to changing circumstances.

1333 Article 6(2)(b) Delaware Compact.

1334 As mentioned below, the States are reluctant toward involvement of the DRBC in land-use regulation and therefore, these standards are minimal.

1335 Or their political subdivisions.

1336 Article 6(2)(a) Delaware Compact.

1337 Similar to the EU, where spatial planning is a topic heavily subject to the subsidiarity principle.

1338 Ibid.

1339 The most recent version of the Flexible Flow Management Program dates from 1 Jue 2016. Delaware River Basin Commission, 'Flexible Flow Management Program' (2016) <https://water.usgs.gov/osw/odrm/ffmp/2016/2016_FFMP_Agreement_Signed.pdf> accessed 13 July 2017.

3.4.8. *Withdrawals and diversions*

382. In the Delaware River Basin, the water allocation regime is a combination of the Supreme Court Decree, the jurisdiction of the Delaware Commission on the basis of the Compact, and state regulations. Indeed, allocation of water has not been determined by federal legislation such as the CWA.[1340] In general in the US, the quantification of the right to use waters in interstate water bodies is inchoate until it has been provided for through the various means available in the water-law landscape.[1341] Allocation of water forms part of the Delaware Compact, which stipulates that the Commission has the power to allocate the waters of the basin to and among the states and among their political subdivisions and subsequently impose conditions, obligations and release requirements.[1342] This is done on the basis of the doctrine of equitable apportionment. The doctrine of equitable apportionment implies that interstate waters must be shared between the co-riparian states, resulting in percentages of use for each State. This doctrine also underpins the 1954 Supreme Court Decree which initially allocated the waters, by giving New York City the right to withdraw a certain daily amount of water and providing for compensating releases to meet a flow objective in New Jersey, also determined by the Decree.[1343]

There are certain limits to the exercise of allocation powers by the Commission. The first limitation relates to the 1954 Supreme Court Decree, in that the Commission may not influence the diversion, obligations, compensating releases, and so forth, stipulated by the Court except with the unanimous consent of the parties. An exception to this rule is if the Commission declares a state of emergency resulting from a drought or catastrophe, in light of which it may direct increases or decreases in diversions, releases and flow targets, without consent of the parties to the decree. However, in preparation of an extreme drought, and with the unanimous consent of the decree parties, the Commission has adopted regulations that provide for phased reductions in diversions, flow targets and reservoir releases in the event that New York City's Delaware Basin reservoir storage diminishes during an extended dry period.

Unlike in the Western United States, allocations decided on by the Commission and its member states do not constitute prior appropriations of waters of the basin. These DRBC and state allocations are not transferable except by the permitting authority and do not result in a hierarchy of rights to the allocated waters. Rather, the right to use of the water is based on current and projected needs and is granted by permits for limited periods of time. Upon renewal, an allocation may be

1340 George William Sherk, 'The Management of Interstate Water Conflicts in the Twenty-First Century: Is It Time to Call Uncle?' (2005) 12 New York University Environmental Law Journal 764.

1341 I.e. Supreme Court litigation, apportionment by Congress, interstate Compacts.

1342 Article 3(3) Delaware Compact.

1343 See *New Jersey v New York*, 283 US 336 (1931) and *New York v New Jersey*, 347 US 995 (1954).

extended, increased or reduced, again depending on the demonstration of need by the applicant.

The powers of the Commission are especially broad when it comes to emergency situations or to special protection measures. For example, the Commission may designate protected areas where there is a threat of water shortage. In these protected areas, entities may not divert or withdraw water except as defined in a permit issued by the Commission or by a signatory State prior to enactment of the Compact.[1344] Thus, the authority to issue permits is with the states, if they have a permit system for water use in place.[1345]

In general, users that withdraw more than 100,000 gallons a day on average over 30 days from basin waters fall under DRBC jurisdiction and need to obtain a permit from the Commission.[1346] State laws refer to this jurisdiction in their respective water allocation regulations. For example, the water allocation rules of the State of Delaware stipulate that: "Compliance with these regulations does not exempt those who make water withdrawals within the Delaware River Basin from the requirements of the Delaware River Basin Commission (DRBC). All water withdrawals having a substantial effect on the water resources of the basin – i.e., averaging more than 100,000 gallons per day over any 30-day period – must have DRBC approval. Application for DRBC approval will be forwarded through the Department and in accordance with the provisions specified in DRBC/ Department administrative agreements.[1347]

The DRBC has developed the One Permit Program, which aims to coordinate the permit-issuing activities of the DRBC and competent agencies of the Parties for activities falling under both state and DRBC jurisdiction. This One Permit Program entails that requirements and determinations of both state agencies and the DRBC are incorporated into one permit. The programme provides a streamlined process for the review of eligible projects that are subject to review under the Compact and by a regulatory agency of any of the Commission's Parties. Eligibility is defined in an administrative agreement between the DRBC and the participating signatory party agency.[1348] Until now, regulatory programmes included in agreements have related to water withdrawals from the basin and discharges into basin waters.[1349] In general, applications for approval, renewal or

[1344] Article 10(2) and 10(3) Delaware Compact.

[1345] Dellapenna (n 1262) 846.

[1346] About 378541.18 litres. Article 10(5) and 10(6) Delaware Compact.

[1347] New Jersey Regulations governing the Allocation of Water of 1987, New Jersey Administrative Code 7:19. Delaware Administrative Code: Title 7 Natural Resources and Environmental Control, 7303 Regulations Governing the Allocation of Water.

[1348] See administrative agreements with regard to the One Permit Program between the DRBC and New Jersey (March 2015, <www.state.nj.us/drbc/library/documents/AA/NJ_March2015_incl-demo-project.pdf> accessed 10 June 2017), and between the DRBC and New York (March 2016, <www.state.nj.us/drbc/library/documents/AA/NYSDEC_march2016.pdf> accessed 10 June 2017).

[1349] § 401.42 of the Administrative Manual.

revision of projects under the One Permit Program are filed only with the agency of the signatory State. The most remarkable stipulation in the context of diversions and withdrawals, is that the regulations supersede any laws of the states that may conflict with the control and regulation exercised by the Commission.[1350]

383. All projects that entail an importation or exportation of water into or out of the basin should be submitted to the Commission for approval, and for a review in terms of coherence with the Comprehensive Plan.[1351] The Water Code stipulates which elements should be included in the application for review by the Commission. The exportations from the basin are subject to consumptive water charges payable to the Commission.[1352]

384. On the basis of Article 10(4) of the Compact, when circumstances such as a drought event might cause a shortage of available water supply, the Commission may declare a state of water-supply emergency, following an obligatory public hearing. During this state of emergency, no withdrawals or diversions may take place beyond the quantities allowed by the Commission.[1353] New York City's reservoirs, New York City's and New Jersey's out-of-basin diversions, and flow targets established by the Commission are automatically placed under "normal", "watch", "warning" and "drought" operations as reservoir levels diminish. In November 2016, the Commission put the basin in "drought watch" status. By unanimous vote, the Commission approved a special permit with the goal of enhancing coordination of operations related to water storage and release in the basin.[1354]

385. Relevant in the context of the allocation of water in the United States, is the existing water rights system. The riparian rights model prevails in the East, whereas the prior appropriation doctrine prevails in the West. Three out of the four states sharing the Delaware basin have a regulated riparian statute. Regulated riparianism is an "improved" form of riparianism, because through this system, water rights are granted for a number of years, and the permits set conditions safeguarding reasonable utilisation, thereby also protecting the environment of the basin and avoiding negative impact of the water use for downstream activities.[1355] The permit-issuing policy of the Delaware River Basin Commission with respect to the use of water is based on such regulated riparianism. As mentioned,

---

1350 Article 10(8) Delaware Compact.

1351 In line with Article 3(8) of the Compact, and Article 2(30)4 of the Delaware River Basin Commission Water Code.

1352 Article 2(30)5 Delaware River Basin Water Code.

1353 Article 10(4) Delaware Compact.

1354 See DRBC News Release.

1355 League of Women Voters of Pennsylvania Citizen Education Fund Common Ground Project, 'Water Use and Water Rights in Pennsylvania' (1998) <http://wren.palwv.org/documents/WaterUse_000.pdf> accessed 14 July 2017.

three out of the four states have adopted this approach. The Commonwealth of Pennsylvania has not empowered any state agency with comprehensive authority to allocate water. This means that the Commission is the primary permitting authority, not only within a large protected area designated by the Commission in South-eastern Pennsylvania, but throughout the Pennsylvania portion of the basin. In the other three states, the Commission acts as coordinator with regard to water-allocation permits, and its requirements are included in permits issued by the states.[1356]

3.4.9. Limits to the power of the Commission

386. The Commission does not have unlimited competences. A general limit on the powers of the Commission and on the binding character of the Compact in general relates to the various disclaimers issued by the federal government with regard to the federal authority. As discussed above, there is a delicate balance and complicated dynamics between the states and the federal level with regard to interstate water resources management. In general in the United States, water allocation is left to the states, whereas water quality remains subject to both federal and state law. However, due to the federal government's exclusive authority over interstate commerce and navigation, it retains certain overriding powers. Federal reservations of rights in interstate water Compacts, such as the Delaware Compact, are not unusual.[1357]

The federal government's approval of the Delaware River Basin Compact was issued with multiple reservations. From the moment a project is to involve funds from the federal level, the Commission must submit a number of information elements to the federal government before it can be put into motion, for example, a complete plan and estimate for the project is necessary to carry out an engineering evaluation.[1358] Moreover, in case of national emergency, the powers of the Commission cannot restrict the executive powers of the President in any way. Another important reservation from the federal perspective is that the Compact cannot touch on the constitutional authority of the United States. If any phrase of the provision related to the powers of Congress to regulate interstate commerce or navigable waters or its power to withdraw the federal government from the Compact is deemed as unconstitutional by a competent court of last instance, the federal government will no longer be bound by the Compact and will cease to be a party to the Compact.[1359] In addition to the federal government, the states have submitted their issues to the commission for consent. For example, the State of New York has stipulated that no agency or officer of the State can exercise the

---

1356 Joseph Dellapenna (n 1262) 846.

1357 Zachary Lorne McCormick, 'The Use of Interstate Compacts to Resolve Transboundary Water Allocation Issues' (Oklahoma State University 1973) 412.

1358 Article 15(1) Delaware Compact.

1359 Article 15(1)(t) Delaware Compact.

power to enter into an agreement with the Commission regarding the regulation and control of withdrawals of surface waters and groundwater from the basin.[1360]

387. The limits with respect to flow management and allocation have already been mentioned in the section above: the Supreme Court allocation must not be overridden with the exception of emergency scenarios during which the Commission may decrease or increase the allocations, diversions and releases permitted by the Court. Whereas generally decisions are taken through majority voting, this has to be done through unanimity voting. Furthermore, the Commission may not issue individual permits or other entitlements to the waters in the basin. This permit-issuing competence remains with the individual states.[1361]

3.4.10. Enforcement and dispute resolution

388. The Commission has the authority to issue enforceable rules and regulations, and provide for the actual enforcement.[1362] A key provision of the Compact is Article 3(8), which provides that all projects that have a substantial effect on the water resources in the basin should be submitted to the Commission for approval. The main parameter in reviewing this project, is the question whether the project would substantially impair or conflict with the Comprehensive Plan. This provision allows the Commission to enforce the provisions included in the Comprehensive Plan. Furthermore, in general, when a violation of a rule or a permit has been observed, a notice of violation can be served. In this event, the Commission can organise a hearing and appoint a Hearing Officer in accordance with the Compact. The Administrative Manual stipulating the rules of practice and procedure provides further details with regard to objections related to administrative actions and decisions of the Commission's Executive Director in the context of water quality. In this case, a decision which would limit the discharge of certain substances into the basin may give rise to a request for a hearing by the discharger in question.[1363] This request should be directed to the Secretary of the DRBC within thirty days following the determination. The Executive Director may then request the submission of a technical report by the objector prior to the organisation of a hearing.[1364] For objections in the context of project reviews in light of conformity with the Comprehensive Plan, the DRBC may also organise hearings at the request of the person in question.[1365] Hearings are conducted by one or more members of

1360 Within the meaning of Article 10(1) Delaware Compact and on the basis of reservation no. 637 of the State of New York.
1361 Marilyn O'Leary and George William Sherk, 'Reinventing the Interstate Water Compact: A New Model' (Proceedings of the Rocky Mountain Mineral Law Institute', Chapter 21, 2006).
1362 Article 14(2) Delaware Compact.
1363 § 401.73 Administrative Manual. E.g. in the context of waste-load allocation.
1364 Ibid. The technical report should relate to a description of the operational characteristics of the waste treatment, chemical and toxicological characteristics, etc.
1365 Subpart F of the Administrative Manual.

the Commission, by the Executive Director or by another Hearing Officer, and in the case of waste-load allocation, the hearing board should consist of two persons, one of which should be appointed by the water pollution control agency of the State in which the discharge originates.[1366] During the hearing the objector has the right to be represented by counsel, examine witnesses, and so forth. The Hearing Officer draws up a report, following which the Commission takes a final decision which is in turn open to appeal by the objector with an appropriate federal court in accordance to Article 15.1 of the Compact.

3.5. SUB-BASIN INITIATIVES IN THE DELAWARE BASIN

3.5.1. Introduction

389. Whereas in the European Union, designation of sub-basins is done by the Member States with no interference from the level of the EU or the joint entity designated as the competent authority at the level of the International River Basin District, the designations of watersheds are partly carried out by federal agencies such as the United States Geological Survey and the National Resources Conservation Service. The participation of the states and local governments in cooperation mechanisms to govern the shared transboundary waters is authorised by legal instruments at the levels of the states. For example, the Constitution of Pennsylvania authorises municipalities to cooperate in exercising any power or responsibility, or to delegate or transfer such power or responsibility to other units of power, such as governmental units, the federal government, another State or a newly created governmental unit.[1367] The Upper Delaware Council[1368], for example, operates within the limits set by the intergovernmental cooperation laws included in the relevant states, i.e. Pennsylvania and New York.[1369]

390. Several sub-basins have been identified in the Delaware River Basin, which are also referred to as "watersheds". These sub-basins have governance and planning structures in place. In order to promote the link with the more local hydrological scale, the Delaware River Basin Commission provides a list of local watershed associations in its basin, but many of these have a more private character. There are watershed coalition organisations that combine several of these smaller

---

1366 § 401.83(b)(1) of the Administrative Manual.

1367 This is further implemented in the Intergovernmental Cooperation Act of 1972, 53 P.S. Sections 483-485 (1972). See Article 9, Sect. 5 of the Pennsylvania Constitution. Similar provisions exist in the other states.

1368 See *infra*.

1369 Conference of Upper Delaware Townships, 'River Management Plan for the Upper Delaware Scenic and Recreational River' (1986) <www.upperdelawarecouncil.org/wp-content/uploads/Upper_Delaware_River_Management_Plan_1986_updated_2015.pdf> accessed 14 July 2017.

scale governance structures, to represent their common interest in protecting the Delaware basin. Parts of the Delaware River Basin are subject to specific conservation programmes triggering coordination between the federal, basin and local levels. Drawing the comparison to the EU mechanism for support at the sub-basin, i.e. regional, scale, reference can be made to the INTERREG programme, which can support, among other things, cooperation activities between regions in the EU. A selection of these sub-basin structures will be discussed below, mainly to gain insight into the interaction between the sub-basin and basin governance levels.

3.5.2. Examples of sub-basin schemes in the Delaware Basin

391. Watershed partnerships in the context of the United States have been defined as "self-directed and locally-focused collection of parties, usually featuring both private and intergovernmental representatives, organised to jointly address water-related issues at the watershed level, [...], and typically reliant on collaborative mechanisms of group interaction characterized by open debate, creativity in problem and solution definition, consensus decision-making, and voluntary action".[1370] Taking a side step, this definition best fits the operation of the river contracts in the Walloon Region in Belgium.

392. An example of such watershed partnership in the US is the Partnership for the Delaware Estuary. The Delaware Estuary is the tidal area of the basin, i.e. the part of the basin which flows into the sea. The Estuary runs through Delaware, New Jersey and Pennsylvania and with approximately 18,000 km^2, it is one of the largest estuaries of the country, covering a significant part of the Delaware Basin. The Delaware Estuary has been included in the federal National Estuary Program, which was established by the federal Environmental Protection Agency on the basis of the CWA, with the aim of protecting the ecological integrity of estuaries of national interest. Within the realm of the National Estuary Program the estuaries establish a governance structure with the watershed as the point of departure and they issue a Comprehensive Conservation and Management Plan.

In addition to the Delaware River Basin Commission, the main steward of the Estuary is the Partnership for the Delaware Estuary, which was created in 1996 as a non-profit organisation to safeguard a healthy water system in the tidal Delaware River. The difference with the DRBC is that the Partnership for the Delaware Estuary is a non-profit organisation that deals with non-regulatory activities, whereas the DRBC is a regulatory agency focussing on issues regarding water quantity and water quality throughout the entire basin.[1371] The lead federal

---

[1370] Mark Lubell, 'Collaborative Environmental Institutions: All Talk and No Action?' (2004) 23 Journal of Policy Analysis and Management 549, 551.

[1371] E-mail correspondence with representative of the Partnership for the Delaware Estuary, January 2017.

agency for the Partnership is the Environmental Protection Agency.[1372] Within the realm of the Partnership for the Delaware Estuary, several committees operate: The Steering Committee provides input on strategy for the estuary from a short-term perspective, whilst the Estuary Implementation Committee provides guidance to the Partnership with the goal of realising its long-term plan, and the Science & Technical Advisory Committee provides advice on technical issues. The plan refers to inventories that have been executed with regard to the estuary, illustrating the disconnect between land-use regulations and the ecosystem perspective.[1373] There are approximately 500 municipalities connected to the estuary, which have their proper zoning and planning regulations.[1374] The Comprehensive Conservation and Management Plan for the Delaware Estuary refers to the Delaware River Basin Commission multiple times.[1375] It is emphasised that the DRBC will support the implementation of its members' actions based on priorities from the basin's perspective.[1376] Reference is made to the potentially significant role of the DRBC with regard to land-use planning, i.e. through its authority with regard to water quality standards and the extraction and use of water.

The plan refers to the need for a watershed planning strategy regarding land-use planning that takes into account the sustainable development vision presented by the Brundtlandt report, i.e. a regional sustainable development strategy sprouting from a consensus-building process. The plan identifies several actions, the related resource needs and the entities that should take the lead in their implementation. Importantly, the plan focuses on implementation by identifying "Implementation Teams" for each action and designating "Leads and Partners", with three annual conferences to further the implementation of the plan.[1377] The Estuary Implementation Committee is considered as a crucial factor in the success of the Partnership.[1378] For several of these implementation actions referred to above, the Delaware River Basin Commission has been identified as the responsible authority, e.g. with regard to the promotion and implementation of water-conservation programmes. Indeed, there is a clear interaction between the Partnership for the Delaware Estuary and the Delaware River Basin Commission. The DRBC forms part of the management structure of the Partnership, with a seat on its steering committee, Estuary implementation committee and science

---

1372 Recall that the lead federal agency associated with the DRBC is the Army Corps of Engineers.

1373 Partnership for the Delaware Estuary, 'Comprehensive Conservation and Management Plan for the Delaware Estuary' (1996) 83 <https://s3.amazonaws.com/delawareestuary/pdf/CCMP.pdf> accessed 4 May 2017.

1374 This is because the States of Pennsylvania, New Jersey and Delaware have authorised local governments to issue such regulations.

1375 More than 150 references to the DRBC are included in the plan.

1376 Ibid.

1377 Partnership for the Delaware Estuary (n 1373) 291.

1378 E-mail correspondence with representative of the Partnership for the Delaware Estuary, January 2017.

and technical advisory committee, and vice versa, the Partnership has seats on committees run by the DRBC as well.[1379] Within the realm of the Partnership for the Delaware Estuary, smaller-scale organisations are active for parts of the Estuary that have been granted special protection statuses and grants. For example, the Environmental Protection Agency gave the Partnership for the Estuary a $1.5 million grant in the context of the Targeted Watershed Initiative Grant with the goal of protecting the Schuylkill River Valley.[1380]

393. Another example of cooperation at the sub-basin scale is the Upper Delaware Council, which oversees the management of the Upper Delaware Scenic and Recreational River. The official acreage of the Upper Delaware Scenic and Recreational River is 55,574.5.[1381] However, this acreage is based on technologies from the 1980s, and is currently being revised through a project developed by the National Park Service at the request of the Upper Delaware Council. The project aims to designate a boundary of the River on the basis of GIS software to enhance planning processes within the corridor.[1382]

This Council was created following the publication of the management plan prepared for this river in 1986.[1383] The Delaware River Basin Commission has reviewed the plan in order to check the consistency with its comprehensive plan for the basin.[1384] The Upper Delaware Council formally includes representatives from the various towns, the two states involved, the federal National Park Service and the Delaware River Basin Commission. Characteristic of the management of the Upper Delaware Scenic and Recreational River is the high level of involvement of local stakeholders, as it favours local-level land and water stewardship.[1385] The Upper Delaware Scenic and Recreational River was categorised as a national wild and scenic rivers system in 1978, on the basis of the Wild and Scenic Rivers Act.[1386] This federal Act aims to preserve the free-flowing condition of rivers that possess remarkable scenic, recreational, wild, geologic, historic, cultural

1379 E-mail correspondence with representative of the Partnership for the Delaware Estuary, January 2017.

1380 For the Schuylkill River watershed, the US EPA, the Delaware River Basin Commission, the Pennsylvania Department of Environmental Protection and the Philadelphia Water Department formed the Schuylkill Action Network, also including local watershed organisations, municipal governments and other local stakeholders. See Philadelphia Water Department, 'The Schuylkill River Watershed Source Water Protection Plan Philadelphia Water Department' (2006).

1381 As reported to the United States Congress. See Upper Delaware River Management Plan, *infra*.

1382 E-mail correspondence with resource specialist of the Upper Delaware Council, December 2016.

1383 The operations of the Upper Delaware Council are underpinned by several federal laws: the Code of Federal regulations (C.F.R.) Title 36- National Park Service lands and activities regulations, Public Law 90-542, 16 U.S.C. 1271- Wild and Scenic Rivers Act, 1968, Public Law 95-625, 16 U.S.C. 1274- Upper Delaware added to Wild and Scenic Rivers System, 1978.

1384 Conference of Upper Delaware Townships (n 1369).

1385 E-mail correspondence with resource specialist of the Upper Delaware Council, December 2016.

1386 Public Law 90-542; 16 U.S.C. 1271 et seq. § 1274 reads: "The segment of the Upper Delaware River from the confluence of the East and West branches below Hancock, New York, to the existing railroad bridge immediately downstream of Cherry Island in the vicinity of Sparrow

or similar characteristics. Rivers can (partially) be designated as wild, scenic, or recreational river areas. Once designated as such in the national system, the wild and scenic river needs to be managed so that the characteristics triggering the designation are maximally safeguarded. Several parts of the Delaware River Basin have been designated as wild and scenic. For rivers that have been classified as such, comprehensive river-management plans need to be drawn up. When the National Park Service was tasked with developing the management plan for the river, instead of acquiring land for the creation of a natural park, it went about it another way, by working alongside local governments and community groups.[1387] The first two versions of the plan faced protest of local residents, mainly out of fear of over-regulation by the federal government. The final version of the plan established the Upper Delaware Council, among other things, to safeguard control by the local governments.[1388] For example, the federal power of eminent domain is restricted and the conditions to be fulfilled in order to exercise such power are defined in the river-management plan. The conditions include, for example, a new land use that poses a "clear and direct threat" to the protection of the objectives of the Upper Delaware Scenic and Recreational River Act.

The Land and Water Use Guidelines are a component of the Upper Delaware River Management Plan, and seek to protect water quality, safeguard water resources, and so forth. The Upper Delaware Council reviews proposed plans and legislation in the two states connected to the river, i.e. New York State and the Commonwealth of Pennsylvania, for conformity with the River Management Plan and the Land and Water Use Guidelines. The River Management Plan is not a federal or state law, but it is a document signed by the relevant levels of government and can as such be enforced. Indeed, New York and Pennsylvania have issued legislation directing their respective agencies to act consistently with the Upper Delaware River Management Plan.[1389] The resource specialist of the Council first reviews the applications and submits it to the Project Review Committee of the Council, and the decision is ultimately made by the National Park Service.

Bush, New York, as depicted on the boundary map entitled "The Upper Delaware Scenic and Recreational River".

1387 Upper Delaware Council, 'Partnering to Protect the River' (2008) <www.upperdelawarecouncil.org/wp-content/uploads/partnering_to_protect_river_supplement.pdf>. The approach was to limit federal ownership, i.e. mainly only with regard to surface water, whilst most lands remained privately owned.

1388 Conference of Upper Delaware Townships (n 1369).

1389 New York Executive Order No. 169 stipulates that all agencies need to act consistently with the Upper Delaware River Management Plan and Pennsylvania Code Subchapter MM provides that: "to the maximum extent permitted by law, all administrative departments, independent administrative boards and commissions, and other State agencies shall act consistently with the goals, policies and objectives of the Upper Delaware Scenic River Management Plan and provide for adequate consideration of the National interest involved in planning for and siting of facilities and the use of resources in the Upper Delaware Scenic River corridor when the planning, siting and uses are of more than local interest."

Local governments in the Upper Delaware Council participate on a voluntary basis, and at the time of writing, thirteen of the fifteen local governments are involved. The 1986 plan acknowledges this voluntary character of participation, but stresses that towns, through participation in the Upper Delaware Council, are able to retain maximum authority by being included in the decision-making processes of the Council. The "stick" in this regard, relates to the consideration that, if the local governments fails to participate, the National Park Service will execute its decision-making power on the basis of Section 704 of the Wild and Scenic River Act, e.g. with regard to the acquisition of land if the conditions required for the NPS to do so have been met, without a strict necessity for local input.[1390]

Whereas the two states and the towns involved are voting members, the DRBC and the National Park Service are members on a non-voting basis. The participation of the Delaware River Basin Commission is authorised on the basis of Article 3.9 of the Compact, which allows the Commission to promote the coordination of programmes with regard to water resources in the basin between federal, state, municipal and private agencies. The river management plan explicitly states that the Upper Delaware Council should be the primary entity through which the various layers of governance involved, i.e. local, state and federal, resolve their differences and agree on joint actions to be taken.[1391] The Council, however, does not have regulatory authority and therefore relies more strongly on voluntary enforcement, in addition to enforcement by the NPS. As the DRBC has the responsibility to the parties that have signed the Delaware River Basin Compact to plan, develop and control the water resources of the basin as a whole, the Special Provisions for the Upper Delaware in the context of its designation as a Scenic and Recreational River identified the Commission as a "full partner in the cooperative intergovernmental planning effort".[1392] According to representatives of the Upper Delaware Council, information flows back and forth between the Council and the Delaware Commission continuously, e.g. through e-mail.[1393] Moreover, a representative from the DRBC attends the full Upper Delaware Council meeting on a quarterly basis, which municipalities, state representatives, a representative from the National Park Service also attend and, in turn, members of the Upper Delaware Council sit on DRBC Committees.[1394] The DRBC has included the provisions related to the designation of the Upper Delaware as Scenic and Recreational River in its Comprehensive Plan.[1395]

1390 Conference of Upper Delaware Townships (n 1369) 37.

1391 Ibid.

1392 Ibid 34.

1393 E-mail correspondence with resource specialist of the Upper Delaware Council, December 2016.

1394 E-mail correspondence with resource specialist of the Upper Delaware Council, December 2016.

1395 Conference of Upper Delaware Townships (n 1369) 17.

To encourage public participation, the Upper Delaware Scenic and Recreational River legislation has established the Upper Delaware Citizens Advisory Council, which is responsible for enhancing public participation, e.g. by providing a platform for individual concerns to be heard, by providing public forums, by providing advice on the River Management Plan and Land and Water Use Guidelines, and so forth. The legislation specifies the membership of the Advisory Council, i.e. seventeen members, which should include members from each of the directly affected states, members appointed by each governor of a directly affected State and a member appointed by the Secretary of the Interior.[1396] A major enabler in successful public participation in the Upper Delaware relates to education, of municipalities and citizens, through the website, outreach appearances, annual reports and biennial newsletter.[1397]

394. Two other examples are the Coalition for the Delaware River Watershed and the Christina Basin Clean Water Partnership. The Coalition for the Delaware River Watershed was created in 2012 and unites approximately 80 non-governmental organisations, enabling these organisations to have their joint priorities in protecting the basin's water resources. The Coalition for the Delaware River Watershed played an important role in lobbying for the adoption of the Delaware River Basin Conservation Act, which, as explained above, aims to establish a non-regulatory restoration programme for the Delaware basin. The Coalition for the Delaware River Watershed is important to refer to, as its geographical scope covers the entire basin, i.e. all non-governmental organisations in the several sub-basins are eligible to become members of the Coalition. The Christina Basin, which spans the territories of Delaware and Pennsylvania, is under the governance of the Christina Basin Clean Water Partnership, whose Policy Committee includes the departments of environmental protection of the two states involved, the EPA and, importantly, the Delaware River Basin Commission.[1398] It has received a $1 million federal grant from the Environmental Protection Agency as part of its Targeted Watershed Initiative.

3.5.3. Conclusion on sub-basin and basin level interaction

395. The governance at hydrological scale levels at the sub-basin and sub-sub-basin levels in the Delaware can therefore be summarised as a combination of scattered and private initiatives and more prominent partnerships and groupings of the smaller associations. The Delaware River Basin Commission is the coordinative body with authority over the entire basin, and in this sense,

---

[1396] See Section 704(f)(2) of the Wild and Scenic Rivers Act.

[1397] E-mail correspondence with resource specialist of the Upper Delaware Council, December 2016.

[1398] Local coordinators involve e.g. water resources authorities from the respective states.

has a clear basin-wide perspective and ability to act to the benefit of the basin as a whole. The more prominent sub-basin governance structures have formal connections to the Commission, for example through non-voting membership of the DRBC of their board of directors' or committees' meetings. The plans drawn up for these sub-basin scales refer to the DRBC as the competent basin-wide authority and to measures taken under the auspices of the Commission. The work of the regional sub-basin commissions and the DRBC is complementary, as the DRBC is a regulatory authority that operates throughout the entire basin and the sub-basin commissions do not have regulatory authority and focus more strongly on bottom-up processes. The DRBC does not actively intervene in sub-basin coordination initiatives *per se*, although Article 3(8) of the Compact applies, i.e. the necessity for all projects that may substantially influence water resources in the basin to be submitted for review and approval by the Commission.

4. CONCLUSION: EVALUATION OF THE DELAWARE RIVER BASIN MECHANISM

4.1. EQUITABLE AND REASONABLE UTILISATION AND NO-HARM

396. In the United States, each state has the right to an equal share of waters shared with other states.[1399] Water allocation at the basin level and the use of shared waters are prominent features of the Delaware Compact, and in fact, the main reasons for launching coordinative efforts in the first place.[1400] The Compact does not include a specific allocation method but empowers the Commission to allocate the waters among the parties and subsequently impose conditions, obligations and release requirements on the basis of the equitable apportionment doctrine.[1401] The doctrine of equitable apportionment entails the right of all riparian states sharing a watercourse to fairly share the benefits of the use of the watercourse.[1402] Specifically, equitable apportionment prevents an upstream State from using more than its share of the waters to the detriment of the downstream

1399 Lauren D Bernadett, 'Equitable Apportionment in the Supreme Court: An Overview of the Doctrine and the Factors Considered by the Supreme Court in Light of Florida v. Georgia' (2014) 29 Environmental Law and Litigation 511, 513.

1400 Emily Jeffers, 'Creating Flexibility in Interstate Compacts' (2009) 36 Ecology Law Quarterly 209.

1401 See for a comparison of allocation formula in interstate Compacts in the US: George William Sherk, 'Compact Review Summaries' (2005). Article 3.3 Delaware Compact. The Commission does need to take into account the Supreme Court appropriation promulgated in the 1954 Judgment *New Jersey v New York*.

1402 Robert Haskell Abrams, 'Interstate Water Allocation: A Contemporary Primer for Eastern States' (2002) 25 UALR L. Rev. 155. See e.g. *Nebraska v Wyoming*, 325 US 589 (1945) for an application of equitable apportionment.

states. This doctrine has been applied numerous times by the Supreme Court and is also reflected in the *New Jersey v New York* decree. The Commission must not deviate from the allocation applied in this judgment, bar the unanimous consent of the parties in certain situations.[1403] The Compact and Commission ensure that adequate amounts of freshwater flows down the river so that the Estuary's salt water is not advanced upstream and influences drinking water and industrial intakes along the river.[1404]

397. There is a clear link between international water-law principles and United States water law: the former is derived from the latter.[1405] Whereas equitable apportionment is the guiding principle in US water-sharing issues, international water law has coined the term equitable utilisation, which has a broader scope and does not put as much weight on the factor of "prior use".[1406] In line with the UN Watercourses Convention, which lists several factors that need to be taken into account in determining equitable and reasonable utilisation of a shared watercourse, certain factors have been applied in the Supreme Court's equitable apportionment cases. However, the Court has also acknowledged that the doctrine is a flexible one, and that endless discussions over certain formulae to apply should be avoided.[1407] Factors that have been applied by the Supreme Court include a comparison of the harms and benefits of the states in question, the efficiency of water use, the protection of economies, the size of the respective state parts of the basin, and the available water supplies.[1408]

In applying equitable apportionment, the "reasonable beneficial use" principle should be upheld. In US riparian law, the meaning of reasonable utilisation is similar to that in the context of international water law.[1409]

As mentioned, the applicable water doctrine in place also has an impact on the equitable apportionment, i.e. prior appropriation in the West and riparianism in the East, although the Court can deviate from the applicable state-law doctrine in

---

[1403] Bar certain exceptions, i.e. through unanimous consent of the parties and in case of droughts/ emergencies. There is also the Delaware River Master, who has been appointed by the Supreme Court to administer the provisions of the Court's Decree. The River Master works under the supervision of the United States Geological Survey. States are represented in the River Master's Advisory Committee.

[1404] Delaware River Basin Commission Press Release, 23 November 2016, "DRBC Approves Drought Management Special Permit – Basin Placed in "Drought Watch" Stage Effective Immediately".

[1405] A Dan Tarlock and Patricia Wouters, 'Are Shared Benefits of International Waters an Equitable Apportionment' (2007) 18 Colorado Journal of International Environmental Law and Policy 523, 525.

[1406] Ibid 526.

[1407] *New Jersey v New York*, 283 US 336, 343 (1931).

[1408] Douglas L Grant, 'Interstate Allocation of Rivers Before the United States Supreme Court: The Apalachicola-Chattahoochee-Flint River System' (2004) 21 Ga.St.UL Rev 401, 413.

[1409] Margaret J Vick, 'The Law of International Waters: Reasonable Utilisation' (2014) 12 Chicago-Kent Journal of International and Comparative Law 141.

the interstate water dispute case.[1410] The regulated riparianism regime applies in the Delaware basin, which is different from traditional riparianism in the sense that water (re-)allocation forms part of a collective decision-making process through agencies based on the reasonable character of use.[1411] In a regulated riparianism model, a permit system controls substantial withdrawals of surface and groundwater, where data collection on the use of water is important and environmental concerns are addressed.[1412]

As part of its allocation powers, the Commission has broad powers to regulate withdrawals and diversions from surface waters and groundwaters of the basin and may designate protected areas following a public hearing.[1413] For example, the Commission may designate protected areas where there is a threat of water shortage. In these protected areas, entities may not divert or withdraw water in excess of the quantities stipulated by the Commission, except when they have been granted a permit by the Commission or pursuant to the applicable laws in the existing states.[1414] This means that the permit issuing authority is left with the states, if they have a permit system for water use in place.[1415] In general, when circumstances, e.g. a drought, might cause a shortage of available water supply, the Commission may declare a state of water-supply emergency, following an obligatory public hearing. During this state of emergency, no withdrawals or diversions may take place beyond the quantities allowed by the Commission.[1416] In this regard, a priority of use applies, where first priority must be given to uses which sustain human life, health, and safety, and second priority to water needed to sustain livestock. The Delaware River Basin Water Code further defines the details of the scheduled reductions in times of drought in terms of maximum allowable diversions and minimum compensation releases.[1417]

398. The initial reason for negotiating a joint governance scheme as extensive as the Delaware Compact related to the fact that states wished to avoid costly and lengthy litigation with regard to questions such as the use of water and the protection of the whole basin. The no-harm rule in the Delaware basin mechanism is reflected in the fact that the states have transferred parts of their sovereignty by

---

1410 This has been explained in this chapter. Lauren D Bernadett, 'Equitable Apportionment in the Supreme Court: An Overview of the Doctrine and the Factors Considered by the Supreme Court in Light of Florida v. Georgia' (2014) 29 Environmental Law and Litigation 511, 513.

1411 Joseph W Dellapenna, 'The Evolution of Riparianism in the United States' (2011) 95 Marq. L. Rev 53, 55.

1412 Dan Tarlock, 'Reconnecting Property Rights to Watersheds' (2000) 25 William & Mary Environmental Law and Policy Review 69, 90.

1413 Article 10 Delaware Compact.

1414 Article 10(2) and 10(3) Delaware Compact.

1415 Dellapenna (n 1262) 846.

1416 Article 10(4) Delaware Compact.

1417 These were issued through regulations, which have been consolidated into the Code. Section 2(5)(3) of the Delaware River Basin Water Code.

giving the Commission a broad mandate to protect the inter-jurisdictional waters of the basin. For example, the DRBC has a legal mandate to "Do No Harm" to Special Protection Waters of the basin that have been designated as wild and scenic under federal law, in which it must safeguard good water quality. The preamble to the Compact explicitly states that the comprehensive plan administered by an agency with a basin-wide overview provides the most effective disaster (flood) risk reduction, conservation, watershed management, and so forth. Because the basin is viewed as a whole, the question of "no harm" relates more to the consideration of harm to the basin than harm caused by one of the jurisdictions in the basin to another. The project-review authority of the Delaware Commission on the basis of Article 3(8) is relevant in this regard.

4.2. INTEGRATED RIVER BASIN MANAGEMENT AND NESTED GOVERNANCE

- *General: A model Compact with certain drawbacks*

399. In literature, the Delaware River Basin Commission is deemed to have been largely effective in attaining the goals set in the Compact in terms of river basin management.[1418] Indeed, the Delaware River Basin Compact is often referred to as a model Compact, on which other Compacts have effectively been based.[1419] In 1981, the US Government Accountability Office (GAO) evaluated the effectiveness of the Delaware River Basin Compact in a report to Congress. The GAO referred to successful management practices by the Commission, namely in managing a regional drought and the establishment of basin-wide planning policies. The report concluded, among other things: "GAO believes that the commissions[1420] are worthwhile and achieve results – such as managing a basin wide drought – attainable only by joint cooperation and action. Their progress has been slow in some areas, but they have been dealing with complex and politically sensitive issues. They can continue to make positive contributions if all members give them

1418 Ibid 144. Examples are: settling of disputes over allocation issues, interstate coordination, etc.

1419 The Susquehanna River Basin Compact is based on the Delaware River Basin Compact. For the reference to the Delaware River Basin Compact as a best practice, see, amongst others, Lynn A Mandarano and Robert J Mason, 'Adaptive Management and Governance of Delaware River Water Resources' (2013) 15 Water Policy 364; Mandarano, Featherstone and Paulsen (n 1121); Bruce Hooper, 'River Basin Organization Performance Indicators: Application to the Delaware River Basin Commission: Supplementary File' (2010) 12 1; Gerald Kauffman (n 1119); Gerald J Kauffman and others, 'Water Quality Trends in the Delaware River Basin (USA) from 1980 to 2005.' (2011) 177 Environmental monitoring and assessment 193; Emily Jeffers, 'Creating Flexibility in Interstate Compacts' (2009) 36 Ecology Law Quarterly 209.

1420 The GAO also evaluated the Susquehanna River Basin Compact, which is a Compact based on the Delaware River Basin Compact.

adequate encouragement and support".[1421] Indeed, rightly so, the GAO stresses that the effectiveness of an inter-jurisdictional joint body is dependent on the commitment of all parties involved to fully participate in the mechanism and support it, "even at the cost of occasionally sacrificing their prerogatives".[1422]

The main advantage of the Commission is that opportunities that arise across the boundaries of the States in question can be seized on a systematic and regular basis. The Commission has certainly been faced with difficulties and challenges along the way. Indeed, as explained in the historical overview of the Delaware basin governance, the process of achieving this type of broadly empowered joint commission such as the Delaware Commission, has been wrought with hiccups such as court proceedings, failed negotiations, unsuccessful agreements, and so forth. Over the years, since the creation of the Compact and Commission in 1960, the Commission has realised impressive basin-wide accomplishments and successes. For example, water quality has improved tremendously under its governance – the Delaware basin was once referred to as one of the most polluted areas of the US, and is now seen as a best practice for integrated river basin management in the country.

– *Difficulties in federal representation and state involvement*

400. Whereas Parties to the Compact have, in general, been supportive of what the Commission embodies and its actions in achieving basin-wide planning, New York State, which is the upstream State in the Delaware Basin, has, at times, revealed itself to be rather sceptic of the efforts of the Commission, taking the stance that the Commission duplicates efforts within the competence of the State and has been hesitant to relinquish state sovereignty with regard to water management.[1423] However, all states have relied on the Commission as coordinator and mediator in the face of emergency situations, such as impending floods or droughts.[1424] Related to this issue of signatory party support are the two main bottlenecks and pressing concerns the Delaware River Basin Compact and its Commission struggle with: the difficulties with regard to the financial contributions both by the states and the federal level, and the federal-state synergies over the years.[1425]

One of the difficulties of river basin management in the US relates to the highly fragmented character of the constellation of rules and actors applicable in

1421 Comptroller General of the United States, 'Federal-Interstate Compact Commissions: Useful Mechanisms for Planning and Managing River Basin Operations' (1981).

1422 Comptroller General of the United States, 'Federal-Interstate Compact Commissions: Useful Mechanisms for Planning and Managing River Basin Operations' (1981) ii.

1423 In the run-up to the GAO report, New York paid less than the financial support agreed on the basis of the Compact. Comptroller General of the United States, 'Federal-Interstate Compact Commissions: Useful Mechanisms for Planning and Managing River Basin Operations' (1981) 17.

1424 See below.

1425 See, e.g., Kauffman (n 1119) 88.

this context, both from the horizontal perspective – the variety of agencies and authorities competent regarding the various aspects of water resources management – and the vertical perspective – the division of competences and power dynamics between the states and the federal level.[1426] The intention of the 'founding fathers' of the Compact was to have (i) one representative from the federal level[1427], (ii) representing – with authority – all the relevant federal agencies. However, the latter part of this intention did not materialise. In theory, the Delaware Compact has been applauded for having one federal authority representing the relevant stakeholders of the federal government. However, in practice, this attempt to centralise the interests of the federal government has been viewed as rather unsuccessful. This is mainly due to the fact that the federal representative has not managed to speak or decide authoritatively on behalf of the federal level.[1428]

Research evaluating the Delaware River Basin Compact in the 1970s revealed that the role of the federal government in the Commission could be described as reluctant, and mainly there to point out the different views of the various relevant federal agencies vis-à-vis the actions of the Commission.[1429] The involvement of the federal government further diminished with the cut in funding in 1997, as explained above.

– *Volatility in financial contributions and participation*

401. In general, one of the most pressing concerns in the functioning of the DRBC, relates to the decrease in government funding, both from the federal and the state levels. As explained in Section 3.4.3, federal and state contributions have been diminishing over the past years. The withholding or decreasing of contributions by one of the parties has a domino effect on the contributions by the other parties. The commitment of the governments also depends on the political situation in question, for example, in an anti-spending, anti-regulation era within a certain jurisdiction, be it state or federal, an interstate agency protecting shared water resources will gain less appeal for financial support. However, the Commission continues to have sufficient financial resources through water-supply charges and other sources of funding. The Commission has thus found a way to ensure its continued viability irrespective of the financial support of the Parties. Linked to the issue of funding, is the fact that, as opposed to the earlier years following the adoption of the Compact, the Governors themselves no longer attend the Commission meetings, but send their political appointees or their deputies.[1430]

1426 Adler and M Straube (n 1118) 1.
1427 Article 2.2 Delaware Compact.
1428 George William Sherk (n 1340) 825.
1429 Jeffrey P Featherstone (n 1159) 281.
1430 Dave Feldman, *Water Policy for Sustainable Development* (The John Hopkins University Press 2007).

– *Integrated river basin management at the hydrological scale?*

402. All in all, the Delaware River Basin Compact can be considered as a good example of integrated river basin management at the hydrological scale and as such shows the traits of integrated water resources management.[1431] The Commission is able to keep a basin-wide perspective, can step up in times of emergency, and can reap mutual benefits from water quantity and quality management. Its legal mandate is clear and transparent and has been defined in a detailed manner in the Delaware River Basin Compact. Agreements between the states under the auspices of the Commission have the force of law in the respective states. The Commission has more than 40 fulltime staff members and independent financial resources through e.g. projects.[1432]

From a vertical integration perspective, in the United States, whereas water quality management is governed by a federal umbrella framework, i.e. the Clean Water Act, through the cooperative federalism approach, water quantity management is governed differently in the different states, because water quantity is a competence of the states. From a horizontal integration perspective, there is fragmentation because of the large number of government agencies responsible for different elements of water management. Moving down from the federate or quasi-federate scale to the river basin scales in question, importantly, the Delaware River Basin in the United States remedies this discrepancy between water quality and water quantity management by integrating the two aspects in the federal-state agreement, thus softening both the vertical and horizontal integration issues.[1433]

Examples of the operationalisation of the integrated approach are plentiful. Reference can be made to the designation of a protected area covering 76 sub-basins in Pennsylvania, mainly aiming at preserving groundwater resources and balance competing uses.[1434] In this regard, an Integrated Resources Planning

---

1431 Brenda Bateman and Racquel Rancier, 'Case Studies in Integrated Water Resources Management: From Local Stewardship to National Vision' (American Water Resources Association 2012) 25 <http://awra.org/committees/AWRA-Case-Studies-IWRM.pdf> accessed 14 July 2017.

1432 Earlier research has also identified several of these elements as good practices in river basin management in Belgium. See Hannelore Mees, Cathy Suykens and Ann Crabbé, 'Evaluating Conditions for Integrated Water Resource Management at Sub-Basin Scale. A Comparison of the Flemish Sub-Basin Boards and Walloon River Contracts' (2017) 27 Environmental Policy and Governance.

1433 E.g. in the context of flood risk management, the Delaware Compact grants explicit authority to the Commission. For example, the Delaware Compact enables the Commission to make a categorisation of lands along the Delaware based on their susceptibility to floods. The Scheldt River studied in the EU, however, does so only to a limited extent. Although the 2002 Scheldt Treaty refers to integrated water resources management, it focuses vaguely on the promotion of water quality management and not on the use of water resources. Moreover, the body governing the basin does not have the competences to achieve integrated water resources management in a coherent manner throughout the basin. See Chapter III of this research.

1434 Bateman and Rancier (n 1431) 27.

has been developed by the Commission, in which the connection between the availability and the quality of water should be considered. Furthermore, the Commission has issued a Water Resources Plan, which explicitly emphasises an integrated approach, in terms of the water supply-water quality link and the surface-groundwater interrelationship.[1435] Key Results Areas have been developed with associated Desired Results, and annual evaluation and progress reports are drawn up.[1436] These Key Results Areas relate to sustainable use and supply, linking land and water resource management and institutional coordination and cooperation. Following the above-mentioned floods of 2004 and 2006, the Interstate Flood Task Force established by the DRBC identified 45 recommendations to realise a more comprehensive flood risk management approach in the basin.[1437] Studies are being conducted by the DRBC to put in place an ecological flow regime in the basin.[1438]

4.3. MONITORING AND LEARNING OPPORTUNITIES

403. Monitoring and the gathering and use of scientific data form a substantial part of the work of the Delaware Commission. Advisory committees support the DRBC regarding a range of issues related to water management, e.g. the regulated flow advisory committee, the water quality advisory committee, and so forth. There is a committee specifically focussing on monitoring activities of several programmes.[1439]

404. Where the main factor of adaptability in the Scheldt lies in the cyclical updates of the RBMPs and FRMPs, adaptability also underpins several aspects of the work of the DRBC. Reference can be made to its key document, the Comprehensive Plan, which plays a major role in the governance of the river basin. It relates to short-term and long-term development and usage of the basin and its water resources, and should be subject to consultation with water users

1435 Which is a continuation of the "flowing towards the future plan" circa 1999. The Delaware River Basin Commission, 'Water Resources Plan for the Delaware River Basin' (2004) <www.nj.gov/drbc/library/documents/BasinPlan_Sept04.pdf> accessed 4 May 2017.

1436 Ibid 12.

1437 Delaware River Basin Commission, 'Statement of Carol Collier before the Subcommittee on Water Resources and Environment Committee on Transportation and Infrastructure United States House of Representatives on Comprehensive Watershed Management and Planning' (2008) <http://nj.gov/drbc/library/documents/crc_testimony062408.pdf> accessed 4 May 2017.

1438 Michele DePhilip and Tara Moberg, 'Ecosystem Flow Recommendations for the Delaware River Basin' (The Nature Conservancy 2013).

1439 Resolution 2014-4 establishing the Monitoring Advisory and Coordination Committee to serve the Commission and the Partnership for the Delaware Estuary <http://nj.gov/drbc/library/documents/Res2014-04_MACCestablishment.pdf> accessed 4 May 2017.

and interested public bodies.[1440] In general, neither federal agencies nor state and local agencies may authorise the construction, acquisition or operation of a project or facility if it has not been included in the Comprehensive Plan.[1441] The Comprehensive Plan has been designed to be continuously updated, and since its initial adoption, it has regularly been amended.[1442] Indeed, Article 13 of the Compact states that the Commission may review and revise this plan from time to time. Furthermore, the Commission drafts a yearly Water Resources Program based on the Comprehensive Plan, which details the needs of the basin in terms of water quantity and quality and the existing and proposed projects in this regard. This document is adopted each year and the annual current expense and capital budget drafted by the Commission should be consistent with this document.

405. The DRBC thus adopts a centralised approach toward the adoption of key plans. This centralised approach to the development of the plan stems from the fact that, at the time that the Compact was concluded, water planning at state levels was still premature.[1443] This top-down approach in planning followed by the Delaware Compact has been criticised for being outdated, as, in terms of planning processes in water bodies, the individual states usually have a better overview of local water needs and projects to be developed in comparison to the centralised competent authority.[1444]

This approach to the development of the Comprehensive Plan, however, has not impeded adaptive management practices with respect to the basin's resources. Indeed, over the past ten years, the Delaware Basin Compact and Commission mechanism has adopted an adaptive management approach, which is considered as a key factor in achieving resilience in governance.[1445] This adaptive management approach is, among other things, illustrated by the adoption of the Flexible Flow Management Plan in the context of flood risk management. In managing the reservoirs, not only water supply, but also other aspects of water management are taken into account, e.g. flood risk management, where decisions are based

1440 It weighs more on policies of the states than the RBMPs/FRMPs and their umbrella plans in the Scheldt district.

1441 Article 11(1) Delaware Compact.

1442 It was codified in 2001 (all actions taken since its initial adoption in 1962). Delaware River Basin Commission, 'Comprehensive Plan' (2001).

1443 Jerome C Muys, George William Sherk and Marilyn CO Leary, 'Utton Transboundary Resources Center Model Interstate Water Compact' 47 *Natural Resources Journal* 17, 86.

1444 Ibid 68.

1445 As discussed in Chapter I. Brian C Chaffin, Hannah Gosnell and Barbara A Cosens, 'A Decade of Adaptive Governance Scholarship: Synthesis and Future Directions' (2014) 19 Ecology and Society; Sally J Priest, Cathy Suykens, Marleen van Rijswick and others, 'The European Union Approach to Flood Risk Management and Improving Societal Resilience: Lessons from the Implementation of the Floods Directive in Six European Countries' (2016) 21 Ecology and Society; Carl Folke and others, 'Adaptive Governance of Social-Ecological Systems' (2005) 30 Annual Review of Environment and Resources 441.

on monitoring and scientific research.[1446] The Flexible Flow Management Plan includes an adaptive release schedule and "consistency with adaptive management principles" is an explicit criteria in the revision of the plan.[1447]

406. In addition to the adaptive character of the various plans and programmes, the adaptability of the DRBC mechanism lies in the fact that the Commission has the tools and flexibility to respond to changing circumstances and needs based on the science and information available.[1448]

407. Studies have shown that the Delaware Compact, due to its adaptive nature, which allows tuning into new and unexpected developments, is an appropriate tool to face, among other things, future climatic pressure.[1449]

4.4. MEANINGFUL PUBLIC PARTICIPATION

408. Public participation requirements are interwoven throughout the Delaware Compact. Vertical and horizontal communication is safeguarded by bi-monthly meetings with the commissioners and information dissemination toward stakeholders such as the citizens and NGOs.[1450] The Compact provides that water users and public bodies should be consulted with regard to the development of the Comprehensive Plan, and that the adoption and any revision of the plan, as

---

1446 Lynn A Mandarano and Robert J Mason, 'Adaptive Management and Governance of Delaware River Water Resources' (2013) 15 Water Policy 364, 370.

1447 "The decree parties and the DRBC will consider the following criteria in reviewing proposed modifications to the FFMP: (i) decree party equity; (ii) net benefits and costs to environmental and economic resources; (iii) source and sustainability of water available to support modification and the environmental or economic resource(s); (iv) habitat types – with naturally occurring habitats receiving consideration over man-made habitats; (v) scientific basis for modification; (vi) impacts to drought management, water supply and flood mitigation, including but not limited to: (1) frequency, duration and seasonal timing of the various levels of drought; and (2) frequency, duration, levels of storage, diversions, releases and flows; (vii) extent to which the diversions and the Montague minimum basic rate of flow provided in the decree are met; (viii) potential impacts to water quality, existing National and State Pollution Discharge Elimination System permits and the assimilative capacity of the Delaware River; (ix) ease and practicability of operation; (x) consistency with adaptive management principles; (xi) applicability and implementation of water conservation practices; and (xii) impacts to salinity."

1448 Delaware River Basin Commission, 'Statement of Carol Collier before the Subcommittee on Water Resources and Environment Committee on Transportation and Infrastructure United States House of Representatives on Comprehensive Watershed Management and Planning' (2008) <http://nj.gov/drbc/library/documents/crc_testimony062408.pdf> accessed 4 May 2017.

1449 Watermark Initiative, 'US Water Stewardship: A Critical Assessment of Interstate Watershed Agreements' (2009) <http://watermarkinitiative.com/wp-content/uploads/2013/10/WMI_Report_09-13.pdf> accessed 10 May 2017.

1450 Gerald Kauffman (n 1119) 84.

well as the water resources plan and annual capital and current expense budgets should be preceded by public hearings.[1451] Moreover, the Commission should publish annual reports, commenting on its programmes, operations and finances and these should be distributed to the Parties and the public.[1452] Advisory committees supporting the work of the DRBC consist of representatives of the public.[1453] All meetings of the DRBC are open to the public, and information with regard to these meetings should be widely distributed to the public, as well as the minutes of these meetings, which are public records.

Every adoption of rules and regulations for the effectuation of Compact provisions should be preceded by public hearings.[1454] The Compact further specifies which initiatives are subject to public hearings. For example, this is the case with respect to the approval of each project related to water supply[1455], an initiative related to pollution control – e.g. when the Commission aims to determine standards of treatment of sewage[1456] – the designation of areas prone to floods[1457], granting contracts with regard to the development of hydroelectric power[1458], the designation of protected areas[1459], and so forth. As mentioned, one of the major responsibilities of the Commission relates to guiding the basin states through times of emergency, especially with respect to drought. Before the Commission can determine an area of shortage to which the emergency regime applies, a public hearing must be held.[1460]

409. In addition to these Compact-specific public-participation requirements, relevant provisions are included in federal and state laws. Public involvement in environmental decision-making has become institutionalised in the United States since the adoption of the National Environmental Policy Act (NEPA) in 1969, which is a procedural statute regulating the environmental impact of certain actions and the associated public-participation requirements.[1461] Federal agencies are obligated to conduct environmental statements for actions significantly affecting the environment and as such, rules have been issued which declare the NEPA applicable to the Delaware River Basin Commission.[1462] This has, however, been the subject of discussion over the years, specifically with

1451 Article 3(2)(a) in conjunction with Article 13.1 Delaware Compact.
1452 Article 14(2) Delaware Compact.
1453 Article 3(10) Delaware Compact.
1454 Article 14(2) Delaware Compact.
1455 Article 4(4) Delaware Compact.
1456 Article 5(2) Delaware Compact.
1457 Article 6(2)(b) Delaware Compact.
1458 Article 9(4) Delaware Compact.
1459 Article 10(2) Delaware Compact.
1460 Article 10(4) Delaware Compact.
1461 42 U.S.C. § 4321 et seq. (1969).
1462 See the Code of Federal Regulations, which refers to the relationship between the DRBC and NEPA. 18 C.F.R. § 404.

regard to the question whether the DRBC is a federal agency within the meaning of NEPA, subject to requirements such as environmental impact statements.

This discussion is beyond the scope of this book, but one specific example will be given to illustrate the difficulties in this regard. The entire non-tidal part of the Delaware Basin has been designated as "Special Protection Waters", in which the DRBC aims to maintain high water quality. However, within these Special Protection Waters, the Marcellus Shale is located, a natural gas bearing formation. The tension between development and environmental protection is thus particularly present.[1463] In 2009, the Commission issued draft regulations prohibiting natural gas extraction within the "Special Protection Waters" of the basin. A broad consultation process was organised and approximately 69,000 public comments were taken into consideration.[1464] Environmental advisory groups requested the DRBC for an extension of the public-participation deadline for regulations on drilling wells that influence the water quality of the upper part of the Delaware River, which it did.[1465] However, criticism resulted from the fact that the Commission had not conducted a cumulative environmental impact analysis of the fracking activities on the water system, and hundreds of letters were submitted to the Commission in this regard.[1466] To date, final regulations with regard to hydrofracking have not yet been issued because it is difficult for the DRBC Commissioners to reach agreement in this regard.[1467] Until such regulations are adopted, a *de facto* moratorium on shale gas drilling is in place in the basin, as the Commission does not process such applications without regulations being in place. Legal proceedings initiated by a corporation wishing to drill a natural gas well on private property and challenging the DRBC's authority to demand prior approval for such activity are currently ongoing.[1468] The question whether an action of the DRBC could be a major federal action requiring an environmental impact assessment remains unanswered.[1469]

1463 See, for a discussion of shale gas and environmental and energy regulation in the EU and US and the DRBC case, Leonie Reins, *Regulating Shale Gas: The Challenge of Coherent Environmental and Energy Regulation* (Edward Elgar Publishing 2017).

1464 These are available on the website of the DRBC: <www.nj.gov/drbc/programs/natural/draftregs-dec2010_comments.html> accessed 6 May 2017.

1465 Environmental Council for Abington Township, Letter to the Delaware River Basin Commission regarding the Natural Gas Development Guidelines <www.state.nj.us/drbc/library/documents/NGC/Officials/PAAbington041511.pdf> accessed 5 May 2017.

1466 See e.g. <www.catskillcitizens.org/learn_one.cfm?t=15&c=14> accessed 5 May 2017. See Complaint filed by the State of New York v the Army Corps of Engineers before the United State District Court, Eastern District of New York, 31 May 2011, <www.eenews.net/assets/2011/06/01/document_gw_01.pdf> accessed 8 May 2017. After environmental impact analysis, New York also issued a ban on fracking.

1467 The website of the DRBC provides updates on the state of play: <www.state.nj.us/drbc/programs/natural/> accessed 8 May 2017.

1468 At the time of writing, May 2017.

1469 This also has an impact on voting behaviour by the federal member of the DRBC. A majority of 3 votes is needed to pass a certain DRBC action, and the question whether federal acts such as

4.5. EFFECTIVENESS AND ENFORCEMENT

410. The Compact enables the Commission to enforce the measures and regulations that it has issued.[1470] For these rules to be effective, they should be preceded by a public hearing and filed in accordance with the laws of the Parties.[1471] The Commission may appoint an officer of the Commission as a peace officer to act as investigator or watchman.[1472] The Compact provides for a cascade sanctions mechanism in case of violation of the Compact provisions or rules or regulations issued by the DRBC. The violation should be punished on the basis of the implementing legislation issued by the states. If such legislation does not provide for a penalty, the violator is liable to a penalty ranging from $50 to $1,000 for each offence. The DRBC does not have the authority to impose fines and needs to go through the courts of competent jurisdiction.[1473] However, as most entities prefer settlements over court rulings confirming their violations, e.g. with regard to permit conditions, these issues are often settled under the auspices of the Commission.[1474] Generally, these settlements relate to the smaller violations whereas the major violations such as spills resulting in substantial water quality violation are handled through state proceedings.[1475]

The Compact also provides for enforcement through Article 3(8) of the Compact, which stipulates that projects that have a substantial impact on water resources in the basin should be submitted to the Commission for a check of compliance with the Comprehensive Plan. This provision enables proactive enforcement of measures and initiatives from the basin-wide perspective.

411. Because the Delaware Commission has been created in order to avoid another bout of Supreme Court litigation between the states sharing the basin, such as the *New Jersey v New York* case, the states make the effort to resolve conflicts through the neutral DRBC. The Delaware Compact itself does not include an extensive dispute settlement mechanism. However, states have relied on the Commission as coordinator and mediator in emergency situations, such as impending floods or droughts.[1476]

NEPA apply has a different outcome depending on which members have voted. Interview with senior staff member Delaware River Basin Commission, 15 December 2016.

1470 Article 14(2) Delaware Compact.

1471 Article 14(2)(a) Delaware Compact.

1472 Article 14(2)(b) Delaware Compact.

1473 Article 14(7) Delaware Compact.

1474 Telephone interview with senior staff member Delaware River Basin Commission, 15 December 2016.

1475 Ibid.

1476 Cf. *infra*.

CHAPTER V
RECOMMENDATIONS

1. INTRODUCTION

412. The bottleneck analysis presented in Chapter II and the evaluation of two specific river basins in Chapters III and IV allow a more normative approach in the present chapter.

Chapter I has explained that the primary focus of this book is on the European legal framework for the governance of International River Basin Districts, from the viewpoint of water quantity management. Based on the data presented in Chapters I through to IV, stemming from the review of international, European and national legislation, jurisprudence and literature, Critical Success Factors (CSF) for transboundary river basin management have be identified. These constitute the groundwork for the specific recommendations which are geared to the EU legal framework in its capacity as facilitator of IRBD management and also to IRBDs as institutional mechanisms. The CSFs will be explained in section 1.1. The next step is to apply the lessons learned to International River Basin Districts, and the underlying EU legal framework. The recommendations associated with this application can be divided into two broad categories: addressing the EU primary law approach to water management and enhancing the cooperation in International River Basin Districts. The latter aspect can be further divided into the EU secondary law and the operationalisation at IRBD levels.

The goal of this book is to improve river basin management in International River Basin Districts, using water quantity as an example. To this extent, this chapter will address the primary EU-law dichotomy between water quantity and water quality, because this is a prerequisite for achieving integrated river basin management. In contrast, the chapter will not present an exhaustive set of recommendations for amendments to water quantity provisions in EU secondary law, because this goes beyond the scope of this book. The recommendations associated with EU secondary law focus on the manners in which the EU legal framework can further facilitate cooperation in and management of International River Basin Districts.

2. CRITICAL SUCCESS FACTORS

2.1. EVALUATIVE COMPARISON

413. Before explaining the Critical Success Factors, it is relevant to recap the evaluation exercise and provide a schematic overview of the river basin management regimes analysed in this book. This schematic overview brings together Chapters I to IV as it presents the application of the resilience benchmarks, which have been explained in Chapter I, to the Scheldt and Delaware regimes, analysed in Chapters III and IV, and, with respect to the Scheldt, the underlying EU legal framework discussed in Chapter II. Moreover, the "five pillars" presented in Chapter II are interwoven throughout the scheme.[1477]

414. The box below presents the questions that are addressed in the table.[1478]

– *Integrated river basin management at the hydrological scale + nested governance*
1. Scope (Pillar I):
1a: What is the territorial scope of the regime?
1b: To what extent is groundwater integrated into the regime?
1c: To what extent have the qualitative and quantitative aspects been integrated?
2. Type of agreement: What type of agreement? (e.g. agreement with basic provisions, and large discretionary powers for the individual States, or basin-wide, centralised authority mechanism) (all pillars)
3. How have cooperation requirements been implemented stemming based on US federal / EU law? (Pillar IV)
4. Relationship between different levels: sub-sub-basin, sub-basin, basin – and e.g. municipal, provincial, regional, national? (Pillar IV)
5. Type of entity (Pillar IV):
5a. Legal personality?
5b. Responsibilities of the body?
5c. (Un)supportive legal framework?
5d. (In)adequate financial and human resources?
– *ERU and No Harm (Pillar II):*
6a. Does the agreement follow the principle of equitable and reasonable use?
6b. How is the no-harm rule translated in the basin?
– *Meaningful public participation (Pillar IV):*
7. Is there a mechanism for basin-level public participation?
– *Monitoring and learning (Pillar IV):*
8a. How are monitoring results evaluated and taken into account?
8b. How are adaptability and climate-change factors incorporated into the regime?
– *Effectiveness, enforcement and dispute resolution (Pillar V):*
9a. How is compliance monitored?
9b. By whom?
9c. What are the consequences in case of non-compliance?
9d. What is the procedure in case of disputes?

1477 These are: Pillar I: scope; Pillar II: substantive provisions; Pillar III: procedural provisions; Pillar IV: institutional mechanisms; and Pillar V: dispute resolution.

1478 These are based on the analysis conducted in Chapters III and IV as well as the literature review in Chapter I, such as the "ten building blocks" method developed by Van Rijswick and others. Marleen van Rijswick and others, 'Ten Building Blocks for Sustainable Water Governance: An Integrated Method to Assess the Governance of Water' (2014) 39 Water International 725.

415. The questions listed above are answered in the table below.

Table 11. Summary of Findings from the Evaluative Comparison of the Scheldt and Delaware River Basin Mechanisms

Resilience: Evaluative Comparison			
Factors	Q	Scheldt: 2002 Scheldt Treaty & International Scheldt Commission	Delaware 1961 Compact & Delaware River Basin Commission
Integrated RBM at the hydrological scale	1a	Area of land and sea, designated in conformity with the WFD, comprising the Scheldt basin, associated basins and the groundwaters and coastal waters.	The area of drainage into the Delaware River and its tributaries including the Delaware Bay
	1b	Reference to sustainable development & implementation WFD – water use not addressed	Integration of quality, quantity & risk. Indicators of IWRM include: water resources plan, flexible flow management plan
	1c	Both groundwater & surface water	Both groundwater & surface water
	2	Best efforts, wide discretionary powers for the States	Basin-wide, centralised regulation
	3	References to cooperation scattered in national frameworks Notification & information exchange requirements in place	Cooperation through DRBC & Compact implemented in state legislation Notification through DRBC
	4	Limited links with sub-sub-basin scales through national levels	Vertical integration through DRBC – links with sub-sub-sub-basin levels present but not priority for DRBC
	5a	Legal personality for competences assigned by the Treaty	Legal personality
	5b	Advice, information exchange and coordination "Competent authorities" within the meaning of WFD and FD are the national authorities (not ISC) Unanimity voting (decisions are non-binding)	Regulatory authority for all aspects of water management Competent authority is the DRBC Majority voting, with exceptions (decisions are binding)
	5c	Treaty lists intentions of Parties, legal mandate ISC is limited	Broad empowerment of Commission. In practice: works best with consensus
	5d	Financial resources provided by parties based on allocation key No other resources Limited staff (<5) ISC Secretariat	Financial resources provided by parties based on allocation key + independent resources through projects 40+ full-time staff members in DRBC Parties' contributions not always reliable

Resilience: Evaluative Comparison			
ERU and no-harm principle	6a& 6b	ERU and no-harm principle apply as Scheldt Treaty implements UNECE Convention + parties are Parties to UN Watercourses Convention Both not explicitly provided for; no agreement on water use FD "Solidarity clause" implemented in national frameworks	Equitable apportionment applies + regulated riparianism Agreements on water use through Supreme Court Decree & Compact No-harm principle reflected in broad mandate Commission
Meaningful public participation	7	Public participation based on Article 14 WFD and EU Aarhus implementation No basin-wide public participation through ISC (limited to publication of roof report on website)	Basin-wide public participation through DRBC: for plans & programmes + public hearings precede adoption of rules & regulations Discussion on application of NEPA
Monitoring and learning	8a & 8b	Monitoring requirements included in WFD Monitoring at basin level (water quality) & at national levels – sound scientific basis Working groups at ISC level Cyclical learning (6 years)	Sound scientific basis Monitoring of various aspects of water management Advisory Committees DRBC has the tools to respond to changing circumstances Long-term planning for climate change is lacking.
Effectiveness, enforcement & dispute resolution	9a, b, c &d	At national levels, enforcement through national authorities Provisions Treaty not enforceable as no obligations of result included Disputes are settled through negotiation	Regulatory authority + follow-up of compliance Proactive enforcement through Article 3(8) Compact No explicit dispute resolution mechanism

2.2. CRITICAL SUCCESS FACTORS

416. Based on the data presented in Chapters II, III and IV, and the review of literature presented throughout this book, Critical Success Factors (CSF) can be identified that support the implementation of integrated river basin management in transboundary waters.

417. It is noteworthy that these Critical Success Factors should not be seen as self-standing, conclusive and exhaustive evaluation criteria for integrated river basin management. These CSFs have not been drawn from a comprehensive study of *all* IRBDs in Europe. The CSFs are mainly drawn from the Scheldt and Delaware studies. However, several of these conclusions included in the Critical Success Factors coincide with and confirm what has been stated in the relevant literature with regard to transboundary water management over the years. The formulation of the Critical Success Factors has been enabled through the resilience benchmarks adopted in this book. The Critical Success Factors are complementary to the resilience benchmarks explained in Chapter I and applied in Chapters III and IV because the Factors are more specific and primarily geared to the institutional mechanisms for integrated river basin management. The Critical Success Factors thus mainly relate to the first evaluation benchmark "integrated river basin management at the hydrological scale and nested governance" and represent a sub-categorisation of this benchmark, as explained in Chapter I.[1479] In the discussion of each individual Critical Success Factor, the possible overlap with and / or confirmation of literature discussed in Chapter I will be emphasised. Principles of broader "water governance" such as the OECD's water principles and the "ten building blocks for sustainable water governance" are relevant in this regard and should be taken into account as well.[1480]

Figure 16. Graphic Overview of Critical Success Factors

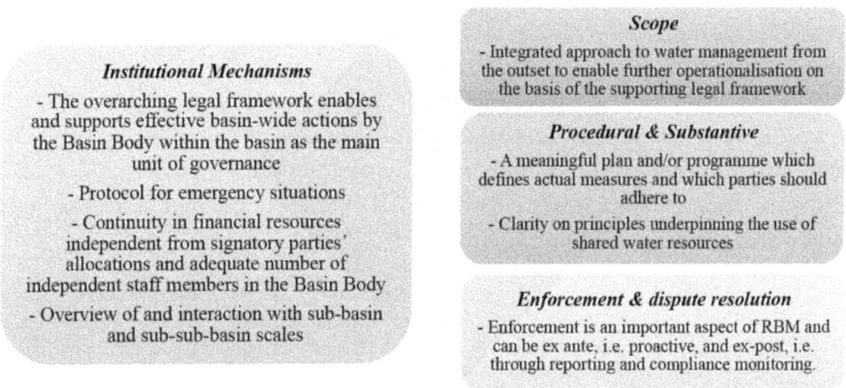

2.2.1. CSF 1: Integrated approach to water management from the outset to enable further operationalisation on the basis of the supporting legal framework

418. The first CSF relates to the scope of the agreement embodying the framework for cooperation between the States and the regions sharing the body of water in question, and underpinning the actions of the joint body.

The concept of integrated water resources management and the importance of addressing quantity and quality in an integrated manner as well as groundwater

[1479] Which is why the figure visualising the resilience benchmarks has highlighted the first benchmark "integrated river basin management at the hydrological scale and nested governance".

[1480] OECD, 'OECD Principles on Water Governance' (OECD 2015) <www.oecd.org/cfe/regional-policy/OECD-Principles-on-Water-Governance-brochure.pdf> accessed 17 July 2017. Marleen van Rijswick and others, 'Ten Building Blocks for Sustainable Water Governance: An Integrated Method to Assess the Governance of Water' (2014) 39 Water International.

and surface water have been established throughout this book.[1481] Integrated water resources management, i.e. addressing the management of water, land and related resources in a coordinated manner, has been widely advocated internationally for several decades.[1482] The two case studies have also confirmed that it is important to also consciously include the risk aspect of water management into the scope. Reciprocal benefits arise from addressing these various aspects associated with water resources in a coordinated manner. Also, the integrated management of groundwater resources and surface water resources is of paramount importance. For example, if administrative requirements for the abstraction of surface water are more strict and cumbersome than those applicable to the abstraction of groundwater, this can result in negative effects, i.e. over-abstraction of groundwater so as to avoid the administrative requirements related to the abstraction of surface water.[1483] These recommendations have long been established by the relevant literature, and confirmed by the two case studies presented in this research. This Critical Success Factor emphasises the importance of including the integrated approach *in the base agreement* governing the transboundary water body to enable integration in the *operationalisations.*

419. The overwhelming level of fragmentation in water management, both from the vertical and the horizontal point of view, has been a leading theme in this book.[1484] This is the case both in the EU and the US. Whereas the EU-based river mechanism examined in Chapter III, however, has not remedied this

---

1481 Amongst others: Ann Crabbé, 'Integraal Waterbeleid in Vlaanderen: van Fluïde naar Solide' (PhD Thesis, Antwerp University 2008); Barbara Cosens, 'Transboundary River Governance in the Face of Uncertainty: Resilience Theory and the Columbia River Treaty' (2010) 30 Journal of Land Resources and Environmental Law 229; Jonas Ebbesson, 'The Rule of Law in Governance of Complex Socio-Ecological Changes' (2010) 20 Global Environmental Change 414; Olivia O Green and others, 'EU Water Governance: Striking the Right Balance between Regulatory Flexibility and Enforcement?' (2013a) 18 Ecology and Society; Jonas Ebbesson and Ellen Hey, 'Introduction: Where in Law Is Social-Ecological Resilience?' (2013) 18 Ecology and Society 1; Marleen van Rijswick and others, 'Ten Building Blocks for Sustainable Water Governance: An Integrated Method to Assess the Governance of Water' (2014) 39 Water International 725; Cathy Suykens, 'EU Water Quantity Management in International River Basin: Crystal Clear?' (2015) European Energy and Environmental Law Review 134; Sally J Priest and others, 'The European Union Approach to Flood Risk Management and Improving Societal Resilience: Lessons from the Implementation of the Floods Directive in Six European Countries' (2017) 21 Ecology and Society.

1482 Hannelore Mees, Cathy Suykens and Ann Crabbé, 'Evaluating Conditions for Integrated Water Resource Management at Sub-Basin Scale. A Comparison of the Flemish Sub-Basin Boards and Walloon River Contracts' (2017) 27 Environmental Policy and Governance 42. Jerome C Muys, George William Sherk and Marilyn CO Leary, 'Utton Transboundary Resources Center Model Interstate Water Compact' (2006).

1483 Kerstin Mechlem, 'Groundwater Governance: A Global Framework for Country Action – Legal and Institutional Frameworks' (2012) <www.groundwatergovernance.org/fileadmin/user_upload/groundwatergovernance/docs/Thematic_papers/GWG_Thematic_Paper_6.pdf>.

1484 Marleen Van Rijswick and Herman Havekes, *European and Dutch Water Law* (Europa Law Publishing 2012); Monika Ambrus, Herman Kasper Gilissen and Jasper JH Van Kempen, 'Public Values in Water Law: A Case of Substantive Fragmentation?' (2014) 10 Utrecht Law Review 8.

fragmentation at basin level, the Delaware mechanism has.[1485] The Compact addresses water quality, water supply, sharing of the resources in terms of use, emergency situations such as droughts, flood risks, and so forth. The fact that these are integrated at the outset, i.e. in the basic legal framework, is a necessary first step which allows taking into account the relevant aspects at the operational levels. Indeed, the Delaware Commissioners, i.e. the governors of Delaware, Pennsylvania, New York, New Jersey and the federal representative, adopted the Water Resources Plan for the basin in 2004, which explicitly addresses integrated water resources management.[1486] This plan was adopted to serve as a guidance policy document for the next thirty years, based on the principles of integrated management, i.e. acknowledging that water supply and water quality should be considered jointly instead of separately and that groundwater and surface water are interrelated.[1487] The plan connects the associated "Key Result Areas" in the context of water management[1488] in order to promote a systematically integrated approach by the stakeholders in the Delaware basin and Commission.[1489] Furthermore, the Flexible Flow Management Plan allows the DRBC to reap mutual benefits from the management of reservoirs both in terms of water quantity and water quality.

420. The analysis of the legal frameworks and governance mechanisms in the countries and regions sharing the Scheldt District has shown that alignment of different strategies in water management, specifically in the context of flood risk management, is crucial to the success of these frameworks. For example, the recovery strategy, notably insurance mechanisms, has an important role to play with respect to adaptive building and preventing constructions in flood-prone areas, which in turn are associated with the risk prevention and mitigation strategies.[1490] Another example is the Flemish water test, which allows the assessment of the impact of a spatial-planning decision, be it a building permit or a spatial plan, on the water system. Importantly, this impact relates to both the quality of the water resources and the quantity and safety aspects.[1491] This integrated approach is enabled by the fact that these various aspects of water

---

1485 This book does not evaluate the governance mechanism of the Scheldt Estuary, which should be seen as separate from the International Scheldt Treaty and Commission.

1486 The Delaware River Basin Commission, 'Water Resources Plan for the Delaware River Basin' (2004) <www.nj.gov/drbc/library/documents/BasinPlan_Sept04.pdf>.

1487 Ibid 11.

1488 These are: sustainable use and supply, waterway corridor management, linking land and water resource management, education and involvement for stewardship and institutional coordination and cooperation.

1489 Guy Pegram and others, 'River Basin Planning: Principles, Procedures and Approaches for Strategic Basin Planning' (2013) UNESCO <http://unesdoc.unesco.org/images/0022/002208/220875e.pdf> accessed 17 July 2017.

1490 Cathy Suykens and others, 'Dealing with Flood Damages: Will Prevention, Mitigation and Ex-Post Compensation Provide for a Resilient Triangle?' (2016) 21 Ecology and Society.

1491 See Article 8 of the Decree Integrated Water Policy.

management are included in a single legal framework, the Decree Integrated Water Policy. The outcome of the water test should be coherent with the goals of the Decree, which, in turn, automatically promotes a comprehensive protection of the water system in all its elements. The basic framework, the DIWP, thus enables operationalisation of the integrated approach. Whereas the Scheldt countries show traits of integration, this is still lacking at the level of the IRBD itself. This CSF therefore mainly relates to national policies, in contrast to the Delaware, where the integration is carried through to the basin level.

This means that adopting an integrated approach in the basic cooperation agreement is crucial in order to achieve integration in subsequent plans, programmes and instruments.

2.2.2. *CSF 2: The overarching legal framework enables and supports effective basin-wide actions by the Basin Body within the basin as the main unit of governance*

421. There are several components to this CSF: (i) the basin should be seen as the main unit of governance, and (ii) the legal framework should be sufficiently supportive and clear.[1492] One of the reasons that the Delaware Commission has been able to make progress in the basin, resulting in basin-wide successes such as managing droughts, water quality, and protection programmes, relates to the fact that the Delaware basin as a whole is the main unit of governance. This is reflected in the Compact by requiring the submission of projects to the DRBC in order to check compliance and coherence with the Comprehensive Plan and the basin-wide overview. State interests are represented through representation in the Commission and the decision-making procedures, but the DRBC is the "go-to" competent authority with regard to issues affecting the quantity or quality of the basin. This has been illustrated by the fact that citizens' organisations have directed their complaints to the DRBC, as opposed to the individual states. As explained in Chapter II of this book, the main unit of governance in the context of the WFD and FD is the national River Basin District. On the basis of the EU legal framework, the IRBD level is the sum of measures taken at RBD levels, with a loose coordinative framework at the basin level, which is reflected in the Scheldt river basin mechanism.

422. Furthermore, the legal mandate of the entity should be sufficiently clear and robust, and rooted in enforceable legislation, whilst ensuring its accountability and striving for broad support.[1493] Creating a broad material

[1492] The importance of clarity in the mandate of the joint entity has been referred to in literature (*infra*), see, e.g., Bruce Hooper, 'River Basin Organization Performance Indicators: Application to the Delaware River Basin Commission: Supplementary File' (2010) 12 1, 10.

[1493] The importance of clarity in the legal mandate has already been stressed by numerous authors, e.g. Ellen Hey and Marleen van Rijswick, 'Transnational Water Management' in Oswald Jansen

scope of competences with regard to all aspects of water management to the joint entity is linked to the first Critical Success Factor[1494] and supports the development of integrated water resources management for the whole basin.[1495] Associated with the clarity of the mandate of the entity, is the importance of public participation processes at the level of the basin.[1496] Meaningful public participation has been identified and applied as a "resilience benchmark" in this book. Hearings of the Delaware River Basin Commission are open to the public, and reports of meetings and news updates are regularly posted on its website.

In terms of decision-making procedures, the Delaware Compact provides both for majority voting and for unanimity voting. Whereas the fact that the DRBC has majority voting on the bulk of issues helps the Commission to move things along, decisions and regulations which have the broad support of the Commissioners have turned out to be the most successful and the Commission continuously seeks such support.[1497] Binding decision-making power therefore does not necessarily equal effective management. Moreover, the effectiveness of regulations is contingent on implementation by the various parties involved.[1498] In the EU, as has been established in Chapter II, the unanimity procedure included in the EU Treaty with respect to quantitative management of water resources has hampered an integrated approach to water management.[1499] The Scheldt Treaty is geared more strongly to listing obligations of best efforts to be strived for by the Parties, than to providing an operating basis and mandate for the ISC. This is translated in the functioning of the Commission as a good-will discussion and data-exchange platform.

---

and Bettina Schöndorf-Haubold (eds), *The European Composite Administration* (Intersentia 2010); Bruce Hooper, 'River Basin Organization Performance Indicators: Application to the Delaware River Basin Commission' (2010) 12 Water Policy 461. The importance of clear allocation of roles and responsibilities is also emphasised in the OECD Water Principles and by Van Rijswick and others in Ten Building Blocks: OECD, 'OECD Principles on Water Governance' (OECD 2015). van Rijswick and others (n 1109).

1494 CSF 1: "integrated approach to water management from the outset to enable further operationalisation on the basis of the supporting legal framework".

1495 Muys, Sherk and Leary. Bruce Hooper, Integrated River Basin Governance: Learning from International Experience (IWA Publishing 2005).

1496 Meaningful public participation has also been explained in Chapter I, and applied in Chapters III and IV.

1497 Telephone interview with senior staff member Delaware River Basin Commission, 15 December 2016.

1498 Indeed, the Commission may issue rules and regulations to apply the provisions of the Compact and develop instruments to enforce these rules and regulations. Before these can be applied, a public hearing needs to be organised and the states need to integrate them into their respective legal frameworks. Article 14.2 Delaware Compact.

1499 See Section 3 of this chapter. Cathy Suykens, 'EU Water Quantity Management in International River Basin: Crystal Clear?' [2015] European Energy and Environmental Law Review 134.

2.2.3. *CSF 3: A meaningful basin-wide plan and/or programme which defines actual measures and links with the sub-basin scales*

423. Both in the EU and in the US, there are a number of procedural requirements that the parties should adhere to in the transboundary water bodies. These procedural requirements mainly relate to the development of plans and programmes. Determining the balance between substantive and procedural requirements deserves vigilant attention. For example, when shifting from substantive obligations toward a focus on procedural rules in EU water law, especially with regard to flood risk management, issues such as access to justice of citizens become more apparent.[1500] However, plans and programmes may constitute useful additions to the more static basic legal framework, which in turn further enables adaptive governance. This is important in light of the various uncertainties the basin as a whole is subject to.

424. The Delaware Commission has to fulfil certain procedural obligations. The Commission has three main tasks in this regard: (i) to develop a Comprehensive Plan, (ii) to draft a Water Resources Program, and (iii) to issue an annual current expense and capital budget. The parties sharing the Scheldt district are obligated, on a five-year basis, to issue River Basin Management Plans (RBMPs) and Flood Risk Management Plans (FRMPs) pursuant to the WFD and FD. An important difference between the Scheldt RBMP and FRMP and the Delaware Comprehensive Plan is the level at which these plans are developed. In the Delaware basin, it is the Delaware Commission that is responsible for developing the Comprehensive Plan on the basis of the Compact, which the states should then comply with. In the Scheldt basin, the States are responsible for developing the RBMPs and FRMPs, with a best-effort obligation to coordinate the plans with the aim of adopting one single RBMP and FRMP, but the possibility to develop plans for the part of the IRBD situated on their territories if such efforts fail.[1501] The process in the Delaware basin is top-down, whereas in the Scheldt it is a more of a bottom-up process. In the IRBD Scheldt, each of the States and regions involved have adopted RBMPs and FRMPs for the currently applicable reporting period, i.e. 2016-2021, and have agreed on one RBMP roof report and one FRMP roof report for the whole IRBD. This means that for the implementation of the WFD, six plans have been issued by the parties, i.e. the Belgian federal level, Brussels-Capital Region, the Flemish Region, the Walloon Region, France and the Netherlands and roof reports issued under the auspices of the ISC. The Scheldt roof reports are more of an informative nature, providing an overview of important data for the district, such as population density, quality and abstraction levels of groundwater

---

1500 Sally J Priest and others, 'The European Union Approach to Flood Risk Management and Improving Societal Resilience: Lessons from the Implementation of the Floods Directive in Six European Countries' (2016) 21 Ecology and Society; Van Rijswick and Havekes.

1501 Article 13(2) WFD.

volumes, natural characteristics, and so forth. The plans do not include specific measures to be implemented by the parties.[1502] In contrast, the Comprehensive Plan developed by the Delaware Commission plays a major role in the governance of the river basin and as such is more strongly determined by policies of the states than the RBMPs/FRMPs and their umbrella plans in the Scheldt district. It relates to short-term and long-term development and usage of the basin and its water resources, and should be subject to consultation with water users and interested public bodies. In general, neither federal agencies nor state and local agencies may authorise the construction, acquisition or operation of a project or facility if it has not been included in the Comprehensive Plan.[1503]

425. In terms of the top-down versus bottom-up planning process, in line with the subsidiarity principle in EU law, the top-down approach, where the joint body develops a plan which the parties should then comply with, does not seem feasible in the EU institutional setting. Moreover, this top-down approach in planning followed by the Delaware Compact has been criticised for being outdated, as, in terms of planning processes in water bodies, the individual states usually have a better overview of local water needs and projects to be developed compared to the centralised competent authority.[1504] However, coordination as such is necessary, and plans should either be reviewed by the competent authority in light of coherence and the maximisation of mutual benefits or states should have more incentive to produce one single plan, which should exceed the informative character of the roof reports for the Scheldt district for the period 2016-2021, for example.

426. It is important to provide links between hydrological scales throughout the basin.[1505] The main actors operating at the sub-basin scales in the Delaware basin, such as the Lower Delaware Wild and Scenic Council, the Upper Delaware Council and the Partnership for the Delaware Estuary have a dynamic link with the DRBC. There is vertical cross-fertilisation between these entities and their roles have been clearly defined and are complementary. The plans issued by the entities at the sub-basin scales take into account the DRBC's Comprehensive Plan.

427. In short, the emphasis of this CSF is not on the top-down versus bottom-up development of the plan, but on the consideration that the plan should be meaningful, entailing that (i) the plan should be the plan of reference for the basin, (ii) the plan

---

1502 International Scheldt Commission, 'Umbrella Part of the River Basin Management Plan for the International River Basin District Scheldt 2016-2021' (2014) <www.isc-cie.org/images/Documents/ODB2-PFPG2_RAPPORT_NL-FR_VDEF.pdf>.

1503 Article 11.1 Delaware Compact.

1504 Jerome C Muys, George William Sherk and Marilyn CO Leary, 'Utton Transboundary Resources Center Model Interstate Water Compact' (2006) 68.

1505 Sally J Priest and others, 'The European Union Approach to Flood Risk Management and Improving Societal Resilience: Lessons from the Implementation of the Floods Directive in Six European Countries' (2016) 21 Ecology and Society.

would ideally cover the entire basin or consist of a roof report which is the main document containing basin-wide data, risk assessments and measures and sub-plans for the different sub-basins accessory to the main plan, (iii) the Basin Body plays a key role in the development of and the implementation of the measures contained in the plan, and (iv) there is a link with the plans issued at the sub-basin scales.

2.2.4. CSF 4: Clarity regarding principles underpinning the use of shared water resources and protocol for emergency situations

428. This CSF relates to the benchmark "equitable and reasonable utilisation and no-harm principle", which has been discussed in Chapter I. The WFD and FD do not regulate the use of shared waters. The problems associated with this lack of reference to the use of shared waters have already been identified by Van Kempen with respect to water quality management.[1506] The Scheldt Treaty does not provide for an agreement on such use, nor a reference to the principles which constitute the basis for the water sharing. The Delaware Compact and its subsequent instruments, in particular the Code and the Comprehensive Plan do provide for allocation mechanisms. Moreover, the legal instruments governing the Delaware basin have provided for emergency situations, such as impending shortage of available water supply. Such an emergency situation is declared by the DRBC, following a public hearing, on the basis of which special conditions apply.[1507] In this context, uses are prioritised, where first priority is given to uses sustaining human life, health and safety and the use of water needed to sustain livestock has second priority. The Code details the action plan in terms of implementing scheduled reductions in times of drought in terms of maximum allowable diversions and minimum compensation releases.[1508] The tools available to the Commission in light of emergency situations, and the protocol in place, have been important in its successful management of the river. Indeed, one of the success factors of the Delaware mechanism in practice has been its ability to step up in times of impending calamities. The DRBC has been able to efficiently take the necessary measures to manage a basin-wide drought, which would have had less favourable outcomes had the parties to the Delaware Compact each addressed the issue separately.[1509] The DRBC gained traction and trust because of these "shock" events, which leveraged the ability of this hydrological scale entity to continue making a contribution to river basin management.

1506 Jasper Van Kempen, *Europees Waterbeheer: Eerlijk Zullen We Alles Delen?* (Boom Juridische Uitgevers 2012).

1507 Article 10.4 Delaware Compact.

1508 These reductions have been defined in regulations, which have been consolidated into the Code. Section 2.5.3 of the Delaware River Basin Water Code.

1509 Telephone interview with senior staff member Delaware River Basin Commission, 15 December 2016. For an confirmation of this success factor for the DRBC, see the GAO report of 1981: Comptroller General of the United States, 'Federal-Interstate Compact Commissions: Useful Mechanisms for Planning and Managing River Basin Operations' (1981).

The added value of controlling emergency situations and changes in water variability in general, which are entirely inherent to water management, in transboundary river basins in combination with the fact that this is often overlooked in the relevant agreements, has been emphasised in literature as well.[1510] It is also the reason why the "ten building blocks method" has a separate block for conflict prevention and resolution.[1511]

2.2.5. CSF 5: Continuity in financial resources independent from Parties' allocations and adequate number of independent staff members in the Basin Body

429. This CSF is relevant to the basin level as well as for the sub-basin levels. The Delaware basin has shown that securing other means of funding in addition to the Parties' contributions is conducive for the continued viability of the entity in question. The DRBC has been able to function adequately despite diminishing and volatile contributions by the states and the federal level. Having a dedicated number of staff available to work exclusively under the auspices of the Basin Body is a success factor in river basin management. In the Scheldt, comparative research between the Flemish sub-basin boards and the Walloon river contracts has shown that means that are fully allocated to the functioning of the hydrological scale entity and the autonomous management of its own budget are success factors. Sub-basin boards are staffed with personnel from the Flemish water managers who have been assigned to perform the task, but often combine these tasks with other tasks. Their Walloon equivalents do have several full-time staff members, in part because these river contracts are in control of their own budgets and can use them for hiring purposes.[1512]

2.2.6. CSF 6: Enforcement is an important aspect of RBM and can be ex ante, i.e. proactive, and ex post, i.e. through reporting and compliance monitoring

430. Implementation of decisions made is an important element of river basin management, as has been established in Chapter I and is widely acknowledged across the board.[1513] Characteristic to the Delaware mechanism, and, as such a

---

1510 Olivia O Green, Barbara A Cosens and Ahjond S Garmestani, 'Resilience in Transboundary Water Governance: The Okavango River' (2013) 18 Ecology and Society. Shlomi Dinar and others, 'Climate Change and State Grievances: The Resiliency of International River Treaties to Increased Water Variability' (2010) 3 Insights 1.

1511 Marleen van Rijswick and others, 'Ten Building Blocks for Sustainable Water Governance: An Integrated Method to Assess the Governance of Water' (2014) 39 Water International 725.

1512 Hannelore Mees, Cathy Suykens and Ann Crabbé, 'Evaluating Conditions for Integrated Water Resource Management at Sub-Basin Scale. A Comparison of the Flemish Sub-Basin Boards and Walloon River Contracts' (2017) 27 Environmental Policy and Governance 42.

1513 As also explained in the context of the discussion of the resilience benchmark "effectiveness and enforcement". See e.g. Marleen van Rijswick and others, 'Ten Building Blocks for Sustainable Water Governance: An Integrated Method to Assess the Governance of Water' (2014) 39 Water International 725, 737. Olivia O Green and others, 'EU Water Governance: Striking the Right

success factor is the fact that the Compact provides through proactive enforcement. Article 3(8) of the Compact provides that projects that have a substantial impact on water resources in the basin should be submitted to the Commission for a check of compliance with the Comprehensive Plan. In addition to implementing decisions, monitoring and reporting mechanisms, which are associated with implementation in general, this provision enables proactive and basin-wide enforcement of measures and initiatives. This proactive implementation has promoted coherence as projects are in line with the goals and provisions of the Comprehensive Plan.

3. APPLICATION TO THE EU LEGAL FRAMEWORK

3.1. TOWARD A MORE COMPREHENSIVE APPROACH TO IRBD MANAGEMENT IN THE EU

431. Chapter II has demonstrated the difficulties associated with the bifurcation that exists between water quality and water quantity in the EU primary law framework in terms of integrated river basin management. The following section aims to present legal concepts and amendments to the existing legal framework that might enable a more encompassing approach.

3.1.1. River Basin District Security as an overarching tool

Figure 17. River Basin District Security

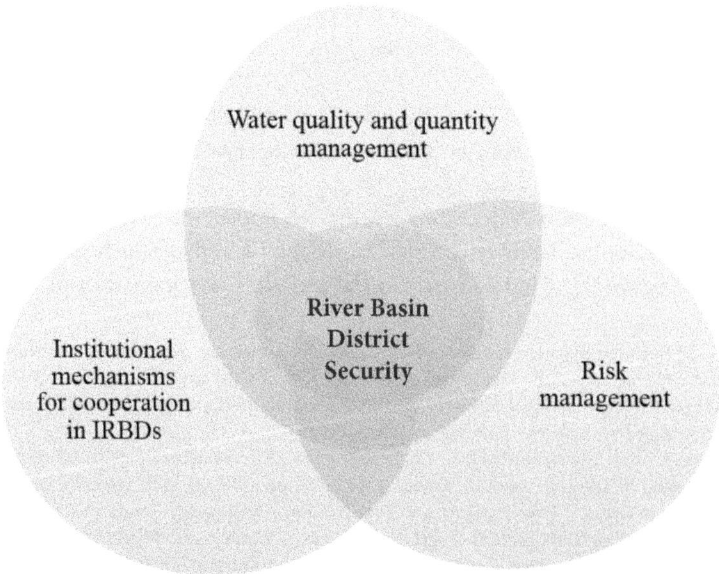

Balance between Regulatory Flexibility and Enforcement?' (2013) 18 Ecology and Society art10. For more literature, see this benchmark as explained in Chapter I.

432. This book presents the notion of "River Basin District Security" as a comprehensive assessment tool to be used in river basin districts across the EU. The notion "River Basin District Security" encompasses three main components: (i) quality and quantity, (ii) risk management, and (iii) legal and institutional mechanisms for cooperation in IRBDs. It is derived from the more commonly known notion of "water security", the general importance of which will be explained below.

433. Water, in its relationship with security and peace issues, was discussed as an independent topic for the first time under the auspices of the UN Security Council on 23 November 2016. This has heightened the importance of water security as a global issue.[1514] Moreover, water security has been marked as one of the UN Sustainable Development Goals (SDGs).[1515] As explained in Chapter I, the planetary boundary research has shown that the planetary boundary for freshwater use is substantially lower than originally thought.[1516] The amounts of water to safeguard environmental flow requirements have been appropriated in many water bodies throughout the world.[1517] More precisely, the water systems where environmental flows are compromised cover around 15% of the land surface globally, which makes addressing water insecurity a pressing issue.[1518] Water security has also been mentioned in relation to circular water management, with regard to which the World Business Council for Sustainable Development has recently published a business guide. Circular water management can be seen as a good business case, where the reuse and recycling of water is seen as the cost-effective option and as beneficial to water security because of the quantities of freshwater that are preserved.[1519]

Magsig has interpreted water security as follows: "a community is water secure when it has sustainable access to freshwater of sufficient quantity and quality, or to the benefits derived therefrom; and the ability to minimize water-related risks and its various repercussions to an acceptable level – without compromising the

1514 Even a global peace issue. Bjorn-Oliver Magsig, 'Water Security as an Evolving Paradigm: Local, National, Regional and Global Considerations' in Alistair Rieu-Clarke, Andrew Allan and Sarah Hendry (eds), *Routledge Handbook of Water Law and Policy* (Routledge 2017).

1515 SDG6 aims to ensure availability and sustainable management of water and sanitation for all.

1516 2015 in comparison to 2009. See Will Steffen and others, 'Planetary Boundaries: Guiding Human Development on a Changing Planet' (2015) 347 Science.

1517 Dieter Gerten and others, 'Towards a Revised Planetary Boundary for Consumptive Freshwater Use: Role of Environmental Flow Requirements' (2013) 5 Current Opinion in Environmental Sustainability 551, 553.

1518 Vladimir Smakhtin, Carmen Revenga and Petra Döll, 'A Pilot Global Assessment of Environmental Water Requirements and Scarcity' (2004) 29 Water International 307 <www.tandfonline.com/doi/abs/10.1080/02508060408691785>.

1519 World Business Council for Sustainable Development, 'Business Guide to Circular Water Management: Spotlight on Reduce, Reuse and Recycle Contents' (2017) <www.wbcsd.org/contentwbc/download/3437/44956> accessed 13 June 2017.

supporting ecosystems".[1520] This interpretation of water security comprises four components: "availability", which relates to water quality and quantity issues; "access", which relates to the use of shared waters; "adaptability", which pertains to the ability of a legal framework to respond to changing circumstances; and "ambit", which is geared toward defining the territorial and personal scope of the governance scheme in question.[1521] Unlike food security and energy security, key in the notion of water security is its twofold meaning as it relates both to the adequate quantity and quality of water and to the management of the risks that are associated with water as a resource, e.g. flood events. The majority of natural disasters are associated with water, which justifies a strong incorporation of the risk-related aspect of water management.

434. In the EU, the concept has not been used in the context of the WFD or FD. Water security has been referred to in relation to the supply of drinking water, where it has been proposed that water utilities could develop Water Security Plans to assess their vulnerability and enhance the manner in which it addresses drinking-water safety.[1522] This means that this is specifically geared toward water utilities, and should not be confused with the recommendations made regarding IRBD management. Water security within the meaning of this book has a broader, more encompassing relevance with certain nuances. Because of the focus of this book, i.e. river basin management, the water security concept is tailored to (international) river basin districts.

The concept of water security is not mutually exclusive with concepts such as integrated water resources management, integrated river basin management and environmental flow. On the contrary, these concepts fit under the overarching umbrella notion of water security. For example, the environmental flow can be used as a bridge between water quantity and quality as the question of good quality water resources forms part of the water security assessment. Using this umbrella concept also allows the evaluation of the institutional mechanisms created at the different basin levels, i.e. ranging from the sub-sub-basin to the international river basin district levels. Governance and legal mechanisms are key factors in determining whether a certain regime, e.g. the transboundary water regime, is water secure and to what extent.

1520 Bjørn-Oliver Magsig, International Water Law and the Quest for Common Security (Routledge 2015) 134.

1521 Bjørn Oliver Magsig, 'Introducing an Analytical Framework for Water Security: A Platform for the Refinement of International Water Law' (2009) 20 Journal of Water Law 61, 67-68.

1522 Andreas Weingartner and Jordi Raich-montiu, 'Proposal for a Water Security Plan to Improve the Detection of Threats in the Distribution Network Affecting Drinking Water Quality' (Joint Research Centre, European Union 2015) <https://erncip-project.jrc.ec.europa.eu/sites/default/files/JRC101135_Proposal%20for%20a%20water%20security%20plan%20to%20improve%20the%20detection%20of%20threats%20in%20the%20distribution%20network%20affecting%20drinking%20water%20quality.pdf> accessed 13 June 2017.

435. The security concept within the meaning of this book excludes the component of the human right to water, as it is mainly meant as a tool to enhance inter-jurisdictional water cooperation. Also, this chapter will not address the possible precedent value of the judgment in India and the Bill in New Zealand granting specific rivers a human right.[1523] River Basin District Security throughout the EU requires a re-evaluation of both EU primary and EU secondary law. The concept is likely to trigger a leap toward more integrative governance. It is relevant both at the level of primary and secondary EU law, meaning that it provides justification to moving away from the bifurcation between water quantity and quality management in the TFEU, and enables a more balanced treatment of quantity and quality. Focusing on water security as an umbrella concept and ultimate goal to be reached in river basin districts instead of separately focusing on water quality, water quantity and water safety, allows a more comprehensive approach to water management in the EU.

3.1.2. Remedying the primary law bifurcation between quantity and quality

3.1.2.1. Article 192(2) under scrutiny

436. This section will address the scope of "quantitative management of water resources or affecting, directly or indirectly, the availability of those resources", as stipulated in Article 192(2)(b)(ii). For the sake of clarity, the current text of Article 192 TFEU is copied in the box below.

1. The European Parliament and the Council, acting in accordance with the ordinary legislative procedure and after consulting the Economic and Social Committee and the Committee of the Regions, shall decide what action is to be taken by the Union in order to achieve the objectives referred to in Article 191.
2. By way of derogation from the decision-making procedure provided for in paragraph 1 and without prejudice to Article 114, the Council acting unanimously in accordance with a special legislative procedure and after consulting the European Parliament, the Economic and Social Committee and the Committee of the Regions, shall adopt:
(a) Provisions primarily of a fiscal nature;
(b) Measures affecting:
  i. Town and country planning
  ii. Quantitative management of water resources or affecting, directly or indirectly, the availability of those resources,
  iii. Land use, with the exception of waste management
(c) Measures significantly affecting a Member State's choice between different energy sources and the general structure of its energy supply.

1523 The implications of these initiatives will be assessed in another research project, currently in the conception phase. Mohd. Salmi v State of Uttarkhand and others (writ petition (PIL) n 126 of 2014) 20 March 2017. Te Awa Tupua (Whanganui River Claims Settlement) Act 2017 (2017/7), 20 March 2017.

437. In moving towards a more comprehensive approach to water management, the EU primary-law approach to water quantity and quality management should be re-evaluated. As has been illustrated throughout the previous chapters, these aspects are inextricably linked. The need for a more advanced integration of water quantity and quality at EU primary and secondary law levels has been established. The question arises how this can be done, taking into account the current constitutional limitations hindering such integration. Because of the difference in decision-making procedure in the TFEU, there is an *a priori* discrepancy between water quantity and water quality.

438. Moreover, Chapter II has exposed the confusing scope of "quantitative management of water resources" in the different versions of the EU Treaty. Article 192(2) had not yet served as a legal basis, which heightens the degree of uncertainty regarding the scope of its provisions.[1524] The Water Framework Directive and Floods Directive are both based on Article 192(1) TFEU, despite the fact that both instruments show clear traits of water quantity management. For the Water Framework Directive, the process toward its adoption has had some hiccups along the way and several amendments proposed by Parliament were dismissed by the Commission in order to maintain the focus on water quality management. The Floods Directive, dealing with water quantity and safety issues, has also circumvented the Treaty provisions associated with quantitative management and land management. Chapter II has illustrated that this recourse is somewhat justifiable as the FD does not contain any tangible, substantive provisions but only requires Member States to fulfil a series of procedural steps.

Finally, the need for the EU to better address the quantity aspect of water management was also referred to by stakeholders in the context of a public consultation procedure in light of the fitness check of EU water policy conducted in 2012.[1525]

439. There are two ways to go about revisiting Article 192(2) TFEU: one for the short term, i.e. not requiring Treaty revision, and one for the long term, which requires revising the provision. Before proposing amendments to the provision related to "quantitative management of water resources or affecting, directly or indirectly, the availability of those resources", several preliminary remarks should be made.

In this regard, it should be mentioned that this book does not intend to promote a fully-fledged transfer of all aspects of water management to the EU

[1524] This has led authors to suggest the removal of the reference to the unanimity requirement for energy and environmental measures altogether. See Leonie Reins, *Regulating Shale Gas: The Challenge of Coherent Environmental and Energy Regulation* (Edward Elgar Publishing 2017) 172.

[1525] Commission, 'Public Consultation to Support the Fitness Check of EU Freshwater: Policy Analysis of the Responses' (2012) <http://ec.europa.eu/environment/water/blueprint/pdf/public%20consultation%20report.pdf> accessed 11 June 2017.

level. First, a distinction should be made between substantive water-related provisions requiring Member States to take specific actions, e.g. to reach a certain target, construct infrastructure, and limit abstraction, and provisions requiring Member States to cooperate regarding such actions without defining them at the EU level. This distinction is important because it is linked to EU competence and the related limits pursuant to Article 192(2) TFEU. Second, the intention is to create an enabling environment for a more integrated approach to water management by removing some of the obstacles rooted in EU primary law. At the same time, several aspects associated with the quantitative management of water resources are inherently local and should remain local. For example, zoning requirements and building permits are mainly the competence of municipal authorities. Such is the case in the EU and in the US. Upscaling all these competences to the EU level is not a prerequisite for attaining an integrated approach to water management in IRBDs. Again, techniques and mechanisms bridging water management and spatial planning, such as the Dutch and the Flemish water test and the French *Barnier* Fund bridging flood risk prevention and flood recovery, explained in Chapter III, are useful in terms of cooperation exchange. The EU level could, for example, be relevant in providing a catalogue with such techniques.[1526] Third, the relationship with associated areas within and beyond Article 192 TFEU should be taken into account.[1527] Water-management measures can affect two or three indents of Article 192(2) simultaneously. This would be the case, for example, if an EU instrument were to require the construction of a hydro-electric dam. Indeed, such a requirement would touch on the "control of water" and the "sources of energy supply".[1528] The intersection between town and country planning and quantitative management of water resources is especially relevant in the area of flood risk management, e.g. EU requirements regarding legal safety norms. An example of an area beyond Article 192 is the issue of ex-post compensation of damages following flood events, and more specifically, insurance policies forming part of a comprehensive flood risk management regime.[1529]

1526 Especially as the deficient link between water management and other sectors has been identified by stakeholders as the main barrier to water management in the context of a public-consultation procedure in light of the EU fitness check of water legislation. European Commission, 'Public Consultation to Support the Fitness Check of EU Freshwater: Policy Analysis of the Responses' (2012) <http://ec.europa.eu/environment/water/blueprint/pdf/public%20consultation%20report.pdf> accessed 11 June 2017.

1527 See e.g. Kurt Deketelaere, 'The Use of Fiscal Instruments in European Environmental Policy: Review Essay' (1999) 10 European Environmental Policy 146.

1528 See also Opinion of AG Léger in Case C-36/98, *Spain v Council of the European Union* [2001] ECLI:EU:C:2000:246, Opinion of AG Léger, para 100.

1529 Other provisions, such as Article 114 TFEU and 345 TFEU are relevant, the latter limiting EU law making with respect to property ownership systems. Article 345 TFEU states that: "The Treaties shall in no way prejudice the rules in Member States governing the system of property ownership". There is also an important "services" component to water management, which will not directly be addressed in this chapter. Water is explicitly excluded from the

440. Both options will be presented below. A central element in both suggested routes is the use of bridging mechanisms to allow competent authorities to link the aspects of water management from a legal and policy perspective. With or without a Treaty revision, there is a need for mechanisms bridging the legal and administrative gap between these hydrologically linked aspects of water management. In literature, bridging mechanisms have been defined as instruments to link actors, sectors and policy instruments in order to remedy fragmentation. For example, in the context of flood risk management, different types of fragmentation were distinguished by Gilissen and others.[1530] These types of fragmentation are linked to the five flood risk management strategies, which have been explained in Chapter III: i.e. risk prevention, mitigation, protection, preparation, and recovery.[1531] These touch on different sectors, e.g. spatial planning, water management, emergency management and environmental protection. Fragmentation occurs (a) where different actors operate in different sectors in the pursuit of different flood risk management strategies, (b) where different actors operate in the same sector applying the same strategies, (c) where different actors apply the same flood risk management strategy but from a different sector, and (d) where different actors work in the same sector but pursue different strategies.[1532] Extrapolating this concept of bridging mechanisms to the dichotomy between water quality and quantity in the EU Treaty, its relevance lies in bridging the gap at the outset, in line with CSF 1. Adopting an integrated approach from the outset, in this case in EU primary law, is likely to limit fragmentation down the road.

441. One bridging mechanism in particular will be presented below, i.e. the environmental flow.[1533]

3.1.2.2. Environmental flow: A useful bridge

442. The concept of environmental flow was introduced in Chapters I and II and makes a re-appearance in this chapter as an important bridge between water quality, water quantity and risk management. The environmental flow refers to the amount of flow that the river needs in order to maintain a good ecosystem state.[1534]

Concessions Directive. Directive (EU) 2014/23 of the European Parliament and of the Council on the award of concession contracts [2014] OJ L 94.

1530 Herman Kasper Gilissen and others, 'Bridges over Troubled Waters: An Interdisciplinary Framework for Evaluating the Interconnectedness within Fragmented Domestic Flood Risk Management Systems' (2016) 25 Journal of Water Law 12, 15.

1531 They together add up to form a flood risk management policy of a certain country, region, province, and so forth.

1532 Herman Kasper Gilissen (n 1530) 12.

1533 Also referred to as "ecological flow".

1534 Will Steffen and others, 'Planetary Boundaries: Guiding Human Development on a Changing Planet' (2015) 347 Science. Permanent Court of Arbitration, Indus Waters Kishenganga Arbitration (Pakistan v. India), Final Award, 20 December 2013, <http://archive.pca-cpa.org/

It links the quality of the water body with the quantity and also the risk elements of water resources management. With regard to the latter link, between flood risk management and environmental protection, reference can be made to nature-based solutions in flood risk management, such as floodplain 'leaky barriers' and planting riparian buffer zones. Environmental flow can thus play a role in linking the scope of the Water Framework Directive with the scope of the Floods Directive.[1535]

443. The question arises to what extent the notion of environmental flow, despite the primary EU law dichotomy between quantity and quality, is already anchored in the EU legal regime. The importance of the environmental flow has been acknowledged by the EU Commission and is partly linked to the attainment of the environmental objectives included in Article 4 of the WFD.[1536] However, this link is implicit, and as mentioned it is partial, rather than explicit.[1537] The EU Water Directives do not determine flow regimes or methodologies in this regard, or include the requirement to streamline these at IRBD level, but environmental flows are relevant in the implementation of the WFD. The WFD aims to maintain and improve the quality of aquatic ecosystems in the EU. For groundwater

---

PK-IN%20Final%20Award,%2020%20December%202013d770.pdf?fil_id=2471> accessed 25 April 2016; Owen McIntyre, 'The Emergence of an "Ecosystem Approach" to the Protection of International Watercourses under International Law' (2004) 13 Review of European Community and International Environmental Law 1; Owen McIntyre, 'The Protection of Freshwater Ecosystems Revisited: Towards a Common Understanding of the Ecosystems Approach to the Protection of Transboundary Water Resources' (2014) 23 Review of European, Comparative & International Environmental Law 88; Alistair Rieu-Clarke and Christopher J Spray, 'Ecosystem Services and International Water Law: Towards a More Effective Determination and Implementation of Equity?' (2013) 16 Potchefstroom Electronic Law Journal 12; Robert Speed and others, 'Basin Water Allocation Planning: Principles, Procedures and Approaches for Basin Allocation Planning' (2013) <www.adb.org/publications/basin-water-allocation-planning> accessed 14 July 2017.

1535 E.g. when such flow is preserved in flood risk management. European Commission, 'WFD CIS Guidance Document No. 31: Ecological Flows in the Implementation of the Water Framework Directive' (2015) 79.

1536 See Chapter II of this book. Moreover, environmental flows are relevant to the implementation of the Birds and Habitats Directive, as flow regimes are important to protected species and habitats' conservation status. Article 6 (1) states "For special areas of conservation, Member States shall establish the necessary conservation measures involving, if need be, appropriate management plans specifically designed for the sites or integrated into other development plans, and appropriate statutory, administrative or contractual measures which correspond to the ecological requirements of the natural habitat types in Annex I and the species in Annex II present on the sites". European Commission, 'WFD CIS Guidance Document No. 31: Ecological Flows in the Implementation of the Water Framework Directive' (2015). The analysis of EFlows in the Birds and Habitats Directive goes beyond the scope of this book, but it is relevant to mention that environmental flows are implicitly present in various EU legal instruments. Council Directive (EEC) 92/43 on the conservation of natural habitats and of wild fauna and flora [1992] OJ L 206. Directive (EC) 2009/147 of the European Parliament and of the Council of 30 November 2009 on the conservation of wild birds [2009] OJ L 20.

1537 Rafael Sanchez and Guido Schmidt, 'Environmental Flows as A Tool to Achieve the WFD Objectives' 20 <www.pianc.org/downloads/euwfd/Eflows%20discussion%20paper%20draft%20NAVI%20TG%20response%20v2%2010-8-12.pdf> accessed 14 July 2017.

bodies, the quantitative status is an explicit element in the determination of good status, where ecological flow needs are relevant in the context of the applicable tests.[1538] As has been explained in Chapter II, the biological elements and the hydromorphological elements, and the chemical and physical-chemical elements supporting the biological elements are the relevant elements in deciding on the ecological status of surface water bodies.[1539] Hydromorphological quality elements include the hydrological regime, i.e. the quantity and dynamics of the flow, the river continuity and the morphological conditions, e.g. channel patterns and flow velocities. The values of the hydromorphological elements are not generally used to determine the ecological status of a water body, but must be used in the determination of the High Ecological Status.[1540] With respect to good, moderate, poor and bad water status, the hydromorphological quality variable only relates to "conditions consistent with the achievement of the values specified for the biological quality elements".[1541] The same line of reasoning is applicable to the Heavily Modified Water Bodies. In the fourth WFD implementation report issued in 2015, hydromorphological pressure, i.e. changes to the physical shape and flow of water bodies, has been identified as one of the main barriers to achieving good water status in the implementation of the Directive.[1542]

As mentioned, the EU has accepted the importance of the environmental flow in the implementation of the WFD and has issued a Common Implementation Strategy document, which defines the role of the environmental flow in calculating the applicable status of the water bodies in question, possible methodologies, and monitoring requirements.[1543] The RBMPs have also been reviewed in light of environmental flows. From the review of the first-cycle RBMPs, it has become clear that Member States generally only consider the minimum flows that need to be maintained in the summer months, without considering the environmental flows and the elements necessary for well-functioning ecosystems.[1544] The fourth implementation report, which deals with the Programmes of Measures, refers to the fact that there is a major gap in biological assessment methods that take sufficient account of hydromorphological pressures.[1545] From the review of the Programmes of Measures in 2012, it seemed that only one Member State

---

1538 European Commission, 'WFD CIS Guidance Document No. 31: Ecological Flows in the Implementation of the Water Framework Directive' (2015) 26.

1539 See Annex V of the WFD.

1540 Commission, 'WFD CIS Guidance Document No. 3: Analysis of Pressures and Impacts' (2003) 26.

1541 Annex V of the WFD. See Sanchez and Schmidt (n 1537).

1542 Commission, 'The Water Framework Directive and the Floods Directive: Actions towards the 'good status' of EU water and to reduce flood risks' COM (2015) 120 final.

1543 Commission, 'WFD CIS Guidance Document No. 31: Ecological Flows in the Implementation of the Water Framework Directive' (2015) 25.

1544 Commission, 'The Water Framework Directive and the Floods Directive: Actions towards the 'good status' of EU water and to reduce flood risks' COM (2015) 120 final 7.

1545 European Commission, 'Report on the Progress in Implementation of the Water Framework Directive: Programmes of Measures' SWD (2015) 50 final 12.

had actually analysed the level of reduction in abstraction rates necessary to achieve the environmental objectives set in the WFD.[1546] At the time of writing, an assessment of the second-cycle RBMPs was still pending. The roof report of the Scheldt IRBD refers to the fact that "awareness of the ecological minimum flow rate" would be beneficial to the biological quality of the watercourses.[1547]

3.1.2.3. A matter of interpretation

444. A Treaty revision is not a short-term option. In the meantime, more clarity should be obtained regarding the current wording of Article 192(2) TFEU with respect to "quantitative management of water resources". In the absence of Treaty revisions, (i) this scope can be clarified and (ii) the gap can be remedied with the environmental flow concept serving as a supportive interpretative tool.

445. As has been established in Chapter II, the mere fact that provisions included in EU secondary-law instruments influence the use of land or the quantitative management of water resources, does not trigger the unanimity decision-making procedure of the TFEU.[1548] Several CJEU judgments have referred to the "centre of gravity" test with respect to water quantity in relation to water quality and have acknowledged the legitimacy of EU water quantity measures insofar as these are necessary to reach water quality objectives.[1549]

In the context of *Spain v Council*, France supported an interpretation in which the unanimity requirement applies to the physical control of the water, e.g. measures governing the use of available water supplies, flow regulation and infrastructure. Portugal and Finland suggested a similar interpretation, focussing on water use and infrastructure. Reference was also made to the economic exploitation of water in relation to the unanimity requirement.[1550] As explained in Chapter II, the WFD has developed a legal regime for water services in light of the cost-recovery principle, which has been examined by the CJEU. The cost-recovery principle implies that environmental and resource costs need to be taken into account for these services, which link water quantity and environmental protection. Pursuant to the judgment in *Commission v Germany*,

---

1546 I.e. Spain. Ibid 74.

1547 International Scheldt Commission, 'Roof Report of the Management Plan for the International Scheldt River Basin District' (2015) <www.isc-cie.org/images/Documents/ODB2-PFPG2_RAPPORT_ENG_V17.pdf>.

1548 See Chapter II and also see e.g. Petra Lindhout, Cost Recovery as a Policy Instrument to Achieve Sustainable and Equitable Water Use in Europe and the Netherlands (Ph.D. Thesis, Utrecht University 2015).

1549 Case C-36/98, *Spain v Council of the European Union* [2001] ECLI:EU:C:2001:64; Case 525/12 *European Commission v Germany* [2014] ECLI:EU:C:2014:2202; Case C-686/15, *Vodoopskeba i odvodnja d.o.o. v Željka Klafurić* [2016] ECLI:EU:C:2016:927.

1550 I.e. by Portugal.

Member States have broad discretionary powers to interpret the exact meaning of the notion "water services" and are not required to qualify each activity listed in Article 2(38) WFD as such.[1551] Member States may decide whether a certain measure should be subject to cost-recovery.[1552] This has implications for EU water quantity management, in the sense that activities that are not mentioned explicitly as a water service in Article 2(38) WFD, such as flood risk management and impoundment of water for hydroelectric power do not necessarily have to answer to the cost-recovery principle. Indeed, in a restricted interpretation of the notion of water services, approximately 20% of water abstraction activities are considered as a water service.[1553] This is another application of the centre of gravity-test. If the WFD had listed all types of water-related activities constituting water services, in turn subject to the cost-recovery principle, the unanimity provision would have been triggered. In fact, the AG in *Commission v Germany* argued that the Commission had tried to circumvent this provision by subjecting activities such as the above-listed to its pricing provision.[1554] On the other hand, in granting broad discretionary powers to the Member States in terms of (i) which activities are subject to the cost-recovery principle, and (ii) how the pricing policies should be calculated, the "quantitative management of water resources" is not "directly or indirectly" affected. The Court of Justice recently stipulated that it is up to the Member States to determine the methods associated with water-pricing policies, as long as they ensure that the cost-recovery principle regarding water use is applied.[1555]

446. In moving EU water management forward within the current institutional constellation, an interpretative framework is suggested, as illustrated by figure 18 below. The figure does not contain an exhaustive list of measures touching on water management, but is meant to present an indicative tool.

1551 The text of Article 2(38) and (39) WFD is as follows: 38. "Water services" means all services which provide, for households, public institutions or any economic activity:
(a) abstraction, impoundment, storage, treatment and distribution of surface water or groundwater,
(b) waste-water collection and treatment facilities which subsequently discharge into surface water.
39. "Water use" means water services together with any other activity identified under Article 5 and Annex II having a significant impact on the status of water.
This concept applies for the purposes of Article 1 and of the economic analysis carried out according to Article 5 and Annex III, point (b).

1552 Petra Lindhout and others, 'Kostenterugwinning van Waterdiensten in Woelig Water: Bedenkingen Bij Het Arrest C-525 / 12 van Het Hof van Justitie Bezien in Rechtsvergelijkend Perspectief' (2015) 1 Tijdschrift voor Milieurecht 4.

1553 Case 525/12 *European Commission v Germany* [2014] ECLI:EU:C:2014:449, Opinion of AG Jääskinen, para 31.

1554 Ibid para 70.

1555 Case C-686/15, *Vodoopskeba i odvodnja d.o.o. v Željka Klafurić* [2016] ECLI:EU:C:2016:927, para 21.

Figure 18. Quantitative Management: Lower Bound – Upper Bound

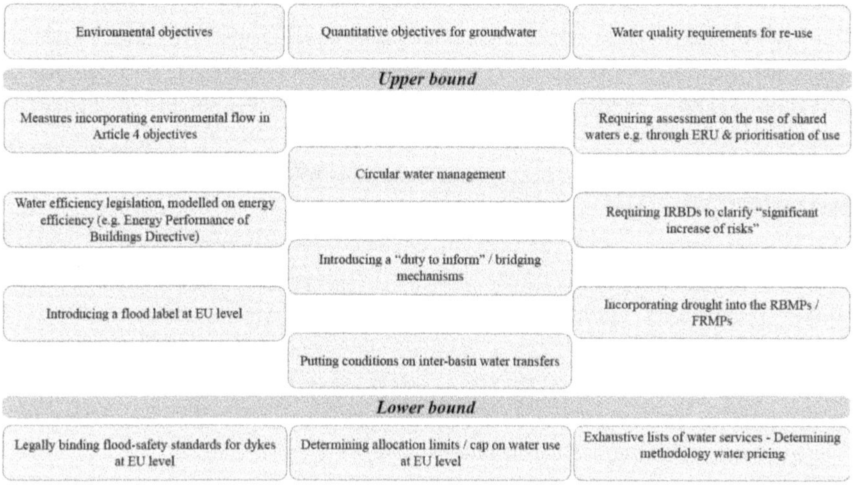

447. This interpretative framework adopts a narrow interpretation of the wording "quantitative management of water resources". The lower bound represents the unanimity decision-making procedure, whereas the upper bound reflects the qualified majority voting procedure. Measures which exceed the lower bound are measures that have a substantive physical impact on the territory of the Member States in river basin districts in national territories. Physical control is central in this regard. This also ties in with the other indents of Article 192(2), in particular town and country planning and land use. Measures exceeding the upper bound are measures where it is currently clear that the ordinary legislative procedure applies. The "grey" area in between the lower and upper bound is subject to interpretation, the closer to the upper bound, the higher the likelihood that the ordinary procedure applies. For example, explicitly incorporating environmental flows into the "good status" objectives included in Article 4 WFD should be subject to the ordinary legislative procedure. Indeed, such measures mirror the intricate link between quality and quantity aspects of water management. Another example relates to water efficiency. By analogy, energy efficiency is elaborately regulated at EU level: reference can be made to the Energy Performance of Buildings Directive.[1556] Incorporating water efficiency more strongly in the EU legal framework would, in the narrow interpretation of Article 192(2) TFEU, not trigger the unanimity requirement because it does not substantially and physically influence the territories of Member States. Another example is in the area of flood risk management.

1556 Directive (EU) 2012/27 of the European Parliament and of the Council of 25 October 2012 on energy efficiency, amending Directives 2009/125/EC and 2010/30/EU and repealing Directives 2004/8/EC and 2006/32/EC [2012] OJ L 315.

Whereas legally binding safety standards for dykes set at EU level would likely require unanimity, other measures are possible within the current constellation that go beyond what is currently included in the Floods Directive. Whereas town and country planning and land use management are the competence of the Member States, bridging mechanisms linking water management with other sectors quintessential to reaching water management objectives can be relevant to consider at EU level. For example, in a narrow interpretation of "quantitative management", instruments raising awareness of citizens such as flood labels and the "duty to inform" could be introduced. The flood label entails an indication of the flood-proof character of the building in question.[1557] Such an instrument already exists in EU energy law: the energy performance certificate.

448. Whereas the interpretation route offers temporary relief, a more structural solution to the current state of confusion is necessary. The following section will suggest a new wording of the relevant provisions of Article 191 and 192(2) TFEU.

3.1.2.4. Revision of the Treaty provision

449. In the context of a possible Treaty revision, the scope of Article 192(2) TFEU could be revisited, as suggested in the table below.

Table 12. Suggested Amendments to Articles 191 and 192 TFEU

Article 191 TFEU

1. Union policy on the environment shall contribute to pursuit of the following objectives:
- preserving, protecting and improving the quality of the environment,
- protecting human health,
- prudent and rational utilisation of natural resources, ***and promoting water security***,
- promoting measures at international level to deal with regional or worldwide environmental problems, and in particular combating climate change.

Article 192(2) TFEU

2. By way of derogation from the decision-making procedure provided for in paragraph 1 and without prejudice to Article 114, the Council acting unanimously in accordance with a special legislative procedure and after consulting the European Parliament, the Economic and Social Committee and the Committee of the Regions shall adopt:

(d) Provisions primarily of a fiscal nature;

(e) Measures affecting:

   i. Town and country planning

   ~~ii. Quantitative management of water resources or affecting, directly or indirectly, the availability of those resources~~

   iii. Land use, with the exception of waste management

(f) Measures significantly affecting a Member State's choice between different energy sources and the general structure of its energy supply

[1557] Thomas Hartmann and Marc Scheibel, 'Flood Label for Buildings – a Tool for More Flood-Resilient Cities', FLOODrisk 2016 – 3rd European Conference on Flood Risk Management (2016).

450. The proposed amendments enable a more comprehensive approach to water management by anchoring it more strongly in Article 191 TFEU on the one hand, and removing the confusion with regard to quantitative water management in Article 192(2) TFEU on the other.

451. The question arises whether it is necessary to maintain the reference to "quantitative management of water resources", which emphasises the contrast with water quality management, where in fact these are two sides of the same coin. The continued reference to "quantitative management of water resources" is unsustainable in light of the need for integrated river basin management, as its scope is too broad. The suggestion is to add water security as one of the objectives in Article 191 TFEU. This allows for a more comprehensive approach to water management, and it solves the lack of clarity related to the applicable decision-making procedure. Including water security as an objective in Article 191 TFEU makes the legal basis for flood risk management at EU level more straightforward.[1558] The FD is based on Article 192(1) TFEU, and there are various reasons justifying this choice of legal basis. One of these reasons relates to the link between the FD and the WFD. However, it equally raises questions, e.g. with respect to its link with quantitative management of water resources and crisis management. By referring to water security, and thus to quality, quantity and also risk management in Article 191, flood risk management initiatives issued at EU level are more clearly enshrined in Article 192(1). Restrictions stemming from associated Treaty provisions should naturally still be taken into account.[1559]

The main rationale behind the proposed amendment relates to the above section related to the interpretation of the provision, with the environmental flow bridge and integrated management in general gravitating toward the ordinary decision-making procedure and aspects substantially linked to land management and the physical control of national territories associated with the special decision-making procedure.

452. It is important to address the existing primary law gap and clear the confusion regarding "quantitative management of water resources" in order to create a more "enabling" environment for IRBD management at EU level. However, the available institutional mechanisms should also be addressed, which will be done in the sections below. When applying River Basin District Security to the EU legal framework applicable to water management, a distinction should be made between the elaboration of the concept as such, which can be done at EU secondary law level, and the operationalisation of the concept, which should be done at the (international) River Basin District level.

1558 As the water security concept embodies the "risk" element.

1559 E.g. Article 345 TFEU related to property ownership systems.

3.2. ENHANCING COOPERATION IN IRBDs

3.2.1. The question of Institutional Mechanisms

453. The tension between the different levels of applicable law in International River Basin Districts has been illustrated in Chapters II and III. Member States make use of international mechanisms based on international treaties to implement the hydrological unit of governance requirements based on the EU Water Framework Directive and the Floods Directive.[1560] These treaties aim to implement both the international law instruments, in particular the UNECE Water Convention, and the EU Directives. The Directives, in turn, also aim to implement these international law instruments, such as the UNECE Water Convention.[1561] IRBDs are subject both to *inter se* Treaties and *inter se cum tertis* Treaties.[1562] Whereas international river commissions, such as the International Scheldt Commission, traditionally served as framework mechanisms through which States define the general principles of cooperation in transboundary rivers, they now mainly serve as implementing tools for the Directives, which is reflected by the structure and content of their working groups.[1563] In this constellation, the Member States retain individual responsibility to comply with the WFD and FD in their accountability toward the EU:

(a) The International River Basin District does not operate as the unit of governance in the implementation of the WFD and FD, as explained in Chapter II;
(b) Cooperation requirements with respect to the procedural provisions of the Directives, e.g. the obligation to submit joint international river basin management plans and flood risk management plans, are a matter of best efforts, not of result.
(c) The "Competent Authorities" may be designated at the level of the IRBD, but are, e.g. with respect to the Scheldt District, designated at the level of the national river basin district. These basins are not governed by one competent

1560 Andrea M Keessen, Jasper JH Van Kempen and Marleen Van Rijswick, 'Transboundary River Basin Management in Europe Legal Instruments to Comply with European Water Management Obligations in Case of Transboundary Water Pollution and Floods' (2008) 4 Utrecht Law Review 35.

1561 Recital 35 WFD.

1562 Ellen Hey and Marleen van Rijswick, 'Transnational Water Management' in Oswald Jansen and Bettina Schöndorf-Haubold (eds), *The European Composite Administration* (Intersentia 2010) 240.

1563 Ellen Hey, 'Multi-Dimensional Public Governance Arrangements for the Protection of the Transboundary Aquatic Environment in the European Union: The Changing Interplay between European and Public International Law' (2009) 6 International Organizations Law Review 191, 207. Ellen Hey and Marleen van Rijswick, 'Transnational Water Management' in Oswald Jansen and Bettina Schöndorf-Haubold (eds), *The European Composite Administration* (Intersentia 2010) 240.

authority for the whole district, but by several competent authorities operating at the respective national levels.

(d) The Scheldt case study has shown that the IRBD level serves as a discussion platform and platonic coordination, but the heart of the decision-making process is entirely at the national levels, where the sub-basin levels and the basin levels are unconnected.

454. The question now arises what could be done to remedy this situation. The following sections will propose suggestions toward remedying this institutional mechanism gap and enhance cooperation in IRBDs. By shifting the institutional mechanisms for IRBD management from the international level to the EU level, the current multi-level governance tensions associated with IRBD management would be remedied to some extent. The question arises whether the EU could move a top-down obligatory recourse of States sharing IRBDs to the institutional mechanism provided by the EU legal framework. Furthermore, would this have to be a new EU-level institutional mechanism or could IRBD States make use of existing instruments, and if so, how could these be 'sculpted' to tailor river basin management and what would be the incentives for initiating such mechanism?

455. Suggestions have been made in literature to create a "transboundary European district water board" through the EU legal framework, which would constitute a type of joint responsibility mechanism for States sharing International River Basin Districts in the European Union.[1564] This entails a twofold recommendation: (i) a move toward an EU institutional mechanism for river basin management instead of those based on international law, and (ii) a stronger emphasis on the competences of the joint entity at the hydrological scale. The Delaware mechanism has indeed shown that integrated river basin management is promoted when there is a basin-wide entity with an adequate number of staff and resources, and whose actions are rooted in a solid agreement.

An impediment to the creation of a top-down *ex ante* joint responsibility mechanism at the EU level is the fact that not all International River Basin Districts are located entirely within the territory of the EU. If the EU legal framework were to impose the creation of joint entities, 'sculpted' on the basis of EU law, as the formal competent authorities for the implementation of the Directives, this would create a tension between Member States in IRBDs located entirely within the European Union and Member States in IRBDs extending beyond the EU. The EU does not have the competence to impose the creation and designation as formal authority of such river commissions on non-EU Member States, and only subjecting the EU Member States in those IRBDs extending beyond the EU to such commissions would go against the idea of hydrological scale governance and the necessity to safeguard basin-wide governance should be taken into account. Chapter IV has

[1564] Hey and van Rijswick (n 1563) 240.

illustrated that the top-down imposed cooperation in inter-jurisdictional waters has proven to be difficult in practice. Reference can be made to the Title II-river basin commissions created on the basis of the Water Resources Planning Act, which were established on the initiative of the federal Water Resources Council and were abolished in 1981 because they were mistrusted by the states and did not succeed in securing financial support by the states.[1565] However, the *status quo* in EU IRBD management should not be maintained.

456. On the basis of the WFD and the FD, Member States should have the possibility to resort to institutional mechanisms on the basis of EU law to manage their shared waters. The provision of alternatives to rely on international law mechanisms for the implementation of the Water Directives can be achieved following a step-by-step approach. Before determining whether it is necessary for the Water Directives to create a proper institutional mechanism which IRBD States can use, it is necessary to look beyond these Directives. Indeed, any such EU-level action would need to be in line with the subsidiarity and proportionality principle.

3.2.2. Enhanced Cooperation Model (ECM)

457. The question whether the EU legal framework already provides for a useful institutional mechanism for IRBD cooperation can be answered in the affirmative, but needs qualifying.

The most apparent available mechanism for adoption in IRBDs is the European Grouping for Territorial Cooperation. This instrument, and its legal framework in the form of an EU Regulation, has been explained in Chapter III. In 2016, the European Parliament issued a report on European Territorial Cooperation, stressing the enormous potential of EGTC in facilitating transboundary cooperation and calling on Member States to increase their efforts in facilitating the creation and existence of EGTCs.[1566] The possibility of using the EGTC in the context of transboundary river basin management has already been suggested by several authors.[1567] In line with the CSFs listed in this chapter, there are several underlying conditions that need to be fulfilled before considering using the

---

1565 Lynn A Mandarano, Jeffrey P Featherstone and Kurt Paulsen, 'Institutions for Interstate Water Resources Management' (2008) 44 Journal of the American Water Resources Association 136, 140.

1566 European Parliament, 'Report on European Territorial Cooperation – best practices and innovative measures' (2015/2280(INI)) (2015) <www.europarl.europa.eu/sides/getDoc.do?pubRef=-//EP//NONSGML+REPORT+A8-2016-0202+0+DOC+PDF+V0//EN> accessed 9 June 2016.

1567 These articles date from the period before the 2013 amending Regulation was issued. The 2013 Regulation makes the EGTC more attractive. Herman Kasper Gilissen, *Internationaal en Regionale Grensoverschrijdende Samenwerking in Het Waterbeheer* (Sdu Uitgevers 2009). Hey and Van Rijswick (n 1563). Marleen Van Rijswick, Herman Kasper Gilissen and Jasper van Kempen,

EGTC as a cooperation tool in IRBDs. The members of the EGTC should be the competent authorities with respect to the IRBD in question. This is a well-known prerequisite for achieving successful river basin management.[1568] An important caveat is that certain hydrological scale entities, such as the Flemish sub-basin boards, cannot be a formal partner in an EGTC, should an EGTC related to river basin management be created, due to their lack of legal personality.[1569] Such entities could be involved indirectly, through the formal water manager, here the Flemish Region, or via participation in a specific working group on water organised under the auspices of the EGTC. There should be clear added value in using the EGTC instead of the existing cooperation mechanisms.[1570] In this regard, importantly, the EU cannot "force" Member States or decentralised authorities to create an EGTC in IRBDs. One of the fundamental characteristics of the tool pursuant to the EGTC Regulation is its voluntary nature. This can be traced back to the difficult adoption process of the Regulation, with Member States fearing that such vehicle would circumvent their authority.[1571] However, the use of these mechanisms can be promoted in other ways, which will be discussed in the following sections.

The arguments in favour of making use of the EGTC Regulation and the limitations of the instrument are described below.

- *Arguments in favour of the EGTC as an appropriate institutional mechanism*

458. Several arguments can be formulated in favour of using the EGTC as a vehicle for IRBD management. First, it will accommodate concerns expressed by several authors with respect to the tension between international law and European law cause by the use of international mechanisms to implement EU legislation in River Basin Districts.[1572] The EGTC will be created on the basis of EU law, and will be governed by EU law. As stated in a study ordered by the Committee of the Regions in 2007: "The EGTC will thus be a legal entity under Community law, and not a legal entity defined by a national legal system to which subnational authorities governed by a foreign legal system are allowed

---

'The Need for International and Regional Transboundary Cooperation in European River Basin Management as a Result of New Approaches in EC Water Law' (2010) 11 ERA Forum 129.

1568 E.g. interview with senior staff member of the Secretariat of the International Commission for the Protection of the Rhine, 27 April 2017.

1569 However, this is an internal law issue rather than an issue associated with the EGTC.

1570 Sabine Zillmer and Maria Toptsidou, 'Potential and Limits of the EGTC Instrument for Enhancing Integration across Borders' (Spatial Foresight 2014) <www.spatialforesight.eu/tl_files/files/editors/dokumente/Brief-2014-5-141111.pdf> accessed 14 June 2017.

1571 This also implies that it would be difficult for the WFD and FD to create an *ex ante* joint responsibility mechanism as the obligatory "go to" mechanism for Member States sharing IRBDs. Elisabetta Nadalutti, *The Effects of Europeanization on the Integration Process in the Upper Adriatic Region* (Springer 2015) chapter 4.

1572 See e.g. Hey and van Rijswick (n 1563); Keessen, Kempen and Rijswick (n 1560); Van Kempen (n 1567).

access by an international instrument".[1573] The EGTC is governed by EU law, the EGTC Convention constituting the basic agreement concluded between the parties involved and, for the aspects not regulated under the Regulation, the national law of the Member State where the EGTC has its office.[1574] Second, using the EGTC as a pilot project for river basin management might have a snowball effect toward other basins and as such increase coherence in IRBD management throughout the EU. In relation to this, although the creation of the EGTC as such does not automatically trigger EU funding, its existence increases this possibility of securing funding. The EGTC may respond to calls for proposals and act as the Managing Authority. Third, the EGTC may include both national governments, regional and local authorities and public undertakings. This resonates with the reality of river basin management and the division of competences, as illustrated in Chapter III. The EGTC Regulation allows for participation of national authorities in the cooperation vehicle, with the underlying rationale that competences that are national in one country are regional in another.[1575] This is important for water management, where in some countries important tasks are within the competence of functionally decentralised entities, such as the regional water authorities in the Netherlands, but where this is not the case in other countries. This is important because the Flemish Region, for example, is a formal water manager but does not have the legal status of "decentralised authority" in Belgium. The "multi-level mismatch" which has been referred to in Chapter III, is thus remedied to some extent. Fourth, the EGTC can be created by EU Member States and non-EU Member States, if at least two EU Member States are involved.[1576] This is rather important, considering the existence of IRBDs extending beyond the EU. Fifth, the EGTC has legal personality and has the obligation and responsibility to draft and manage a budget. The EGTC can act as the managing authority of certain programmes in an autonomous manner.[1577] The Regulation also allows the parties to approve the participation

1573 Nicolas Levrat, 'The European Grouping of Territorial Cooperation' (2007) <http://cor.europa.eu/en/documentation/studies/Documents/The%20European%20Grouping%20of%20Territorial%20Cooperation%20-%20EGTC/The%20European%20Grouping%20of%20Territorial%20Cooperation%20-%20EGTC%20-%20EN.pdf> accessed 13 June 2017.

1574 Article 2 of the EGTC Regulation as amended by Regulation (EU) No 1202/2013.

1575 Recital 7 of the Regulation (EU) No 1302/2013 of the European Parliament and of the Council of 17 December 2013 amending Regulation (EC) No 1082/2006 on a European grouping of territorial cooperation (EGTC) as regards the clarification, simplification and improvement of the establishment and functioning of such groupings, OJ L. 347.

1576 Article 3.2 of the EGTC Regulation as amended by Regulation (EU) No 1202/2013. The added value of using EGTCs in the context of cross-border river basin management as key vehicles in investment projects, e.g. in light of the European Fund for Strategic Investment, has explicitly been referred to. Jürgen Pucher, 'The EGTCs Investing: Implementing EU Funds. Which Role in the European Fund for Strategic Investments ? Which Procurement ?' (Committee of the Regions 2016).

1577 Nadalutti, *The Effects of Europeanization on the Integration Process in the Upper Adriatic Region* (Springer 2015) 110.

of a member in the EGTC, a competence that does not cover all the tasks of the EGTC.[1578]

- *The downside of the EGTC as institutional mechanism*

459. However, there are restrictions to using the EGTC as a vehicle to manage a common river system. The main restriction relates to the fact that the scope of competences of the EGTC is limited. Pursuant to the EGTC Regulation, it cannot receive public authority powers such as police and regulatory powers. This has to do with the reluctant stance of Member States in creating an instrument at EU level creating such a legal cooperation instrument such as the EGTC and the process toward adoption of the Regulation has been very difficult.[1579] The implementation process has also been slow.[1580] The EGTC entity will not be able to enforce its decisions. This is a fundamental difference with, for example, the Delaware River Basin Commission, which is a regulatory agency with enforcement powers. Taking into account the *travaux préparatoires* of the Water Framework Directive and Floods Directive and the Member States' interventions in *Spain v Council* it is doubtful that Member States would be willing to transfer regulatory powers to a joint entity at the IRBD level.[1581] The EGTC Regulation does provide for the enforcement of the rights of third parties that have been disadvantaged by the actions or omissions of the EGTC, which allows for a broad interpretation of its task pursuant to Article 7 of the Regulation.[1582] In this regard, it is relevant to refer to the Benelux Grouping for Territorial Cooperation, which does allow the parties to transfer public regulatory competences. The BGTC can manage projects, has control over its financial means and personnel and can adopt decisions that are binding on the citizens. This type of entity, of all the cooperation instruments studied in Chapter III, would most closely resemble the Delaware River Basin Commission.

460. The EGTC Regulation only provides for the creation of the EGTC entity, in contrast to, e.g. the Benelux Treaty, which provides for three types of cooperation, ranging from "light" to "fully-fledged".[1583] Within a given basin, it would be possible to create a Benelux Grouping for Territorial Cooperation with binding decision-making powers and legal personality, which could be relevant for the basin as a whole, and a joint entity without legal personality as well as an administrative agreement. These three cooperation routes reflect the needs of a

1578 Article 7 EGTC Regulation.
1579 Elisabetta Nadalutti (n 1577) 105.
1580 Implementation is required for the designation of the authorities that should approve the EGTC.
1581 See preceding chapters.
1582 Article 15 Paragraphs 1 and 2 of the 2006 EGTC Regulation as amended by the 2013 Regulation.
1583 See Chapter III.

river basin, where it would not be beneficial to create entities such as EGTCs and BGTCs throughout the basin. Indeed, it would be more appropriate to create one EGTC covering the entirety of the basin, whereas the lower-level hydrological scales require a different type of cooperation instruments, such as the two types explained in Section 3.3 below.

Transboundary river basin management would take place in the context of one main legal framework, which would improve clarity and coherence.

– *A measure beyond the EGTC: proportional?*

461. Creating an instrument at EU level on top of the EGTC to 'sculpt' transboundary river basin management would face the risk of breaching the proportionality principle. As explained in Chapter II, the proportionality entails that the content and form of measures at EU level must not go beyond what is necessary to achieve the objectives of the EU Treaties.[1584] The creation of a joint entity on the basis of the EU legal framework operating in parallel with the EGTC is likely to raise questions with regard to the "necessity" aspect of the proportionality test. Indeed, such an instrument would inevitably show traits similar to the EGTC. However, EU action and guidance is necessary in order to tailor the EGTC to river basin management. As has been illustrated in Chapter III, the instrument has not been carried over to the IRBD and sub-basin levels. The EGTC could be deployed in the context of an Enhanced Cooperation Model, which will be proposed below.

3.2.3. The EU Water Directives

462. In addition to the EU primary-law framework and within and beyond the EU Water Directives, several actions can be taken to increase the facilitating power of the EU in IRBDs.

Two suggestions relate to "scope" and "institutional mechanisms". For reasons explained in Chapter II, and in line with the Critical Success Factors, the International River Basin District should be added as a unit of governance to the WFD and FD by including it explicitly in Article 2. Currently, as explained in Chapter II, the River Basin District is the "main unit for management of river basins".[1585] Stipulating that the International River Basin District is the main unit for management of river basins districts covering the territory of more than one Member State would lay the foundation for improved IRBD governance. Member States have already made the administrative arrangements to assign water bodies to IRBDs on the basis of Article 3(1) of the WFD. One of the aspects that relates to the assignment of water bodies to the IRBD is the designation of the competent

[1584] See Article 5 TEU. See Geert Van Calster and Leonie Reins, *EU Environmental Law* (Edward Elgar Publishing 2017) 21.

[1585] Article 2(15) WFD.

authority. These competent authorities are assigned to river basin districts by the Member States. Pursuant to Annex I of the WFD, Member States have to define more details with regard to the competent authorities related to e.g. their legal status, responsibilities and the international relationships. The Directives could require the Member States to designate the IRBD entity as the primary competent authority. Annex I of the WFD could be reviewed through the comitology procedure, as the WFD allows for an adaptation of this Annex in light of scientific and technical information. In parallel, the Common Implementation Strategy could focus on producing a document specifically targeting IRBD management. The CIS document should focus both on international cooperation models and on territorial cooperation models. The various elements of the WFD and FD where there is a lack of clarity in terms of the impact of provisions on IRBD cooperation identified in Chapter II should be examined.

Furthermore, IRBD management could be enhanced through the applicable procedural requirements. I agree with Van Kempen's suggestion that the qualification of the requirement to promulgate a single River Basin Management Plan in International River Basin Districts should constitute an obligation of results.[1586] The "basin" rationale underlying this consideration is not restricted to water quality management, and is equally important for water quantity and risk management. The upstream-downstream interaction is equally apparent. Therefore, this requirement should not only extend to RBMPs, but to FRMPs as well. More emphasis should be placed on the importance of these single basin-wide plans. An assessment based on the River Basin District Security concept as part of the ECM will be explained below.

463. A survey conducted in the context of the Fitness Check of EU freshwater policy showed that 57.9% of stakeholders identified insufficient finance as one of the barriers to implementing EU water law, along with a lack of capacity in relevant public bodies.[1587] In a first phase, the Enhanced Cooperation Model could be implemented as a pilot project for "integrated river basin management" through EU funding. The joint entity at the level of the basin could be granted financial and human resources and clear competences to achieve coherence throughout the basin. In this first phase, the EGTC framework could be used. Following a WFD / FD planning cycle, in which the ECM is implemented and a River Basin District Security Assessment is carried out, the appropriateness of the EGTC as a tool for river basin management should be evaluated. The Critical Success Factors described in the section above could be used to guide its set-up and scope, as will be discussed in the sections below.

1586 Jasper Van Kempen (n 1567) 389.

1587 Commission, 'Public Consultation to Support the Fitness Check of EU Freshwater: Policy Analysis of the Responses' (2012) <http://ec.europa.eu/environment/water/blueprint/pdf/public%20consultation%20report.pdf> consulted 11 June 2017.

464. Finally, the local hydrological scale levels and their relationship with the IRBD level should not be overlooked in future revisions of the Water Directives. The EU WFD and FD focus on International River Basin Districts and national River Basin Districts, but not on the cross-border, more local parts of the IRBDs. A specific action, for example, relates to the adoption of transboundary sub-basin management plans. Sub-basin aspects are currently integrated in the plans through the national RBD levels. The more local hydrological scale level is discussed in section 3.3.

3.2.4. River Basin District Security assessment

465. As has been established the sections above the "River Basin District Security" concept can be operationalised by requiring an associated assessment. The nuts and bolts of such possible assessment mechanism will be discussed in the sections below. A great deal of information necessary to conduct this assessment already exists based on data gathered in light of the planning and reporting cycles for the Water Directives. For the WFD, these requirements for which data should be collected are included in Articles 4 to 11 and two planning cycles have already been completed.[1588] The same holds true for risk management, where the procedural requirements of the FD have resulted in detailed risk assessments, cartographical information and plans detailing measures to manage flood risks in EU Member States.[1589] The new or innovative aspect of these two components is the more integrated approach in comparison to the situation as is.

466. A component which is not that apparent in current planning cycles relates to the evaluation of institutional mechanisms for cooperation in IRBDs. The European Commission has published a map of International River Basin Districts, pictured below, with an indication of four categories of cooperation. Category 1 reflects the top category and refers to a situation where a cooperation agreement, cooperation body and international RBMP Are in place, Category 2 where a cooperation agreement and cooperation body are in place, but no international RBMP is in place, Category 3 where a cooperation agreement is in place but no cooperation body or international RBMP is in place, and Category 4 where no cooperation has been formalised. This categorisation is depicted in figure 19.

1588 Programmes of measures are set up in the context of achieving the environmental objectives of Article 4, Member States have made economic analyses of water use, have identified the waters used for the abstraction of drinking water, have identified protected areas, have established monitoring programmes for surface waters and groundwaters, and have rolled out cost-recovery programmes in light of water services.

1589 The three phases of the Floods Directive have been explained in Chapter II of this book. Articles 4 to 7 include the data requirements.

Figure 19. EU River Basin Districts Indicating Transboundary Cooperation

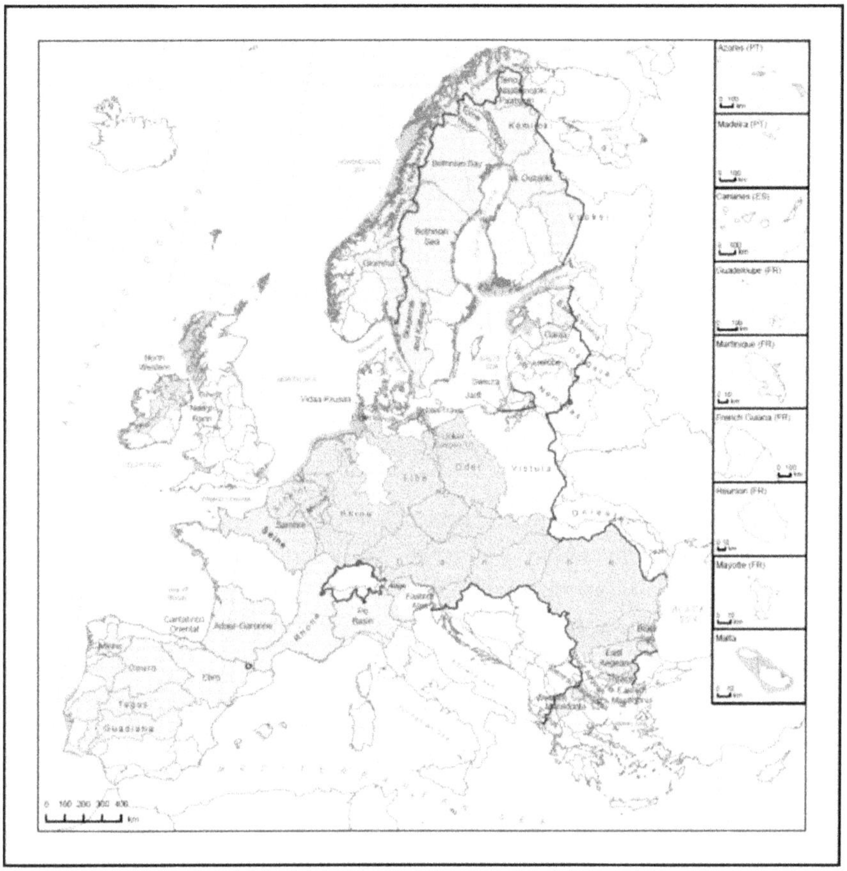

Legend:

- Category 1: Co-operation agreement, co-operation body and international RBMP in place
- Category 2: Co-operation agreement and co-operation body in place, but no international RBMP in place
- Category 3: Co-operation agreement in place but no co-operation body or international RBMP in place
- Category 2/3: Not clear whether both co-operation agreement and co-operation body in place
- Category 4: No co-operation formalised
- Uncategorised
- National River Basin Districts (within the EU)
- International River Basin Districts (outside of the EU)
- Coastal waters
- Country borders
- EU27 extent

Map produced by WRc plc on behalf of the European Commission©, DG Environment, 2012

467. By way of example, the Scheldt River has been assigned to the top "category 1". Chapter III, however, has shown that the institutional mechanism governing the Scheldt River, by way of the International Scheldt Treaty and the International Scheldt Commission, does not consistently foster coherent and integrated river basin management. The mechanism is characterised by several bottlenecks impeding its resilience to the various types of uncertainties described in Chapter I.[1590] The Scheldt is designated as Category 1 cooperation, i.e. a basin with a

[1590] I.e. not only in terms of climate change, but also in terms of infrastructure, economic and political pressures, demographic pressures, urbanisation, and so forth.

cooperation agreement, a cooperation body and an international RBMP. Indeed, one cannot deny that cooperation exists in the Scheldt basin. However, the question arises to what extent this cooperation is adequately robust and resilient to face future water pressures, both from the perspective of too much and too little water. Looking at Category 1, i.e. "top of the class", some comments need to be made: (a) the cooperation agreement does not include any substantive and binding requirements and the definition of its intention is limited to vague; (b) the cooperation body, here the International Scheldt Commission, does not have clear competences; and (c) there is no adequate international RBMP, but only a roof report which contains references to intentions and general measures in a generalist and non-committing manner – the actual measures are included in the separate national plans, which have been drawn up by the respective parties individually.[1591] Within sub-basins, within one country, structural measures are often not coordinated – reference can be made to the sluices in the Dender sub-basin shared by the Walloon and Flemish regions.

468. The categorisation tool used by the EU is rather unrefined and its parameters, "the existence of a cooperation agreement", "the existence of a cooperation body", and "the existence of an international RBMP", are not sufficiently detailed to adequately assess cooperation practice in the IRBDs in question.

469. By including not only technical scientific data on the quality and use of the water systems and the flood risks associated with these systems, but also a thorough assessment of the river basin mechanism governing the basin as a whole, i.e. the legal and governance aspect, IRBD management would likely be promoted. The Critical Success Factors described in the section above could be useful as broad parameters guiding this exercise. The assessment would form part of the planning cycle and would therefore be conducted by the States themselves, ideally through the joint entity governing the IRBD.

3.3. ECM IMPLEMENTED IN IRBDS

470. On the basis of the data gathered in the previous chapters, the evaluation framework defined in Chapter I, and extensive literature review, a step-by-step approach toward reaching an Enhanced Cooperation Model (ECM) is presented. Within the realm of the ECM, the basin-wide agreement is concluded which creates a basin body, where the Critical Success Factors can serve as a guiding tool.

[1591] Commission, 'Commission Staff Working Document Accompanying the Report from the Commission to the European Parliament and the Council on the Implementation of the Water Framework Directive (2000/60/EC) River Basin Management Plans' (2012) SWD(2012) 379 final.

3.3.1. *Institutional set-up and scope of the mechanism*

471. Section 3.2 has established that it would be advisable to use the EGTC in moving toward an institutional mechanism for IRBD management on the basis of EU law, rather than creating a new EU-wide mechanism. In general, the EGTC can be set up for various reasons: it can act as a Managing Authority with the goal to implement a specific programme or project, it can be set up to ensure the management of infrastructure such as road networks, and it can act as a governance mechanism, which has a functional purpose in the sense that certain sectorial areas are subject to the EGTC Convention.[1592] Reference can be made to the EGTC "Eurométropole Lille-Kortrijk-Tournai", discussed in Chapter III, which was created to deal with issues such as transport and tourism.

472. In the case of IRBD management, the most appropriate set-up would be the EGTC as a governance structure, where the joint management of infrastructure would be especially relevant. The EGTC should exceed a project basis. Funding, e.g. stemming from the EU level, may constitute an incentive in launching negotiations toward achieving a joint institutional mechanism but it should not be the main goal of the vehicle.

473. The Convention, i.e. basic agreement, creating the EGTC should stipulate the tasks of the EGTC, in accordance with Article 7 of the EGTC Regulation in its 2013 amended form. The EGTC could be referred to as a "River District Commission". In using the EGTC as a cooperation vehicle, the Critical Success Factors listed above should be taken into account. In accordance with Article 8 (1)(j) of the EGTC Regulation, the EGTC Convention would refer to the WFD and FD as directly relevant legal instruments whose provisions the EGTC aims to implement.[1593]

(a) **CSF 1**: *Integration from the outset to facilitate integrated operationalisation.* The EGTC Convention should include all relevant elements of water management pursuant to the integrated approach as it lays the groundwork for integrated river basin management, including risk management, environmental objectives, environmental flow agreements and the use of water. The Convention should clearly define the territorial and hydrological scope of the legal framework and should explicitly refer to the implementation of the WFD and FD as one of the goals. A forward-looking approach, which takes into account climate change prognoses specific to the region in question,

1592 Alfonso Alcolea Martínez, 'Towards a New Generation of European Groupings of Territorial Cooperation' (2014) 2 European Structural and Investment Funds Journal 89, 90.

1593 Other relevant Directives are the SEA/EIA Directives, the Nitrates Directive, Urban Wastewater Directive. The water scarcity and droughts communication issued by the Commission should also be referred to, see Chapter II.

is paramount. This could be stipulated as a condition in promulgating the basin-wide plans and programmes.

(b) **CSF 2**: *Clear mandate of the joint entity rooted in supportive legal framework.* An EGTC is an entity with legal personality. Its mandate should be rooted in the EGTC Convention and associated Statutes in a clear and straightforward manner. This implies that the tasks of the entity and its tools for responding to changing circumstances should be listed. It could be designated as the primary entity controlling the communications and reporting with the EU Commission. As the EGTC has legal personality, it can participate in projects benefitting the basin. The EGTC River District Commission (RDC) could also be directed toward the coordinated management of infrastructure within the basin. The above sections have already pointed out that the EGTC as an entity is limited in terms of its regulatory and police authority. However, this does not prevent the Parties from agreeing on the submission of projects to the EGTC RDC for a check of coherence with the EGTC Convention and applicable plans. Moreover, rules issued by the Delaware Commission can only enter into effect following implementation by the Parties. The analysis of the Delaware mechanism has shown that binding decision-making power is not a *sine qua non* in achieving hydrological scale successes, as the DRBC is continuously seeking broad support from the states involved. However, it should be clear that the EGTC RDC is the primary entity for issues concerning the basin as a whole. This can be facilitated by giving the RDC the authority to coordinate the single management plan, in combination with the designation of the RDC as the competent authority regarding the requirements of the Water Directives in IRBDs. With reference to CSF 4, enabling the RDC to play a role in steering the basin through impending emergencies, may function as a catalyst to offset further enhanced cooperation through this entity.

(c) **CSF 3**: *Meaningful basin-wide plan with links to the lower-level hydrological scales.* One of the EGTCs could focus on executing the procedural requirements based on these Directives. Importantly, the plans should go beyond the currently formulated "roof reports", e.g. in the Scheldt, and strive toward a more comprehensive, basin-wide plan and programme. In this respect, the EGTC could be created with one of the goals being the preparation and coordination of a single river basin management plan and flood risk management plan during an entire planning cycle. Alternatively, one of the major tasks of the EGTC River District Commission could consist of coordinating the River Basin District Security Assessment, on the basis of data provided by the Parties. As mentioned above, the transboundary governance and legal aspect of the assessment would be the main additional element in comparison to the evaluation that must be submitted to the EU Commission. The RDC should have a clear overview of (a) the competent authorities at the sub-(sub-)basin scales and (b) the formal and informal cross-border connections between these entities, both in terms of institutional mechanisms and of procedural links, the step-by-step approach for which is defined in the following section. The Delaware analysis has shown that local river basin management plans, i.e. for those sections of the river where different entities have joined forces, are reviewed in light of the basin-wide Comprehensive Plan. Finally, public participation should be safeguarded by the EGTC RDC, through regular updates of the website, e-mail notifications to users, widespread dissemination of reports of meetings, involvement of stakeholders early on in the planning process, and so forth. In the context of the River District Commission, the lower-level hydrological scale mechanisms, such as the sub-basin boards, could be useful in raising public awareness and increasing participation at the local level. It would be worthwhile to further explore the "Citizens' Juries" tool that has been explained in Chapter II, as this tool would fit well in a transboundary river basin management mechanism.[1594]

(d) CSF 4: *Clarity on the use of shared waters and emergency protocol.* As part of the River Basin District Security Assessment, the use of shared waters could be expressed as well as the prioritisation of uses in times of emergency. There should be a clear understanding of the principles underpinning the use of shared waters, i.e. in particular the translation of the "equitable and reasonable use" principle in the basin in question. The EGTC Convention could authorise the River District Commission to clarify the meaning of "significant increase of flood risks" for the basin in question.[1595] Under the auspices of the RDC, an emergency protocol could be drawn up which would allow the entity to coordinate actions in times of impending emergencies.

(e) CSF 5: *Continuity in financial resources and independent staff members.* The Statutes of the EGTC need to designate the decision-making procedure in question, the number of representatives of the parties in the bodies of the EGTC should be clarified, and the financial allocations should be arranged.[1596] The EGTC should produce an annual expense with a rubric for operational expenses. The Regulation provides that the EGTC should consist, as a minimum, of an Assembly representing the parties and a Director acting on behalf of the EGTC. The parties can create additional bodies.[1597] The EGTC "Eurométropole Lille-Kortrijk-Tournai" also includes a Board as the executive body and a civil society forum. It is important that the EGTC has its own staff and budget to respond to the needs arising from the basin. This is strongly

1594 E.g. network juries. See Chapter II.

1595 This is the reflection of the solidarity principle enshrined in the Floods Directive, the scope of which requires clarification at basin level.

1596 Article 9 EGTC Regulation.

1597 Article 10(3) EGTC Regulation.

linked to CSF 2 related to the mandate of the entity. The Eurométropole has several motivated, full-time staff members who prepare the strategies and policy course. Funding could be secured with EU support, this being one of the incentives for IRBD States to engage in the Enhanced Cooperation Model.

(f) **CSF 6**: *Proactive enforcement*. The EGTC Convention could provide for proactive reviews of projects substantially influencing the basin by the EGTC RDC. Considering the restrictions in terms of regulatory authority of the entity, it is doubtful that the RDC could make a binding decision. It is useful to note that the water test in Belgium is non-binding, but that permit-issuing authorities need to elaborately motivate the reasons for deviating from the recommendations made by the water manager in question.

474. The focus of cooperation mechanisms in IRBDs should not only be directed toward the basin-wide level, but to the more local levels as well and to achieving coherence throughout. The following section will describe a step-by-step approach to connect the dots within the IRBD in the context of the ECM. This approach is mainly geared toward enhancing scalar links.[1598]

3.3.2. Connecting the dots within the IRBD

475. Section 4 of Chapter III has described the formal territorial cooperation instruments available to the relevant authorities in the Scheldt District, which mainly relate to the EGTC, the Benelux Treaty and the Brussels Agreement. Section 2 of the same chapter also highlighted the instruments available to water authorities in the countries in question, in particular the Walloon river contract, the French river contract and the Dutch water agreement. The analysis of the sub-basin specific parts of the Flemish River Basin Management Plans in the District has shown that sub-basin and sub-sub-basin authorities do not make widespread use of these cooperation instruments, except for the limited use of the EGTC, which is used to address various topics including water management.[1599] The Scheldt case study has shown that there is still significant room for improvement in terms of cooperation at the sub-basin scales.

476. The main message is that the existing internal water cooperation instruments in the States sharing the transboundary basin can be linked with the toolbox based on territorial cooperation. Through the territorial cooperation toolbox, the internal water cooperation instruments used can be transformed to fit into the transboundary

[1598] See Chapter I and e.g. C Folke and others, 'Adaptive Governance of Social-Ecological Systems' (2005) 30 Annu. Rev. Environ. Resourc. 441; Barbara a Cosens, 'Legitimacy, Adaptation, and Resilience in Ecosystem Management' (2013) 18 Ecology and Society.

[1599] E.g. the EGTC Eurométropole " Lille-Kortrijk-Tournai ".

context. Importantly, the use of the territorial cooperation route takes away some of the internal law barriers to transboundary river basin management.

(a) *Stakeholder mapping*: the first step would be to map all the relevant competent authorities active at the basin, sub-basin and sub-sub-basin levels throughout the entire basin. These should also include entities that are active at the hydrological scale but without formal and binding powers, such as the Flemish sub-basin boards. In this regard, the legal status of these entities and their scope of competences should be clarified.

(b) *Identification of the available legal tools:* the available legal instruments should be analysed meticulously. The EGTC Regulation applies throughout the EU, but States have often concluded bilateral or multi-lateral Treaties which facilitate territorial cooperation, i.e. remove the barriers in terms of differences in legal and institutional regimes on the respective sides of the border. Moreover, specific instruments that exist in the countries or regions in question can be relevant in shaping transboundary cooperation. Internal law barriers and opportunities should also be studied, e.g. the 50% cap on participation based on French law is relevant in this respect.[1600]

(c) *Cooperation clarification*: a second step would be to identify the cooperation structures active, or non-active, throughout the basin.[1601] These include both the structures directly involved in water management and the structures that have been created for other purposes. This is important in order to avoid further overlap of efforts in river basin management, e.g. as would be important for the EGTC Eurométropole "Lille-Kortrijk-Tournai" regarding the activities of the Walloon river contract "Escaut-Lys".

(d) *Application of territorial cooperation framework to (sub-)river basins*: once the available range of instruments has been defined, an evaluation exercise can be conducted with a view to identifying the way forward in district basin governance.

3.3.3. *Instruments relevant to the lower level scales in the territorial cooperation route*

477. Two instruments can be presented as appropriate tools in transboundary river basin management, which could be extrapolated to a broader level. These instruments are the Dutch water agreement and the Walloon river contract. These are currently mainly used within national borders, but can be used in

1600 See Chapter III.

1601 Chapter III has performed this exercise for the Scheldt District.

a transboundary setting as well, provided there is an overarching framework which facilitates such cooperation. These two types of cooperation mechanisms have been selected because (i) they have been evaluated positively in their own country, and (ii) they represent two different types of cooperation: a mechanism where an entity is created and a mechanism where existing players in the river basin management scene agree on mutually beneficial actions without creating a joint entity.[1602]

478. The Dutch water agreement is an example of the latter type of cooperation, and is based on the Water Act. These water agreements are available to water managers within a river basin district, who can make agreements on aspects of the water management that go beyond their territorial area of competence, with the goal of achieving coherent and purposeful water management.[1603] The water managers can also involve other public authorities, e.g. municipalities, that are charged with water-related tasks. It is important to note that these water agreements do not create a "new" entity, but are characterised as "competence agreements", which are agreements between public authorities defining the mutual deployment of public competences and obligations in executing certain public tasks.[1604] An example is the water agreement concluded between Rijkswaterstaat "IJsselmeergebied" and the regional water authority "Vallei & Eem", which includes provisions on the relationship between water quantity management and water quality management, and actions to be taken in case of regional water scarcity or when calamities with respect to the water system occur.[1605] Specific agreements are made with respect to the water levels in times of water scarcity.[1606]

479. The river contract model has been explained in Chapter III, and, in the Scheldt District, is used by France and by the Walloon Region. Entities in both areas have also concluded transboundary river contracts, specifically with regard to the sub-basin Semoi.[1607] Several other countries use the river contract as an

---

1602 Gilissen (n 1567); Hannelore Mees, Cathy Suykens and Ann Crabbé, 'Evaluating Conditions for Integrated Water Resource Management at Sub-Basin Scale. A Comparison of the Flemish Sub-Basin Boards and Walloon River Contracts' (2017) 27 Environmental Policy and Governance 42.

1603 Article 3(7) Water Act.

1604 Gilissen (n 1567) 130.

1605 See water agreement "Het gebiedsoverschrijdend water en stoffen tussen Rijkswaterstaat IJsselmeergebied en Waterschap Vallei & Eem" (1 July 2010) <www.rijksoverheid.nl/bestanden/documenten-en-publicaties/convenanten/2010/06/01/waterakkoord-rijkswaterstaat-ijsselmeergebied-waterschap-vallei-eem/waterakkoord-ijg-waterschap-vallei-en-eem.pdf> accessed 8 June 2017.

1606 Article 5 of the above-mentioned water agreement.

1607 The European Regional Development Fund and also funding by authorities in France and the Walloon Region, e.g. municipalities, water agencies and ministries.

instrument in water management, e.g. Spain, Italy, and Luxemburg.[1608] The river contract instrument is a voluntary and participatory management tool which is supported by the applicable legal framework.[1609] A transboundary river contract for sub-basins within a certain transboundary basin could both focus on specific projects with mutual benefits across the border and on the publication of mutual plans or programmes. It is a bottom-up, voluntary tool, but it functions as an entity with control over staff and financial resources. As indicated, this is the main difference with the Dutch version.

480. The EGTC Regulation does not refer to other forms of cooperation in addition to the creation of the EGTC as such. This in contrast to the Benelux Treaty, which provides for the creation of the BGTC, but also refers to the creation of an entity without legal personality and the conclusion of an administrative agreement. Should an EGTC be created with the purposes of river basin management, it would be possible to accommodate instruments such as the river contract and the water agreement within the realms of the basin-wide constellation. In the sub-basin "IJzer", the EGTC "West-Vlaanderen / Flandre – Dunkerque – Côte d'Opale" operates. This EGTC has created a working group on water, which has identified specific necessities, on the basis of which studies have been commissioned, in turn resulting in concrete actions. Once the basin-wide stakeholder-mapping exercise has been conducted, the cooperation gaps and opportunities have been identified, and the territorial cooperation framework has been clarified, sub-basin cooperation initiatives could be promoted by the EGTC River District Commission. A working group could focus on the adoption of a cross-border sub-basin management plan. For Benelux countries, the BGTC could be created as an alternative for the EGTC, with the entity without legal personality and the administrative agreement being geared to the cross-border sub-basins and sub-sub-basins, with a feedback and linking mechanism to communicate with the BGTC. If the Dutch water agreement pursuant to Article 3(7) of the Water Act could be used as a transboundary cooperation tool, the administrative agreement within the meaning of the Benelux Treaty would be the appropriate route. As established in Chapter III, the applicable law would be the law of the country in whose territory the relevant aspect of the water agreement is carried out.

1608 These countries have not been studied in this book. See, for a comprehensive overview of existing river contracts and their underlying rationale Maria Laura Scaduto, *River Contracts and Integrated Water Management in Europe* (2016).

1609 The Walloon river contract tool strongly focuses on its participatory capacity. Francis Rosillon and Jérôme Lobet, 'Transboundary River Contract Semois-Semoy between Belgium (Wallonia) and France' in Patrick Meire (ed), *Integrated Water Management: Practical Experiences and Case Studies* (Springer 2008) 202.

Table 13. Application of CSFs at EU and IRBD levels

Critical Success Factors Applied to:	EU Legal FW	IRBD
Integrated RBM	- Addressing EU primary law dichotomy between quality and quantity through revision or interpretation - Revisions of WFD / FD: IRBD as unit of governance in Article 2 - Amendment of Annex I WFD through comitology procedure - CIS guidance on IRBD management: clarification of cooperation requirements and river basin security assessment - EU funding for Enhanced Cooperation Model project / EU demonstration project	- Application of Enhanced Cooperation Model
Meaningful plan or programme	- Requirement for adoption of single management plan - Requirement for River Basin District Security assessment	- Draft and submit, monitoring, cyclical evaluations and reporting
Clarity in underlying principles regarding shared use & protocol emergency situations	- Requirement for agreement on use of shared waters in WFD - Clarification of "significant increase of risks" through CIS document	- Agreement on use and priorities of use, e.g. based on UN Watercourses Convention criteria - Clear allocation of tasks and responsibilities in basic agreement and subsequent plan
Continuity of financial resources	- EU funding	- Negotiate at IRBD level in the context of ECM
Overview & interaction with sub-(sub-)basin scales	- Inter-scalar links as part of reporting process - CIS guidance: implementing a sub-basin management plan	- Stakeholder mapping in IRBD up to the most local level - Transboundary river contract or transboundary water authority project
Enforcement, including pro-active	- Evaluate mediation provision EU Commission in WFD/FD	- Negotiation at IRBD level: clout of the RDC

4. CONCLUDING REMARKS

481. Examining the legal and governance mechanisms related to two specific river basins, the Scheldt in the EU and the Delaware in the US, has allowed the identification of a number of Critical Success Factors for integrated river basin management. These relate to the institutional set-up of the joint entity and the elements of the basic agreement. It is important for this basic agreement to adopt an integrated approach to water management from the outset, and it is important that plans and programmes are coherent throughout the basin, that there is clarity with regard to the use of shared waters including in times of impending emergencies, and that links with the sub-(sub-)basin scales are in place e.g. through the review of plans.

These CSFs have been translated at the EU level. The discrepancy in decision-making procedures between qualitative and quantitative water management in EU primary law, and the associated confusion with regard to the scope of the latter, are persistent barriers to integrated river basin management in the EU which should be changed. A revision of the relevant Treaty provisions has been proposed. In the meantime, the quantitative water management provision should be interpreted narrowly. The CSFs are also implemented in the EU legal framework through the Enhanced Cooperation Model, which represents an incremental approach toward improving IRBD management. The ECM aims to address the concerns expressed by international scholars and confirmed in this book with regard to the delicate relationship between international and EU law in IRBDs in the current constellation. The EGTC is presented as a EU vehicle for comprehensive river basin management, where its usefulness should be evaluated e.g. following a WFD-FD planning cycle.

The River Basin District Security Assessment has been proposed as a comprehensive tool. The most important additional element in comparison to the current procedural requirements at the EU level and the analysis of the governance and legal framework are specifically tailored to cooperation. Indeed, the current categorisation of IRBDs at EU level is rather unrefined and deserves further study.

482. The need for basin-wide governance can be fulfilled via the EU legal framework through a combination of revised procedural requirements and existing institutional mechanisms beyond the Water Directives. The suggested Treaty revision and more integrated management in IRBDs facilitated by EU law through the Enhanced Cooperation Model are the goals to work toward. However, it is clear that steps forward should be facilitated, as it is widely acknowledged that achieving "good governance" in transboundary water management is an incremental process. This consideration should not be equated with acquiescence in the status quo, which is not sustainable in light of the water insecurity that society is currently facing.

CHAPTER VI
CONCLUSIONS

483. Since the entry into force of the WFD in 2000, the EU river basin approach has triggered an important shift in the environmental law landscape. This river basin approach brings about interesting legal challenges and opportunities. As established in the previous chapters, International River Basin Districts are subject to highly complex and fragmented multi-level governance systems. Member States sharing these International River Basin Districts are subject to EU-law requirements related to water management and need to ensure, to some extent, a coordinated implementation in these districts. This book has addressed the question to what extent the European legal framework warrants cooperation between Member States in International River Basin Districts with regard to water quantity management. The final conclusions of this book will be formulated in the sections below.

1. RESILIENCE THEORY AS A TOOL FOR EVALUATION

484. This book has built on social-ecological resilience theory to identify appropriate benchmarks for the evaluation of transboundary river basin management. Resilience theory is well equipped as a theoretical framework, because it revolves around finding ways for a certain legal or governance mechanism to thrive in the face of changing circumstances. River basins are inherently susceptible to such changing circumstances. Indeed, as was explained in Chapter I, several phenomena and uncertainties are associated with the management of freshwater resources. A major type of uncertainty that should be taken into account relates to the hydrological cycle and the impact of climate change on this cycle, for example changes in annual runoff, decrease of groundwater flow and the intensification of extreme events. However, climate change is not the only type of uncertainty river basin management schemes need to take into account. Other relevant factors include the performance of infrastructure relevant to the basin, economic and political volatility, demographic developments, and increased urbanisation. These phenomena are synergistic – climate refugees are a good illustration of this point.

Six benchmarks were formulated, which constituted a clear theme throughout this book:

(a) Integrated river basin management at the hydrological scale and nested governance;
(b) Equitable and reasonable utilisation and "no harm";
(c) Transfer of information across jurisdictions;
(d) Meaningful public participation;
(e) Adaptability: monitoring and opportunities for learning; and
(f) Effectiveness and enforcement.

485. These benchmarks were applied to the case studies of the Scheldt and the Delaware in Chapters III and IV. More specifically, the EU legal framework was first dissected in Chapter II, and subsequently included in the evaluation conducted in Chapter III in terms of the manner in which it facilitates transboundary cooperation in the International River Basin District Scheldt. The evaluation of the applicable regime in the Delaware River basin has provided valuable lessons learned. The following section will provide a concise overview of the main lessons learned from the EU legal framework analysis.

2. LESSONS LEARNED FROM THE DISSECTION OF THE EU LEGAL FRAMEWORK

486. The "five pillar analysis" has enabled the identification of a number of bottlenecks. The combination of the high threshold for legislation related to quantitative management of water resources and the fact that the provision has been circumvented for the EU Water Directives, has raised questions on its exact scope.[1610] EU secondary law with explicit traits of quantitative management of water resources has been adopted through the ordinary legislative procedure, on the basis of Article 192(1) TFEU.

An analysis of the development of the legal basis for water management through time, relevant case law and the relevant directives and their preparatory works, provided some degree of guidance in this regard.[1611] Measures need to have an impact, although no "significant" impact[1612], on physical territories or the availability

1610 As a reminder, measures that aim to achieve the objectives set in Article 191 TFEU are subject to the ordinary legislative procedure. An exception to this rule exists for "quantitative management of resources or measures, directly or indirectly, impacting the availability of those resources", which is subject to the special legislative procedure and therefore unanimity voting.

1611 The analysis of the preparatory works of the WFD and FD has been described in Chapter II, as well as the development of the Treaty provisions over time. For case law, see, in particular, Case C-36/98, *Spain v Council of the European Union* [2001] ECLI:EU:C:2001:64.

1612 This is in contrast to e.g. the energy provision of Article 192, i.e. Article 192(2)(c) which subjects measures "significantly affecting a Member State's choice between different energy sources"

of water resources. Provisions setting allocation limits for example, would affect the availability of resources. As the Floods Directive clearly illustrates, a purely procedural instrument stipulating obligations such as the issuance of cartographical information, preliminary flood risk assessments, and so forth, does not trigger the unanimity requirement, even when such measures deal with the quantitative management of water resources. Still, some confusion remains. Several amendments which would have promoted integrated management of water resources were rejected in the run-up to the adoption of the WFD to avoid the "centre of gravity" from shifting away from water quality. Water quantity measures form part of the WFD, but rather indirectly and partially, or through procedural requirements. Examples of measures related to water quantity are the provisions, although rather weak, related to water pricing and the requirements with respect to the register of water abstraction.[1613] The management of water use crept into the WFD through the programmes of measures, where, importantly, the discretionary powers with regard to limits of use remain with the Member States. Quantitative aspects constitute a small factor in the determination of the Good Ecological Status of surface water bodies.[1614] The Floods Directive, in the context of a more general trend toward proceduralisation[1615] in the EU legal landscape for water, requires Member States to take phased actions toward mapping and determining their own management strategies for flood risk management.[1616] The Directive does not contain any substantive requirements.

487. Cooperation requirements in International River Basin Districts are a matter of best efforts, and the exact aspects of the Water Directives which warrant cooperation are obvious. The WFD and Floods Directive focus on all water bodies in the EU, not only the transboundary water bodies. In comparison to the UNECE Convention and the UN Watercourses Convention, which are specifically geared to transboundary waters, the EU Water Directives are therefore less specific regarding the aspects related to water sharing and cooperation. In essence, the river basin

---

to the unanimity requirement. The threshold for deciding when substantive measures in this regard trigger the unanimity requirement is therefore rather low in the current constellation. See for an extensive analysis using shale gas as a case study: Leonie Reins, *Regulating Shale Gas: The Challenge of Coherent Environmental and Energy Regulation* (Edward Elgar Publishing 2017).

1613 As explained in Chapter II, the water-pricing obligations leave a great deal of discretionary powers to the Member States.

1614 Especially with respect to the High Ecological Status – for the determination of the good, moderate, poor and bad water status, the hydromorphological quality variable only relates to "conditions consistent with the achievement of the values specified for the biological quality elements".

1615 And the associated shift from "government" toward " governance ". See Saskia van Holten and Marleen van Rijswick, 'The Governance Approach in European Union Environmental Directives and Its Consequences for Flexibility, Effectiveness and Legitimacy' in Marjan Peeters and Rosa Uylenburg (eds), *EU Environmental Legislation: Legal Perspectives on Regulatory Strategies* (Edward Elgar 2014) 16.

1616 William Howarth, 'Aspirations and Realities under the Water Framework Directive: Proceduralisation, Participation and Practicalities' (2009) 21 Journal of Environmental Law 391, 397.

approach is geared to national River Basin Districts. Competent authorities are designated at River Basin District level, which is the main unit of governance for the implementation of the WFD and FD.[1617] For the WFD, there are coordination requirements for the achievement of environmental objectives and with respect to the programmes of measures.[1618] The coordination mechanism included in Article 3(4) WFD was not translated into the FD, which can be explained by the lack of objectives and programmes of measures to be issued on the basis of the FD. Even though there is a general obligation to cooperate, and an obligation of best efforts to coordinate measures throughout the IRBDs, from the joint reading of Article 7(2) and Article 5(2) FD, it seems that Member States are subject to the procedural requirement of setting objectives individually and for their own territories. Due-diligence requirements to coordinate RBMPs and FRMPs exist, but there are no legal consequences if Member States fail to produce a single plan.[1619]

The equitable and reasonable use principle and the "no harm" rule are not prominent features of the EU Water Directives. The main expression of the "no harm" rule is reflected in the FD's solidarity principle, which entails that Member States should not include measures in their plans that would significantly increase risks of flooding in another basin State. The FD does not require Member States to express the meaning of this solidarity principle and its operationalisation at IRBD level. The Scheldt case has shown that riparians find it difficult to agree on this operationalisation. This brings us to the next part of this book: the case studies.

3. EVALUATION OF THE SCHELDT AND DELAWARE MECHANISMS

488. Chapters III and IV assessed the legal instruments and governance mechanisms of two river basins, one situated in the EU and one in the US.

489. Chapter III provided an overview of flood risk management in the Scheldt countries through the five flood risk management strategies: risk prevention, defence, mitigation, preparation, and recovery.[1620] It has illustrated that flood risk management approaches in the different Scheldt countries are rather divergent, which confirms the idea that there is no "one size fits all" in flood risk management. The Dutch have legally binding safety standards, whereas further upstream, Flemish authorities determine these standards on a case-by-case basis following cost-benefit analyses. In Belgium and France, ex-post compensation following

1617 Article 3(2) and 3(2) WFD.
1618 Article 3(4) WFD in conjunction with Article 4 WFD.
1619 Article 13(2) WFD and Article 8(2) FD.
1620 Peter PJ Driessen and others, 'Toward More Resilient Flood Risk Governance' (2016) 21 Ecology and Society.

floods is provided under an insurance scheme, although with varying degrees of solidarity and risk differentiation, whereas in the Netherlands, this aspect of flood risk management is only marginally addressed.[1621] Where similar instruments exist in the Scheldt countries, e.g. the water assessment, these are applied differently. Mutual lessons from the application of these instruments throughout the basin are apparent, as established in Chapter III. There are varying degrees of cooperation that can be associated with the flood risk management strategies, ranging from the exchange of lessons learnt, exchange and coordination of information, and coordination of measures. The most apparent opportunities for cooperation are situated in the defence, risk mitigation and preparation strategies.

490. The multi-level mismatch between the relevant authorities in the Scheldt district is obvious.[1622] In Belgium, hydrological scale entities are advisory bodies, whereas the actual decision-making takes place in the pre-existing structures.[1623] The river basin approach is applied, but has not led to an institutional shift. In the Flemish and Walloon Region in Belgium, water management is in the realm of competences of the regions, provinces and municipalities. Whereas the majority of Flemish municipalities transferred their competences over third-category watercourses to the level of the provinces following the Internal State Reform of 2014, this has not occurred in the Walloon Region, increasing the asymmetry between the water managers in the respective Regions. Moreover, the Walloon river contract and the Flemish sub-basin board, the relevant hydrological scale entities, have a different institutional set-up and role to play in water management. The Netherlands has a hydrological governance system in place, with broad authority for the functionally decentralised regional water authorities. France also has an elaborate river basin management scheme in place.

491. In terms of cooperation in the IRBD Scheldt, the three-dimensional analysis, i.e. national, regional and international, has shed light on how cooperation is integrated into the different layers of governance in the basin. The national levels reflect the due-diligence nature of cooperation requirements based on the EU Water Directives.[1624] In terms of cross-border cooperation between the relevant authorities for water management in the Scheldt District, although there are successful examples of such cooperation, Chapter III has pinpointed that there is

[1621] Cathy Suykens and others, 'Dealing with Flood Damages: Will Prevention, Mitigation and Ex-Post Compensation Provide for a Resilient Triangle?'.

[1622] Tobias Chilla and others, 'On the Territoriality of Cross-Border Cooperation: "Institutional Mapping" in a Multi-Level Context' (2016) 20 European Planning Studies 961.

[1623] Water managers in Belgium are at the level of the respective regions, provinces and municipalities.

[1624] For example, the relevant legislative acts, such as the Flemish Decree Integrated Water Policy, refer to the due diligence coordination requirement set in the WFD and FD related to the adoption of a single plan at IRBD level.

a "double opportunity gap" in this regard. The first gap relates to the discrepancy between theory and practice. Whereas there is a myriad of cooperation frameworks that can be used by water managers to 'sculpt' joint initiatives throughout the basin, these are rarely used in practice. Moreover, water-related cooperation instruments based on internal law, e.g. the Dutch water agreements, have not been linked to these territorial cooperation frameworks or to cross-border initiatives in general. The sub-basin management plans issued at the local hydrological scales in the Flemish Region in light of the second-generation RBMPs indicated the need for more thoroughly developed cooperation across the relevant borders. There is therefore substantial room for improvement in terms of approaching cooperation at the sub-basin and sub-sub-basin levels more systematically.

At the international level, the International Scheldt Commission as the entity active at the IRBD level is not equipped with the right tools to assure basin-wide coherence. It operates as an advisory platform with limited legal personality, and limited financial and human resources. The applicable basic agreement, the 2002 Scheldt Treaty, does not provide a clear legal mandate for the Commission. Coordination efforts do exist, for example with respect to monitoring, the establishment of working groups for various issues of water management, and the adoption of a "roof report" for the EU institutions. The actual measures are included in the national-level plans and programmes. Although progress has been made under the WFD and the FD, in law and in fact, the output of the IRBD mainly consists of the sum of the national legal frameworks and policies with respect to the respective parts of the Scheldt River. The IRBD Scheldt as a hydrological unit is rather flawed. This further supports the conclusion drawn in Chapter II of this book, namely that the EU legal framework for water quantity management in IRBDs is not sufficiently robust but rather a blueprint that needs to be further developed. This conclusion also confirms the assumption presented in Chapter I that the EU legal framework is inadequately equipped to foster effective transboundary river basin management.

492. Chapter IV provided an insight into the legal and governance framework for interstate water management in the United States. It established that water quantity management is mainly within the realm of the individual States, but that the federal government has influence through its power to regulate interstate commerce.[1625] There are three ways in which water quantity management in interstate waters can be addressed in the US: through federal legislation, through equitable apportionment by the Supreme Court, or through a contract with regard to the shared waters. Over time, many interstate water disputes have been brought before the Supreme Court, which has encouraged States to adopt a more cooperative, proactive approach to water sharing.[1626] Over time, several types of interstate

[1625] US Const. Article 1, § 8, cl. 3.

[1626] See Section 2.6 of Chapter IV for a more elaborate overview of the recourse to the original jurisdiction of the Supreme Court in apportionments of shared waters.

agreements and entities have been experimented with.[1627] Chapter IV focussed on cooperation instruments concluded on the basis of the Compact Clause, of which the Delaware River Basin Compact is a prime example. Multi-level governance issues are ubiquitous in the Delaware Basin, with nineteen federal agencies, 14 interstate agencies and 43 state agencies, 38 counties and 838 municipalities having stakes in the basin necessitating conciliation. The Delaware Basin has been subject to important Supreme Court apportionment judgments and several attempts at creating joint institutional mechanisms.[1628] The Delaware is currently governed by the Delaware River Basin Compact, which entered into force in 1961 following its implementation at the federal level and in the four basin States.[1629]

In a fragmented and scattered legal landscape with respect to water quantity management, the Parties to the Delaware Compact have managed to pursue integrated river basin management in the Delaware River Basin. There were several difficulties and hiccups along the way, e.g. the dwindling contributions from the state levels and the lack of federal contribution since 1997, coupled with watered-down representations of the Parties in Commission meetings. Still, the Delaware River Basin Compact has accomplished basin-wide successes and has been evaluated positively on the basis of the resilience parameters. The Delaware River Basin Compact addresses all relevant aspects of water management, from pollution to floods to allocation issues, and clearly defines the rights and responsibilities of the Delaware River Basin Compact. By systematically reviewing projects that have a substantial impact on the water resources in the basin against the provisions of the Comprehensive Plan, the basin-wide perspective is safeguarded.

The following section presents the normative part of this research.

4. CRITICAL SUCCESS FACTORS IN RELATION TO IRBD MANAGEMENT IN THE EU

493. The case studies of the Scheldt and the Delaware have provided the necessary tools to identify Critical Success Factors (CSFs), which have been applied to IRBD management in the European Union. This step completes this book's full circle. These Critical Success Factors are useful additions to the existing body of literature and policy papers related to "good governance" principles for water management. Considering the fact that the Critical Success Factors are mainly based on the two case studies conducted in this book, they are neither conclusive nor exhaustive. However, based on literature review of transboundary water management, and the consideration that

---

1627 E.g. Watershed Councils, single federal authorities, the Regional Authority, Title-II Commissions. See Section 2.7 of Chapter IV.

1628 *New York v New Jersey*, and, in particular, the Interstate Commission on the Delaware River Basin (INCODEL).

1629 Delaware River Basin Commission Compact, Pub. L. No. 82-573, 66 Stat. 738 (1961).

the basins studied in this book are situated on two different continents, broadening the geographical scope, they are relevant beyond application to the Scheldt and the Delaware. The Critical Success Factors are listed in the table below.

Table 14. Overview of the CSFs

CSF 1: Integrated approach to water management from the outset to enable further operationalisation on the basis of the supporting legal framework
CSF 2: The overarching legal framework enables and supports effective basin-wide actions by the Basin Body in the basin as the main unit of governance
CSF 3: A meaningful basin-wide plan and/or programme which defines actual measures and links with the sub-basin scales
CSF 4: Clarity regarding principles underpinning the use of shared water resources and protocol for emergency situations
CSF 5: Continuity in financial resources independent from Parties' allocations and adequate number of independent staff members in the Basin Body
CSF 6: Enforcement is an important aspect of RBM and can be ex ante, i.e. proactive, and ex post, i.e. through reporting and compliance monitoring

494. Chapter V presented a number of recommendations, both geared to EU primary law, with the goal of closing the gap between water quantity management and water quality management, and secondary law, with a view to increasing the capacity of the EU framework to warrant cooperation in IRBDs. Recommendations are further divided between the actions that can be taken at EU level and those that can be taken at IRBD level.

495. The concept "River Basin District Security" was presented, which is derived from the more commonly known "water security", as a comprehensive assessment tool to be used in (International) River Basin Districts. The River Basin Districts Security Assessment encompasses water quality, water quantity, and risk management, but also entails the evaluation of the governance aspect of transboundary water management, i.e. the institutional mechanisms for cooperation in IRBDs. Indeed, currently, this aspect is underdeveloped in the planning and reporting requirements at the EU level. This aspect of the assessment should be specifically geared to the question of hydrological scale management, coherence throughout the basin, the performance of the joint entity, and so forth. The current categorisation and parameters for International River Basin Districts, namely the four categories described in Chapter V, are not sufficiently refined as characterisations.

In line with CSF 1, it is paramount that an integrated approach is adopted from the outset. For International River Basin Districts, this means that the EU primary-law discrepancy between quality and quantity should be bridged. The environmental flow is a useful tool in this regard, as it relates to the amount of flow necessary for the aquatic ecosystem to thrive. By anchoring water security, encompassing quality, quantity and risk management, more strongly into Article 191 TFEU, and limiting the scope of the "quantitative management of water resources" as currently formulated in Article 192(2) TFEU, water management would gain steadier footing in the Treaty in comparison to the current version. Pending such revision, an interpretative framework has been proposed, using lower and upper boundaries which represent the unanimity versus the majority voting procedures.

In order to meet the concerns related to the implementation of EU requirements in shared waters via mechanisms based on international law, an incremental approach is proposed. In first instance, reliance on the existing territorial cooperation instrument, the European Grouping for Territorial Cooperation is suggested. The usefulness of the instrument with respect to IRBDs may be evaluated, e.g. following a full WFD/FD planning cycle. Creating a new instrument without having resorted to the EGTC would likely raise proportionality concerns.

496. In the Enhanced Cooperation Model, a number of revisions could be implemented in the Directives, for example the requirement to adopt a single River Basin Management Plan / Flood Risk Management Plan, and the revision of Annex I of the Directive through comitology to better reflect the International River Basin District as the unit of governance. When implementing the Enhanced Cooperation Model in International River Basin Districts, the EGTC River District Commission may be created, applying the six Critical Success Factors presented in Chapter V. Importantly, the dots in the IRBD should be connected through stakeholder mapping and linking the territorial cooperation framework with existing water-related instruments in internal law.

497. Figure 20 visualises the application of Critical Success Factors to the EU legal framework.

Figure 20. Summary of Findings Related to CSFs and the EU Framework

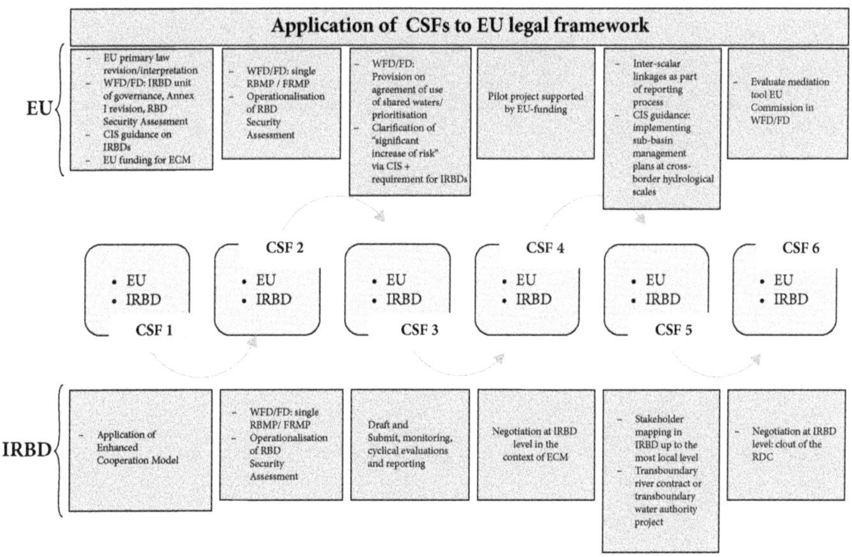

BIBLIOGRAPHY

JOURNAL ARTICLES

Abrams RH, 'Interstate Water Allocation: A Contemporary Primer for Eastern States' (2002) 25 University of Arkansas Little Rock Law Review 155-173.

Acreman MC, 'Environmental Flows and the European Water Framework Directive' (2010) 55 Freshwater Biology 32-48.

Adler R and Straube M, 'Watersheds and the Integration of US Water Law and Policy: Bridging the Great Divides' (2000) 25 William & Mary Environmental Law and Policy Review 1-67.

Ajami NK, Hornberger GM and Sunding DL, 'Sustainable Water Resource Management under Hydrological Uncertainty' (2008) 44 Water Resources Research 1-10.

Akamani K, 'Adaptive Water Governance: Integrating the Human Dimensions into Water Resource Governance' (2016) Journal of Contemporary Water Research & Education 2-18.

Allan A, Loures F and Tignino M, 'The Role and Relevance of the Draft Articles on the Law of Transboundary Aquifers in the European Context' (2011) 8 Journal for European Enviornmental & Planning Law 231-251.

Alreemy Z, Chang V, Walters R and Wills G, 'Critical Success Factors (CSFs) for Information Technology Governance (2016) 36 International Journal of Information Management 907-916.

Ambrus M, Gilissen HK and Van Kempen JJH, 'Public Values in Water Law: A Case of Substantive Fragmentation?' (2014) 10 Utrecht Law Review 8-30.

Baaner L, 'The Programme of Measures of the Water Framework Directive – More than just a Formal Compliance Tool' (2011) JEEPL 82-100.

Beijen B, van Rijswick M and Anker H, 'The Importance of Monitoring for the Effectiveness of Environmental Directives: A Comparison of Monitoring Obligations in European Environmental Directives' (2014) 10 Utrecht Law Review 126-135.

Benvenisti E, 'Collection Action in the Utilisation of Shared Freshwater: The Challenges of International Water Resources Law' (1996) 90 The American Journal of International Law 384-415.

Bernadett LD, 'Equitable Apportionment in the Supreme Court: An Overview of the Doctrine and the Factors Considered by the Supreme Court in Light of Florida v. Georgia' (2014) 29 Environmental Law and Litigation 511-534.

Borg S, 'The Influence of International Case Law on Aspects of International Law Relating to the Conservation of Living Marine Resources beyond National Jurisdiction' (2012) 23 Yearbook of International Environmental Law 44-79.

Bruggeman V, 'A Critical Comparison of the Main Compensation Mechanism for Victims of Natural Catastrophes in Belgium and the Netherlands. With a Law and

Economics Twist' (2011) 8 Journal for European Environmental & Planning Law 46-61.
Brun A, 'Les Contrats de Rivière En France: Enjeux, Acteurs et Territoires' (2010) 51 Les Cahiers de droit 679-704.
Buijze A, 'Promoting Sustainable Water Management in Area Development: A Regulatory Approach' (2012) 24 The Journal of Water Law 166-174.
Elola Calderón T, 'La Politique de L'eau de l'Union Européenne: Vers Une Gestion Quantitative Des Ressources Hydriques ?' (2010) 51 Les Cahiers de droit 859-878.
Carette A and De Smedt P, 'Het Vernieuwde Decreet Integraal Waterbeleid: Sneller En Beter?' (2013) Tijdschrift voor Milieurecht 576-601.
Carter J and Howe J, 'The Water Framework Directive and the Strategic Environmental Assessment Directive: Exploring the Linkages' (2006) 26 Environmental Impact Assessment Review 287-300.
Certini G and others, 'Holocene as Anthropocene Geological Evidence for the Anthropocene' (2015) 349 Science 246-247.
Chaffin BC, Gosnell H and Cosens BA, 'A Decade of Adaptive Governance Scholarship: Synthesis and Future Directions' (2014) 19 Ecology and Society 56.
Chilla T and others, 'On the Territoriality of Cross-Border Cooperation: " Institutional Mapping " in a Multi-Level Context' (2016) 20 European Planning Studies 961-980.
Clemons J, 'Interstate Water Disputes: A Road Map for States' (2004) 12 Southeastern Environmental Law Journal 115-142.
Cosens B, 'Transboundary River Governance in the Face of Uncertainty: Resilience Theory and the Columbia River Treaty' (2010) 30 Journal of Land Resources and Environmental Law 229-265.
Cosens B, 'Legitimacy, Adaptation, and Resilience in Ecosystem Management' (2013) 18 Ecology and Society 3.
Craig RK and Ruhl JB, 'Governing for Sustainable Coasts: Complexity, Climate Change, and Coastal Ecosystem Protection' (2010) 2 Sustainability 1361-1388.
Deason J, Schad T and Sherk GW, 'Water Policy in the United States: A Perspective' (2001) 3 Water Policy 175-192.
Deketelaere K, 'The Use of Fiscal Instruments in European Environmental Policy: Review Essay' (1999) 10 European Environmental Policy 146-210.
Dellapenna J, 'Interstate Struggles over Rivers: The Southeastern States and the Struggle over the 'Hooch' (2005) 12 N.Y.U. Environmental Law Journal 828-900.
Dellapenna J, 'The Customary International Law of Transboundary Fresh Waters' (2001) 1 International Journal of Global Environmental Issues 264-305.
Dellapenna J, 'International Law's Lessons for the Law of the Lakes' (2006) 40 University of Michigan Journal of Law Reform 747-797.
Dellapenna J, 'The Evolution of Riparianism in the United States' (2011) 95 Marq. L. Rev. 53-90.
De Smedt P, 'Towards a New Policy for Climate Adaptive Water Management in Flanders: The Concept of Signal Areas' (2014) 10 Utrecht Law Review 107-125.
De Sadeleer N, 'Principle of Subsidiarity and the EU Environmental Policy' (2012) 9 Journal for European Environmental & Planning Law 63-70.
De Smedt P, 'Towards a New Policy for Climate Adaptive Water Management in Flanders: The Concept of Signal Areas' (2014) 10 Utrecht Law Review 107-125.

Dietz T, Ostrom E and Stern PC, 'Struggle to Govern the Commons' (2003) 302 Science 1907-1912.
Dinar S, Green, O, McNally, A, Blankespoor, B and Kurukulasuriya, P, 'Climate Change and State Grievances: The Resiliency of International River Treaties to Increased Water Variability' (2010) 3 Insights 1-29.
Doorn-Hoekveld W van, 'Compensation in Flood Risk Management with a Focus on Shifts in Compensation Regimes Regarding Prevention, Mitigation and Disaster Management' (2014) 10 Utrecht Law Review 216-238.
Driessen, PPJ, Hegger DLT, Bakker MHN, Van Rijswick MHFMW, and Kundzewicz ZW, 'Toward more resilient flood risk governance' (2016) 21 Ecology and Society 53.
Ebbesson J, 'The Rule of Law in Governance of Complex Socio-Ecological Changes' (2010) 20 Global Environmental Change 414-422.
Ebbesson J and Hey E, 'Introduction: Where in Law Is Social-Ecological Resilience?' (2013) 18 Ecology and Society 25.
Eckstein G, 'A Hydrogeological Approach to Transboundary Ground Water Resources and International Law' (2003) 19 American University International Law Review 201-258.
Eckstein G, 'Commentary on the UN International Law Commission's Draft Articles on the Law of Transboundary Aquifers' (2007) 18 Colorado Journal of International Environmental Law & Policy 537-610.
Estes M, 'The Effect of the Federal Endangered Species Act on State Water Rights' (1992) 22 Environmental Law 1027-1066.
Evrard E, 'The European Grouping of Territorial Cooperation (EGTC): Towards a Supraregional Scale of Governance in the Greater Region SaarLorLux?' (2016) 21 Geopolitics 513-537.
Faure M and Bruggeman V, 'Catastrophic Risks and First-Party Insurance' (2008) 15 Connecticut Insurance Law Journal 1-52.
Featherstone JP, 'Existing Interstate Compacts: The Law and the Lessons' (2001) 4 Tol. J. Great Lakes L. SCI & Pol'Y 271-282.
Featherstone JP, 'Water Resources Coordination and Planning at the Federal Level: The Need for Integration' (2011) 104 Journal of Contemporary Water Research and Education 52-54.
Florestano PS, 'Past and Present Utilisation of Interstate Compacts in the United States' (1994) 24 Publius: The Journal of Federalism 13-25.
Frankfurter F and Landis J, 'The Compact Clause of the Constitution. A Study in Interstate Adjustments' (1936) 45 Yale Law Journal 685-758.
Gerten D and others, 'Towards a Revised Planetary Boundary for Consumptive Freshwater Use: Role of Environmental Flow Requirements' (2013) 5 Current Opinion in Environmental Sustainability 551-558.
Gilissen HK, Alexander M, Beyers J-C, Chmielewski P, Matczak P, Schellenberger T and Suykens C, 'Bridges over Troubled Waters: An Interdisciplinary Framework for Evaluating the Interconnectedness within Fragmented Domestic Flood Risk Management Systems' (2016) 25 Journal of Water Law 12-26.
Glucker A, Driessen P, Kolhoff A and Runhaar H, 'Public Participation in Environmental Impact Assessment: Why, Who and How?' (2013) 43 Environmental Impact Assessment Review 104-111.

Goodrich C, 'the Gallatin Plan after One Hundred and Fifty Years' (1958) 102 Proceedings of the American Philosophical Society 436-441.

Grant DL, 'Interstate Allocation of Rivers Before the United States Supreme Court: The Apalachicola-Chattahoochee-Flint River System' (2004) 21 Ga.St.UL Rev. 401-427.

Green O, Garmestani A, Van Rijswick M and Keessen A, 'EU Water Governance: Striking the Right Balance between Regulatory Flexibility and Enforcement?' (2013) 18 Ecology and Society 10.

Green O, Cosens BA and Garmestani AS, 'Resilience in Transboundary Water Governance: The Okavango River' (2013) 18 Ecology and Society 23.

Grey D and Sadoff C, 'Beyond the River: The Benefits of Cooperation on International Rivers.' (2002) 4 Water Policy 389-403.

Folke C, 'Resilience: The Emergence of a Perspective for Social-Ecological Systems Analyses' (2006) 16 Global Environmental Change 253-267.

Faure M, 'Insurability of Damage Caused by Climate Change: A Commentary' (2007) 155 University of Pennsylvania Law Review 1875-1899.

Faure M, 'Effectiveness of Environmental Law: What Does the Evidence Tell Us?' (2012) 36 William & Mary Environmental Law and Policy Review 294-336.

Faure M, 'The Complementary Roles of Liability, Regulation and Insurance in Safety Management: Theory and Practice' (2014) 17 Journal of Risk Research 689-707.

Folke C, Hahn T, Olsson P and Norberg J, 'Adaptive Governance of Social-Ecological Systems' (2005) 30 Annual Review of Environment and Resources 441-473.

Gilissen HK, Alexander M, Matczak, P, Pettersson M, Bruzzone S, 'A Framework for Evaluating the Effectiveness of Flood Emergency Management Systems in Europe' (2016) 21 Ecology and Society 27.

Gilissen HK, Alexander M, Beyers, J, Chmielewski, P, Matczak, P, Suykens, C, Schellenberger, T, 'Bridges over Troubled Waters: An Interdisciplinary Framework for Evaluating the Interconnectedness within Fragmented Domestic Flood Risk Management Systems' (2016) 25 Journal of Water Law 12-26.

Giordano M, Drieschova, A, Duncan, JA, Sayama, Y, De Stefano, L and Wolf, AT, 'A Review of the Evolution and State of Transboundary Freshwater Treaties' (2013) 14 International Environmental Agreements: Politics, Law and Economics 245-264.

Grad F, 'Federal-State Compact: A New Experiment in Co-Operative Federalism' (1963) 63 Columbia Law Review 825-855.

Hall N, 'Interstate Water Compacts and Climate Change Adaptation' (2010) 5 Environmental & Energy Law & Policy Journal 237-323.

Hall N, 'Toward a New Horizontal Federalism: Interstate Water Management in the Great Lakes Region' (2006) 77 U. Colo. L. Rev. 405-456.

Heikkila T, Schlager E and Davis MW, 'The Role of Cross-Scale Institutional Linkages in Common Pool Resource Management: Assessing Interstate River Compacts' (2011) 39 Policy Studies Journal 121-145.

Heron K, 'The Interstate Compact in Transition: From Cooperative State Action to Congressionnally Coerced Agreements' (1985) 60 St John's L. Rev. 1-25.

Hey E, 'Multi-Dimensional Public Governance Arrangements for the Protection of the Transboundary Aquatic Environment in the European Union: The Changing Interplay between European and Public International Law' (2009) 6 International Organizations Law Review 191-223.

Hill Clarvis M, Allan A and Hannah DM, 'Water, Resilience and the Law: From General Concepts and Governance Design Principles to Actionable Mechanisms' (2014) 43 Environmental Science and Policy 98-110.

Holling CS, 'Resilience and Stability of Ecological Systems' (1973) 4 Annu.Rev.Ecol.Syst. 1-23.

Hooper B, 'River Basin Organization Performance Indicators: Application to the Delaware River Basin Commission' (2010) 12 Water Policy 461-478.

Hooper B, 'River Basin Organization Performance Indicators: Application to the Delaware River Basin Commission: Supplementary File' (2010) 12 Water Policy 1-24.

Hoornbeek, Policy-Making Institutions and Water Policy Outputs in the European Union and the United States: A Comparative Analysis, (2004) 11 Journal of European Public Policy 461-496.

Howarth W, 'Aspirations and Realities under the Water Framework Directive: Proceduralisation, Participation and Practicalities' (2009) 21 Journal of Environmental Law 391-417.

Hudson P, Botzen, WJ, Feyen, L, Aerts, J, 'Incentivising Flood Risk Adaptation through Risk Based Insurance Premiums: Trade-Offs between Affordability and Risk Reduction' (2016) 125 Ecological Economics 1-13.

Huitema D, Mostert, E, Egas, W, Moellenkamp, S, Pahl-wostl, C, Yalcin, R, 'Adaptive Water Governance: Assessing the Institutional Prescriptions of Adaptive (Co-) Management from a Governance Perspective and Defining a Research Agenda' (2009) 14 Ecology and Society 26.

Jaspers FGW, 'Institutional Arrangements for Integrated River Basin Management' (2003) 5 Water Policy 77-90.

Jeffers E, 'Creating Flexibility in Interstate Compacts' (2009) 36 Ecology Law Quarterly 209-236.

Kauffman GJ and others, 'Water Quality Trends in the Delaware River Basin (USA) from 1980 to 2005.' (2011) 177 Environmental monitoring and assessment 193-225.

Keessen A, Vink, M, Wiering, M, Boezeman, D, Ernst, W, Mees, H, Van Broekhoven, S and Van Eerd, M 'Solidarity in Water Management' (2016) 21 Ecology and Society 35.

Keessen A, Van Kempen J and Van Rijswick M, 'Transboundary River Basin Management in Europe Legal Instruments to Comply with European Water Management Obligations in Case of Transboundary Water Pollution and Floods' (2008) 4 Utrecht Law Review 35-56.

Keessen A, Freriks A and van Rijswick M, 'The Clash of the Titans: The Relation between the European Water and Medicines Legislation' (2010) 47 Common Market Law Review 1429-1454.

Keessen A and Van Rijswick M, 'Adaptation to Climate Change in European Water Law and Policy' (2012) 8 Utrecht Law Review 38-50.

Kelley AK, 'Federal Preemption and State Water Law' (1996) 105 Journal of Contemporary Water Research and Education 4-9.

Kenyon W, 'A Critical Review of Citizens' Juries: How Useful Are They in Facilitating Public Participation in the EU Water Framework Directive?' (2005) 48 Journal of Environmental Planning and Management 431-443.

Kimber CJM, 'A Comparison of Environmental Federalism in the United States and the European Union' (1995) 54 Maryland Law Review 1658-1690.

Klein C a, 'The Dormant Commerce Clause and Water Export: Toward a New Analytical Paradigm' (2011) 35 Harvard Environmental Law Review 131-154.
Kundis Craig R, 'Stationarity Is Dead. Long Live Transformation: Five Principles for Climate Change Adaptation Law' (2010) 34 Harvard Environmental Law Review 9-74.
Lankford B, 'Does Article 6 (Factors Relevant to Equitable and Reasonable Utilisation) in the UN Watercourses Convention Misdirect Riparian Countries?' (2013) 38 Water International 130-145.
Lavranos N, 'The MOX Plant and IJzeren Rijn Disputes: Which Court Is the Supreme Arbiter?' (2006) 19 Leiden Journal of International Law 223-246.
Lavranos N, 'Freedom of member states to bring disputes before another court or tribunal: Ireland condemned for bringing the MOX plant dispute before an arbitral tribunal. Grand Chamber decision of 30 May 2006, Case C-459/03, Commission v. Ireland' (2006) European Constitutional Law Review 456-469.
Lindhout P, 'A Wider Notion of the Concept of Water Services in EU Water Law' (2012) 8 Utrecht Law Review 86-101.
Lindhout P, Carette A and Smedt PDE, 'Kostenterugwinning van Waterdiensten in Woelig Water: Bedenkingen Bij Het Arrest C-525 / 12 van Het Hof van Justitie Bezien in Rechtsvergelijkend Perspectief' (2015) 1 Tijdschrift voor Milieurecht 4-21.
Lindhout P and Van Rijswick M, 'The Effectiveness of the Principle of Recovery of the Costs of Water Services Jeopardized by the European Court of Justice – Annotations on the Judgment in C-525 / 12' (2015) 12 Journal for Environmental and Planning Law 80-94.
Lubell M, 'Collaborative Environmental Institutions: All Talk and No Action?' (2004) 23 Journal of Policy Analysis and Management 549-573.
Maes F, 'The Content of the Agreements on the Protection of the Rivers Scheldt and Meuse' (1997) Revue Belge de Droit International 2, 661-681.
Magsig BO, 'Introducing an Analytical Framework for Water Security: A Platform for the Refinement of International Water Law' (2009) 20 Journal of Water Law 61-69.
Mandarano LA, Featherstone JP and Paulsen K, 'Institutions for Interstate Water Resources Management' (2008) 44 Journal of the American Water Resources Association 136-147.
Mandarano LA and Mason RJ, 'Adaptive Management and Governance of Delaware River Water Resources' (2013) 15 Water Policy 364-385.
Martínez AA, 'Towards a New Generation of European Groupings of Territorial Cooperation' (2014) 2 European Structural and Investment Funds Journal 89-100.
McCaffrey S, 'The International Law Commission Adopts Draft Articles on Transboundary Aquifers' (2009) 103 The American Journal of International Law 272-293.
McCaffrey S, 'The International Law Commission's flawed Draft Articles on the Law of Transboundary Aquifers: the way forward' (2011) 36 Water International 566-572.
McDonald J and Styles MC, 'Legal Strategies for Adaptive Management under Climate Change' (2014) 26 Journal of Environmental Law 25-53.
Mcglothlin RM and Slater SS, 'No Fictions Required: Assessing the Public Trust Doctrine in Pursuit of Balanced Water Management' (2014) 17 University of Denver Water Law Review 53-97.

McIntyre O, 'The Emergence of an "Ecosystem Approach" to the Protection of International Watercourses under International Law' (2004) 13 Review of European Community and International Environmental Law 1-14.

McIntyre O, 'The Proceduralisation and Growing Maturity of International Water Law'(2010) 22 Journal of Environmental Law 475-497.

McIntyre O, 'The Protection of Freshwater Ecosystems Revisited: Towards a Common Understanding of the Ecosystems Approach to the Protection of Transboundary Water Resources' (2014) 23 Review of European, Comparative & International Environmental Law 88-95.

Mees H, Crabbé, A, Alexander, M, Kaufmann, M, Bruzzone, S, Lévy, L, Lewandowski, J, 'Coproducing Flood Risk Management through Citizen Involvement: Insights from Cross-Country Comparison in Europe' (2016) 21 Ecology and Society 7.

Mees H, Suykens C and Crabbé A, 'Evaluating Conditions for Integrated Water Resource Management at Sub-Basin Scale. A Comparison of the Flemish Sub-Basin Boards and Walloon River Contracts' (2017) 27 Environmental Policy and Governance 1.

Meijerink S and Huitema D, 'The Challenges and Pitfalls of Decentralisation in Water Resources Management' (2015) 5 Water Governance 16-21.

Muys JC, Sherk GW and Leary MCO, 'Utton Transboundary Resources Center Model Interstate Water Compact' (2006) 47 Natural Resources Journal 17-155.

Nadalutti E., 'Does the 'European Grouping of Territorial Cooperation' Promote Multi-level Governance within the European Union?' (2004) 51 Journal of Common Market Studies 756-771.

Ntegeka V, Decloedt C, Willems P, Monbaliu J, 'Quantifying the Impact of Climate Change from Inland, Coastal and Surface Conditions' (2013) Comprehensive Flood Risk Management 103-111.

Padowski JC, Gorelick S, Thompson B, Kummu M, Ward P, De Moel H, 'Measuring Global Water Security towards Sustainable Development Goals' (2016) 11 Environmental Research Letters 1-13.

Page B and Kaika M, 'The EU Water Framework Directive: Part 2. Policy Innovation and the Shifting Choreography of Governance' (2007) 13 Eur. Env. 6, 328-343.

Percival R V, 'The Clean Water Act and the Demise of the Federal Common Law of Interstate Nuisance' (2004) 55 alabama Law Review 717-747.

Perko A, 'Watershed Management: A Comparison between Efforts in the United States and the European Union' (2013) 16 University of Denver Water Law Review 166-186.

Pettersson M, van Rijswick M, Suykens C, Priest S, Alexander M, 'How Legitimate Is Flood Risk Governance in Europe? Insights from Intra-Country Assessments' (in review) Water International.

Poelmans L and Van Rompaey A, 'Detecting and modelling spatial patterns of urban sprawl in highly fragmented areas: A case study in the Flanders–Brussels region' (2009) 93 Landscape and Urban Planning 10-19.

Poussin JK, Botzen WJW and Aerts JCJH, 'Stimulating Flood Damage Mitigation through Insurance: An Assessment of the French CatNat System' (2013) 12 Environmental Hazards 258-277.

Poussin JK, Wouter Botzen WJ and Aerts JCJH, 'Effectiveness of Flood Damage Mitigation Measures: Empirical Evidence from French Flood Disasters' (2015) 31 Global Environmental Change 74-84.

Priest SJ, Suykens C, Van Rijswick M, Schellenberger T, Goytia S, Kundzewicz Z, Van Doorn-Hoekveld W, Beyers J-C, Homewood S, 'The European Union Approach to Flood Risk Management and Improving Societal Resilience: Lessons from the Implementation of the Floods Directive in Six European Countries' (2016) 21 Ecology and Society 50.

Regent S, 'The Open Method of Coordination: A New Supranational Form of Governance?' (2003) 9 European Law Journal 190-214.

Rieu-Clarke A, A Survey of International Law relating to Flood Management: Existing Practices and Future Prospects, (2008) International Law and Flood Management 649-677.

Rieu-Clarke A and Spray CJ, 'Ecosystem Services and International Water Law: Towards a More Effective Determination and Implementation of Equity?' (2013) 16 Potchefstroom Electronic Law Journal 12-65.

Robison J and Kennedy D, 'Equity and the Colorado River Compact' (2012) 42 Envtl. L. 1157-1209.

Rockart J, 'Chief Executives Define Their Own Data Needs' (1979) 3 Harvard Business Review 81-93.

Rockström J and others, 'A Safe Operating Space for Humanity' (2009) 461 Nature 472-475.

Ruhl J, 'Regulation by Adaptive Management-Is It Possible?' (2005) 7 Minn. JL Sci. & Tech. 21-57.

Ruhl J and Salzman J, 'Climate Change, Dead Zones, and Massive Problems in the Administrative State: A Guide for Whittling Away' (2010) 98 California Law Review 59-120.

Ryan E, 'The Public Trust Doctrine, Private Water Allocation, and Mono Lake: The Historic Saga of National Audubon Society v. Superior Ct.' (2015) 45 Environmental Law 561-640.

Salman SM, 'The Helsinki Rules, the UN Watercourses Convention and the Berlin Rules: Perspectives on International Water Law' (2007) 23 International Journal of Water Resources Development 625-640.

Scanlan MK, Sinykin JH and Krohelski J, 'Realizing the Promise of the Great Lakes Compact: A Policy Analysis for State Implementation' (2006) 8 Vt. J. Envtl. L. 39-99.

Schmeier S, Gerlak AK and Blumstein S, 'Clearing the Muddy Waters of Shared Watercourses Governance: Conceptualizing International River Basin Organizations' (2016) 16 International Environmental Agreements: Politics, Law and Economics 597-619.

Sherk G, 'The Management of Interstate Water Conflicts in the Twenty-First Century: Is It Time to Call Uncle?' (2005) 12 New York University Environmental Law Journal 764-827.

Smakhtin V, Revenga C and Döll P, 'A Pilot Global Assessment of Environmental Water Requirements and Scarcity' (2004) 29 Water International 307-317.

Steffen W, Richardson C, Rockström J, Cornell S, Fetzer I, Bennet E, Biggs R, Carpenter S, de Vries W, de Wit C, Folke C, Gerten D, Heinke J, Mace G, Persson L, Ramanathan

V, Reyers B and Sörlin S, 'Planetary Boundaries: Guiding Human Development on a Changing Planet' (2015) 347 Science 1-15.
Surminski S, Aerts J, Botzen J, Hudson P, Mysiak J and Pérez-Blanco, 'Reflections on the Current Debate on How to Link Flood Insurance and Disaster Risk Reduction in the European Union' (2015) 79 Natural Hazards 1451-1479.
Suykens C, 'Gas Flaring in Developing Countries – Need for Kyoto Mechanisms or Sectoral Crediting Mechanisms ?' (2010) 1 Carbon and Climate Law Review 42-51.
Suykens C, 'Globalisation of the Nuclear Fuel Cycle and Maritime Carriage of Radioactive Materials: Review of the Legal Regime' (2011) 26 The International Jounral of Marine and Coastal Law 385-411.
Suykens C, 'EU Water Quantity Management in International River Basin: Crystal Clear?' (2015) European Energy and Environmental Law Review 134-143.
Suykens C and Priest S, 'Dealing with Flood Damages: Will Prevention, Mitigation and Ex-Post Compensation Provide for a Resilient Triangle?' (2016) 21 Ecology and Society 50.
Tarlock D, 'The Law of Equitable Apportionment Revisited, Undated and Restated' (1985) 56 U. Colo. L. Rev 381-411.
Tarlock D, 'Reconnecting Property Rights to Watersheds' (2000) 25 William & Mary Environmental Law and Policy Review 69-112.
Tarlock D and Wouters P, 'Are Shared Benefits of International Waters an Equitable Apportionment' (2007) 18 Colorado Journal of International Environmental Law and Policy 523-536.
Teclaff L, 'Evolution of the River Basin Concept in National and International Law' (1996) 36 Natural Resources Journal 359-391.
Termeer C, Dewulf A, Van Rijswick M, Van Buuren A, Huitema D, Meijerink S, Rayner T and Wiering M, 'The Regional Governance of Climate Adaptation: A Framework for Developing Legitimate, Effective, and Resilient Governance Arrangements' (2011) 2 Climate Law 159-179.
Toury J and Denys M, 'Vernieuwde watertoets – vergoedingsregeling bij bouw- en verkavelingsverbond' (2012) Nieuw Juridisch Weekblad 82-92.
Unnerstall H, 'The Principle of Full Cost Recovery in the EU-Water Framework Directive – Genesis and Content' (2007) 19 Journal of Environmental Law 29-42.
Utton AE, 'In Search of An Integrating Principle For Interstate Water Law: Regulation versus the Market Place' (1985) 183 Natural Resources Journal 985-1004.
Van Calster G and Deketelaere K, 'Amsterdam, the IGC and greening the EU Treaty' (1998) 7 European Environmental Law Review 12-25.
Van Doorn-Hoekveld W, 'Transboundary Flood Risk Management: Compatibilities of the Legal Systems of Flood Risk Management in the Netherlands, Flanders and France.' (2017) 26 European Energy and Environmental Law Review 81-96.
Van Doorn-Hoekveld, Goytia S, Suykens C, Homewood S, Thuillier T, Manson P, Chmielewski P, Matczak P and Van Rijswick M, 'Distributional Effects of Flood Risk Management – a Cross-Country Comparison of Preflood Compensation' (2016) 21 Ecology and Society 26.
Van Kempen J, 'Countering the Obscurity of Obligations in European Environmental Law: An Analysis of Article 4 of the EuropeanWater Framework Directive' (2012) 24 Journal of Environmental Law 499-533.

Van Rijswick M, Gilissen HK and van Kempen J, 'The Need for International and Regional Transboundary Cooperation in European River Basin Management as a Result of New Approaches in EC Water Law' (2010) 11 ERA Forum 129-157.
Van Rijswick M and Salet W, 'Enabling the Contextualization of Legal Rules in Responsive Strategies to Climate Change' (2012) 17 Ecology and Society 18.
Van Rijswick M and others, 'Ten Building Blocks for Sustainable Water Governance: An Integrated Method to Assess the Governance of Water' (2014) 39 Water International 725-742.
Van Rijswick M, 'Mechanisms for Water Allocation and Water Rights in Europe and the Netherlands: Lessons from a General Public Law Perspective' (2015) 24 Journal of Water Law 141-149.
Vick M, 'The Law of International Waters: Reasonable Utilisation' (2014) 12 Chicago-Kent Journal of International and Comparative Law 141-178.
Watson N, Deeming H and Treffny R, 'Beyond Bureaucracy? Assessing Institutional Change in the Governance of Water in England' (2009) 2 Water Alternatives 448-460.
Weiland PS, 'Federal and State Preemption of Environmental Law: A Critical Analysis' (2000) 24 Harvard Environmental Law Review 237-286.
Wouters J and Vidal M, 'Eindelijk een publiekrechtelijk kader voor gedecentraliseerde grensoverschrijdende samenwerking' (2005) De Franse Nederlanden 219-224.
Zagonari F, 'Implementing a Trans-Boundary Flood Risk Management Plan: A Method for Determining Willingness to Cooperate and Case Study for the Scheldt Estuary' (2013) 66 Natural Hazards 1101-1133.
Zapletal J, 'The European Grouping of Territorial Cooperation (EGTC): A New Tool Facilitating Cross-Border Cooperation and Governance' (2010) 29 Quaestiones Geographicae 15-26.

BOOKS

Anderson J, O'Dowd L and Wilson T, *New Borders for a Changing Europe: Cross-Border Cooperation and Governance* (Psychology Press 2003).
Barlow M and Clark T, *Blue Gold: The Fight to Stop the Corporate Theft of the World's Water* (The New Press 2001).
Black P, *Readings in Soil & Water Conservation* (Irvington Publishers 1974).
Boeve MN and Groothuijse FAG, *Omgevingsrecht* (Europa Law Publishing 2013).
Broun C, Buenger M, McGabe M, Masters R, *The Evolving Use and the Changing Role of Interstate Compacts: A Practitioner's Guide* (American Bar Association 2006).
Bruggeman V, Compensating Catastrophe Victims: *A Comparative Law and Economics Approach* (Wolters Kluwer 2010).
Crabbé A, *Integraal Waterbeleid in Vlaanderen: van Fluïde naar Solide* (Ph.D. Thesis, Antwerp University 2008).
De Feyter K and others (eds), *The Local Relevance of Human Rights* (Cambridge University Press 2011).
de Jong W, *Transborder Governance of Forests, Rivers and Seas* (Earthscan 2010).
de Meij JM and van der Vlies IC, *Inleiding Tot Het Staatsrecht En Het Bestuursrecht* (Kluwer 2004).

de Sadeleer N, *EU Environmental Law and the Internal Market* (Oxford University Press 2014).
Farber D, *Environmental Law in a Nutshell* (9th edn, WestAcademic Publishing 2014).
Feldman D, *Water Policy for Sustainable Development* (The John Hopkins University Press 2007).
Freriks A, *Juridische Bescherming van Beeksystemen* (WEJ Tjeenk Willink 1994)
Gilissen HK, *Internationaal En Regionale Grensoverschrijdende Samenwerking in Het Waterbeheer* (Sdu Uitgevers 2009).
Gerlak A, *'Federalism and US Water Policy', The Use of Interstate Compacts to Resolve Transboundary Water Allocation Issues* (Edward Elgar 2014).
Hays S, *Conservation and the Gospel of Efficiency: The Progressive Conservation Movement 1890-1920* (University of Pittsburgh Press 1999).
Hedemann-Robinson M, *Enforcement of European Union Environmental Law: Legal Issues and Challenges* (Routledge 2007).
Hooper B, *Integrated River Basin Governance: Learning from International Experience* (IWA Publishing 2005).
Husa J, *A New Introduction to Comparative Law* (Bloomsbury 2015).
INBO and GWP, *The Handbook for Integrated Water Resources Management in Transboundary Basins of Rivers, Lakes and Aquifers* (2012).
Jans J, *European Environmental Law* (Europa Law Publishing 2000).
Johnson J, *United States Water Law: An Introduction* (CRC Press 2009).
Kauffman G, *Governance, Policy, and Economics of Clean Water in the Delaware River Basin* (Ph.D. Thesis, University of Delaware 2014).
Kemmis D, *This Sovereign Land: A New Vision for Governing the West* (Island Press 2001).
Krämer L, *EC Treaty and Environmental Law* (Sweet & Maxwell 1998).
Kulesza J, *Due Diligence in International Law* (Brill Nijhoff 2016).
Langlet D and Mahmoudi S, *EU Environmental Law and Policy* (Oxford University Press 2016)
Lavrysen L, *Handboek Milieurecht* (Kluwer 2010).
Leb C, *Cooperation in the Law of Transboundary Water Resources* (Cambridge University Press 2013).
Louka E, *Water Law and Policy Governance without Frontiers* (Oxford University Press 2008).
Lindhout P, *Cost Recovery as a Policy Instrument to Achieve Sustainable and Equitable Water Use in Europe and the Netherlands* (Ph.D. Thesis, Utrecht University 2015).
Magsig B-O, *International Water Law and the Quest for Common Security* (Routledge 2015).
McCaffrey S, *The Law of International Watercourses* (Oxford University Press 2007).
McCormick Z, *The Use of Interstate Compacts to Resolve Transboundary Water Allocation Issues* (Ph.D. Thesis, Oklahoma State University 1973).
McIntyre O, *Environmental Protection of International Watercourses Under International Law* (Ashgate 2007).
McKinley C, *Uncle Sam in the Pacific Northwest* (University of California Press 1952).
Nadalutti E, *The Effects of Europeanization on the Integration Process in the Upper Adriatic Region* (Springer 2015).
Ostrom E, *Governing the Commons: The Evolution of Institutions for Collective Action* (Cambridge University Press 1990).

Piris J-C, *The Future of Europe: Towards a Two-Speed EU?* (Cambridge University Press 2012).
Reins L, *Regulating Shale Gas: The Challenge of Coherent Environmental and Energy Regulation* (Edward Elgar Publishing 2017).
Richardson B, *Local Climate Change Law: Environmental Regulation in Cities and Other Localities* (Elgar 2012).
Rieu-Clarke A, Allan A and Hendry S (eds), *Routledge Handbook of Water Law and Policy* (Routledge 2017).
Sands P and Galizzi P (eds), *Documents in European Community Environmental Law* (Cambridge University Press 1995).
Santbergen L, *Ambiguous Ambitions in the Meuse Theatre* (Eburon 2013).
Scaduto ML, *Comparative Analysis between France and Italy* (Springer 2016).
Scaduto ML, *River Contracts and Integrated Water Management in Europe* (Springer 2016).
Schmeier S, *Governing International Watercourses: River Basin Organizations and the Sustainable Governance of Internationally Shared Rivers and Lakes* (Routledge 2013).
Scott J, *Environmental Protection* (Oxford University Press 2009).
Senecah S, *The Environmental Communication Yearbook* (Routledge 2013).
Seerden R, *Publiekrechtelijke Grensoverschrijdende Samenwerking tussen Decentrale Overheden* (Maklu 1993).
Siems M, *Comparative Law* (Cambridge University Press 2014).
Tanzi A and others (eds), *The UNECE Convention on the Protection and Use of Transboundary Watercourses and International Lakes* (Brill Nijhoff 2015).
Tribe S, *American Constitutional Law* (The Foundation Press, Inc., 1978).
Van Ast J, *Interactief watermanagement in grensoverschrijdende riviersystemen* (Ph.D. Thesis, Erasmus University Rotterdam 2000).
Van Calster G and Reins L, *EU Environmental Law* (Edward Elgar Publishing 2017).
Van Kempen J, *Europees Waterbeheer: Eerlijk Zullen We Alles Delen?* (Boom Juridische Uitgevers 2012).
Van Rijswick M and Havekes H, *European and Dutch Water Law* (Europa Law Publishing 2012).
Verwijmeren J and Wiering M, *Many Rivers to Cross: Cross-Border Cooperation in River Management* (Eburon Delft 2007).

BOOK SECTIONS

Baranyai G, 'The Water Convention and the European Union: The Benefits of the Convention for EU Member States' in Attila Tanzi and others (eds), *The UNECE Convention on the Protection and Use of Transboundary Watercourses and International Lakes* (Brill Nijhoff 2015) 88-100.
Bensoussan A and Farhi N, 'Uncertainties and Risks in Water Resources Management' in Jean-Michel Lasry, Delphine Lautier and Damien Fessler (eds), *The Economics of Sustainable Development* (Economica 2010) 163-179.
Bogaart M, 'The Emergence of the Framework Directive in EU Environmental Policy: An Exploration of Its Function and Characteristics' in Marjan Peeters Marjan and Rosa

Uylenburg (eds), *EU Environmental Legislation Legal Perspectives on Regulatory Strategies* (Edward Elgar Publishing 2014) 48-69.

Borg S, 'The Conservation of Marine Living Resources under International Law: The 1982 United Nations Convention on the Law of the Sea and Beyond' in Algosia Fitzmaurice, and Norman A. Martinez Gutierrez (eds), *The IMLI Manual on International Maritime Law – Volume I: The Law of the Sea* (Oxford University Press 2014) ch 12.

Brun A, 'France's Water Policy: The Interest and Limits of River Contracts' in Schneier-Madanes (ed), *Globalized Water: A Question of Governance* (2014) 139-147.

Brun A, 'France's Water Policy: The Interest and Limits of River Contracts' in Schneier-Madanes (ed), *Globalized Water: A Question of Governance* (2014) 261-276.

De Baere G and Gutman K, 'Federalism and International Relations in the European Union and the United States: A Comparative Outlook' in Elke Cloots, Geert De Baere and Stefan Sottiaux (eds), *Federalism in the European Union* (Hart Publishing 2012) 131-166.

Graziadei M, 'The Functionalist Heritage' in Pierre Legrand and Roderick Munday (eds), *Comparative Legal Studies: Traditions and Transitions* (Cambridge University Press 2003) 100-127.

Hertoghs M, 'De Grens in Je Achtertuin' in Michael Faure and Marjan Peeters (eds), *Grensoverschrijdend recht* (Intersentia 2006) 112.

Hey E and van Rijswick M, 'Transnational Water Management' in Oswald Jansen and Bettina Schöndorf-Haubold (eds), *The European Composite Administration* (Intersentia 2010) 231-253.

Holten S van and Rijkswick M van, 'The Governance Approach in European Union Environmental Directives and Its Consequences for Flexibility, Effectiveness and Legitimacy' in Marjan Peeters Marjan and Rosa Uylenburg (eds), *EU Environmental Legislation: Legal Perspectives on Regulatory Strategies* (Cheltenham 2014) 13-47.

Kaika M, 'Water for Europe: The Creation of the European Water Framework Directive' in Julie Trottier and Paul Slack (eds), *Managing Water Resources Past and Present* (OUP 2004) 88-116.

Keessen A, 'In Search of a European Legislative Approach to Adaptation to Climate Change' in Marjan Peeters and Rosa Uylenburg (eds), *EU Environmental Legislation: Legal Perspectives on Regulatory Strategies* (Elgar 2014) 193-210.

Kuijper PJ, '"It Shall Contribute to ... the Strict Observance and Development of International Law ..."' in Court of Justice of the European Union (ed), *The Court of Justice and the Construction of Europe: Analyses and Perspectives on Sixty Years of Case-law* (The Asser Press 2013) 9-13.

Magsig BO, 'Water Security as an Evolving Paradigm: Local, National, Regional and Global Considerations' in Alistair Rieu-Clarke, Andrew Allan and Sarah Hendry (eds), *Routledge Handbook of Water Law and Policy* (Routledge 2017) ch 28.

Margerum R, 'Integrated Water Resources Management in the United States: The Rogue and Willamette River Cases' in Jeroen Frank Warner, Arwin van Buuren and Jurian Edelenbos (eds), *Making Space for the River* (IWA Publishing 2012) 79-87.

McIntyre O, 'The Principle of Equitable and Reasonable Utilisation' in Attila Tanzi and others (eds), *The UNECE Convention on the Protection and Use of Transboundary Watercourses and International Lakes* (Brill Nijhoff 2015) 146-159.

McIntyre O, 'A Comparative Analysis of the Legal Frameworks That Govern Europe's Transboundary Waters' in Werner Scholtz and Jonathan Verschuuren (eds), *Regional Environmental Law Transregional Comparative Lessons in Pursuit of Sustainable Development* (Edward Elgar 2015) 386-416.

McIntyre O, 'The Principle of Equitable and Reasonable Utilisation' in Attila Tanzi and others (eds), *The UNECE Convention on the Protection and Use of Transboundary Watercourses and International Lakes* (Brill Nijhoff 2015) 146-159.

McIntyre O, 'Substantive Rules of International Water Law' in Alistair Rieu-Clarke, Andrew Allan and Sarah Hendry (eds), *Routledge Handbook of Water Law and Policy* (Routledge 2017) 234-246.

Mees H, Crabbé A and Suykens C, 'Flood risk governance in Belgium. Towards a resilient, efficient and legitimate arrangement?' in Corinne Larue and others, *Facing hydro-meteorological extreme events in Europe: a governance issue* (Wiley In Press 2018).

Moréteau O, 'Policing the Compensation of Victims of Catastrophes: Combining Solidarity and Self-Responsibility' in Willem Van Boom and Michael Faure (eds), *Shifts in Compensation between Private and Public Systems* (Springer 2007) 199-218.

Morgera E, 'Water Management and Protection in the EU' in M Alberton and F Palermo (eds), *Environmental Protection in Multi-Layered Systems: Comparative Lessons from the Water Sector* (Martinus Nijhoff 2012) 265-289.

Nollkaemper A and de Villeneuve C, 'Recht van Internationale Waterlopen' in Nathalie Horbach, René Lefeber and Olivier Ribbelink (eds), *Handboek Internationaal Recht* (T.M.C. Asser Press 2007) 769-799.

Peeters M, 'The Joint Governance of Transboundary River Basins: Some Observations on the Role of Law' in Michael Faure and Song Ying (eds), *China and International Environmental Liability: Legal Remedies for Transboundary Pollution* (Edward Elgar 2008) xx+360.

Pulido-Velazquez M and Ward FA, 'Comparison of Water Management Institutions and Approaches in the United States and Europe – What Can We Learn From Each Other?' in Jadwiga Ziolkowska and Jeffrey Peterson (eds), *Competition for Water Resources* (Elsevier Inc 2016) 423-441.

Rosillon F and Lobet J, 'Transboundary River Contract Semois-Semoy between Belgium (Wallonia) and France' in Patrick Meire (ed), *Integrated Water Management: Practical Experiences and Case Studies* (Springer 2008) 199-206.

Scott J and Holder J, 'Law and New Environmental Governance in the European Union' in G De Burca and Joanne Scott (eds), *Law and New Governance in the EU and the US* (Hart Publishing 2006) ch 8.

Tanzi A and Kolliopoulos A, 'The No-Harm Rule' in Attila Tanzi and others (eds), *The UNECE Convention on the Protection and Use of Transboundary Watercourses and International Lakes* (Brill Nijhoff 2015) 133-145.

Van Hall A, 'Water Stuurt Ruimtelijke Inrichting. Nieuw Denken over Oud Water' in Frank Maes and Luc Lavrysen (eds), *Integraal Waterbeleid in Vlaanderen en Nederland* (Die Keure 2003) 193-222.

Van Holten S and Van Rijswick M, 'The Governance Approach in European Union Environmental Directives and Its Consequences for Flexibility, Effectiveness and Legitimacy' in Marjan Peeters and Rosa Uylenburg (eds), *EU Environmental Legislation: Legal Perspectives on Regulatory Strategies* (Edward Elgar 2014) 13-47.

Van Rijswick M and Keessen A, 'The EU Approach for Integrated Water Resources Management: Transposing the EU Water Framework Directive within a National Context – Key Insights from Experience' in Alistair S Rieu-clarke, Andrew Allan and Sarah Hendry (eds), *Routledge Handbook of Water Law and Policy* (Routledge 2017) 51-65.

Vykhryst S, 'Public Information and Participation under the Water Convention' in Attila Tanzi and others (eds), *The UNECE Convention on the Protection and Use of Transboundary Watercourses and International Lakes* (Brill Nijhoff 2015) ch 18.

Wessel R, 'Reconsidering the Relationship between International and EU Law: Towards a Content-Based Approach?' in Enzo Cannizzaro, Paolo Palchetti and Ramses Wessel (eds), *International Law as Law of the European Union* (Koninklijke Brill NV 2012) 7-34.

Zellmer S, 'United States: The Emergence of Environmental Considerations' in Joseph W Dellapenna and J Gupta (eds), *The Evolution of the Law and Politics of Water* (Springer Science 2009) 205-223.

REPORTS

2030 Water Resources Group, 'Charting Our Water Future: Economic Frameworks to Inform Decision-Making' (2009).

Adviescommissie Water, 'Bruggen Bouwen Nederlands Waterbeheer in Europees En Grensoverschrijdend Perspectief' (2007).

Bates B, Kundzewicw Z, Wu S and Palukitof J, 'Climate Change and Water: IPCC Technical Paper of the Intergovernmental Panel on Climate Change' (IPCC 2008).

Commission, 'WFD CIS Guidance Document No. 3: Analysis of Pressures and Impacts' (2003).

Commission, 'Water Notes on the Implementation of the Water Framework Directive' (2008).

Commission, 'WFD CIS Guidance on Risk Assessment and the Use of Conceptual Models for Groundwater' (2010).

Commission, 'Commission Staff Working Document Accompanying the Report from the Commission to the European Parliament and the Council on the Implementation of the Water Framework Directive (2000/60/EC) River Basin Management Plans' (2012) SWD(2012) 379 final.

Commission, 'Report from the Commission to the European Parliament and the Council on the Implementation of the Water Framework Directive (2000/60/EC) River Basin Management Plans' (2012).

Commission, 'WFD CIS Guidance Document on Public Participation in relation to the Water Framework Directive' (2014).

Commission, 'Report on the Progress in Implementation of the Water Framework Directive: Programmes of Measures' (2015) SWD (2015) 50 final 12.

Commission, 'WFD CIS Guidance Document No. 31: Ecological Flows in the Implementation of the Water Framework Directive' (2015).

Commission, 'The Water Framework Directive and the Floods Directive: Actions towards the 'good status' of EU water and to reduce flood risks' COM (2015) 120 final.

Comptroller General of the United States, 'Federal-Interstate Compact Commissions: Useful Mechanisms for Planning and Managing River Basin Operations' (United States General Accounting Office 1981).

DePhilip M and Moberg T, 'Ecosystem Flow Recommendations for the Delaware River Basin' (The Nature Conservancy 2013).

Echeverria JD, 'The Intersection of Water Law and Takings Doctrine' (Proceedings of the 60th Annual Rocky Mountain Mineral Law Institute, Rocky Mountain Mineral Law Foundation 2014)

Echeverria JD, 'Water and Takings' (32nd Annual Water Law Conference American Bar Association 2014).

Ek K, Pettersson M, Alexander M, Pardoe J, Priest S, Suykens C and Van Rijswick M, 'Strengthening and Redesigning European Flood Risk Practices Towards Appropriate and Resilient Flood Risk Governance Arrangements Design Principles for Resilient, Efficient and Legitimate Flood Risk Governance; Lessons from Cross-Country Comparisons' (STAR-FLOOD Consortium 2016)

Gore R, 'Delaware River Basin Commission Administrative Memorandum: Equitable Apportionment of the Commission Current Expense Budget (General Fund)' (Delaware River Basin Commission 2012)

Hartmann T and Scheibel M, 'Flood Label for Buildings – a Tool for More Flood-Resilient Cities', FLOODrisk 2016 – 3rd European Conference on Flood Risk Management (2016).

Hegger D and others, 'Flood Risk Management in Europe: Similarities and Differences between the STAR-FLOOD Consortium Countries' (STAR-FLOOD Consortium 2013)

Hegger D, Driessen, P.P.J., Bakker, Marloes, Alexander, Meghan, Beyers, Jean-Christophe, Buijze, A.W.G.J., bruzzone, silvia, Chorynski, Adam, Crabbé, Ann, Deketelaere, K., Delvaux, B., Dieperink, C., Hoekveld, W.J., Ek, Kristina, Fournier, M., Ganzevoort, Wessel, Gilissen, H.K., Goytia, Susana, Gralepois, M., Green, C., van Herten, M.L., Homewood, Stephen, Kaufmann, M., Kundzewicz, Zbigniew, Larrue, Corinne, Levy, L., Jakub, Lewandowsky, Liefferink, D., Manson, Corinne, Matczak, Piotr, Mees, Hannelore, Micou, Anna Paula, Parker, Dennis, Pettersson, Maria, Priest, Sally, Raadgever, Tom, van Rijswick, H.F.M.W., Schellenberger, Thomas, Spegel, Elin, Suykens, Cathy, Szwed, Malgorzata, Tapsell, S., Thuillier, Thomas, Tremorin, J.B. and Wiering, Mark, 'A View on More Resilient Flood Risk Governance: Key Conclusions of the STAR-FLOOD Project' (STAR-FLOOD Consortium 2016).

Kaufmann M, Van Doorn-Hoekveld, W., Gilissen HK and Van Rijswick M, 'Analysing and Evaluating Flood Risk Governance in the Netherlands. Drowning in Safety ?' (STAR-FLOOD Consortium 2016).

Larrue C, Hegger D and Trémorin J-B, 'Researching Flood Risk Governance in Europe: Background Theories' (STAR-FLOOD Consortium 2013).

Larrue C, Bruzzone, S, Lévy, L, Gralepois, M, Schellenberger, T, Trémorin, JB, Fournier, M, Manson, C and Thuilier, T, 'Analysing and Evaluating Flood Risk Governance in France: From State Policy to Local Strategies' (STAR-FLOOD Consortium 2016).

Mees H, Suykens C, Beyers JC, Crabbé A, Delvaux B, Deketelaere K, 'Analysing and Evaluating Flood Risk Governance in Belgium Dealing with Flood Risks in an

Urbanised and Institutionally Complex Country' (STAR-FLOOD Consortium 2016).
Matczak P and others, 'Comparing Flood Risk Governance in Six European Countries: Strategies, Arrangements and Institutional Dynamics' (STAR-FLOOD Consortium 2016).
O'Leary M and Sherk GW, 'Reinventing the Interstate Water Compact: A New Model' (Proceedings of the Rocky Mountain Mineral Law Institute' 2006).
Powell JW, 'Report on the Lands of the Arid Region of the United States, with a More Detailed Account of the Lands of Utah. With a More Detailed Account of the Lands of Utah. With Maps.' (1978) H.R. Exec. Doc. 73, 45th Cong., 2d Sess.
Pucher J, 'The EGTCs Investing: Implementing EU Funds. Which Role in the European Fund for Strategic Investments ? Which Procurement ?' (2016).
Rieu-Clarke A and Loures F, 'The Role and Relevance of the UN Convention on the Law of the Non-Navigational Uses of International Watercourses to the EU and Its Member States' (IHP-HELP Centre for Water Law, Policy and Science 2010).
Sherk GW, 'Compact Review Summaries' (2005) Utton Center Model Compact Program.
Suykens C, 'Water Management and Maritime Spatial Planning' (2013) 1 IUCNAEL 59.

ONLINE SOURCES

Able M, 'Floods Dominate Natural Catastrophe Statistics in First Half of 2013' *Munich RE* (2013) 4 <https://www.munichre.com/en/media-relations/publications/press-releases/2013/2013-07-09-press-release/index.html> accessed 17 July 2017.
Adviescommissie Water, 'Bruggen Bouwen Nederlands Waterbeheer in Europees En Grensoverschrijdend Perspectief' (2007) <www.adviescommissiewater.nl/Images/Advies%20over%20Nederlands%20Waterbeheer%20in%20Europsees%20perspectief%20-%2019%20november%202007_tcm308-268502.pdf> accessed 13 July 2017.
Bateman B and Rancier R, 'Case Studies in Integrated Water Resources Management: From Local Stewardship to National Vision' (American Water Resources Association 2012) <http://awra.org/committees/AWRA-Case-Studies-IWRM.pdf> accessed 14 July 2017.
Battelle Memorial Institute, 'Delaware River Basin Comprehensive Flood Risk Management Interim Feasibility Study and Integrated Environmental Assessment for New Jersey' (2015) US Army Corps of Engineers <www.nap.usace.army.mil/Missions/Civil-Works/Delaware-River-Basin-Comprehensive-Study/> accessed 17 July 2017.
Bos-Gorter L, Huitema D, de Kerkhof M, 'Public Participation on its Own Barricades: Citizens' Jury on Water Management from Experiment to Instrument' <www.macaulay.ac.uk/pathconference/outputs/PATH_abstract_5.1.3.pdf> accessed 10 July 2017.
Bovee K and others, 'A Decision Support Framework for Water Management in the Upper Delaware River' (2007) <https://www.fort.usgs.gov/sites/default/files/products/publications/21938/21938_508.pdf> accessed 14 July 2017.
Commission, 'Public Consultation to Support the Fitness Check of EU Freshwater: Policy Analysis of the Responses' (2012) <http://ec.europa.eu/environment/water/blueprint/pdf/public%20consultation%20report.pdf> accessed 11 June 2017.

DePalma A, 'A Fight over Taming the Delaware' *The New York Times* (20 April 2008) <www.nytimes.com/2008/04/20/nyregion/nyregionspecial2/20delawarenj.html?pagewanted=all> accessed 4 May 2017.

Browner C, 'Watershed Approach Framework' (1996) EPA <https://www.epa.gov/sites/production/files/2015-06/documents/watershed-approach-framework.pdf> accessed 14 July 2017.

Circulaire 5 février 1981 relative aux contrats de rivières (1981) <www.gesteau.fr/sites/default/files/Circulaire_contratriviere_19810205.pdf> accessed 13 July 2017.

Circulaire du 24 octobre 1994 relative au plan décennal de restauration et d'entretien des rivières (1994) <www.gesteau.fr/sites/default/files/circulaire_19941024_cr.pdf> accessed 13 July 2017.

CIW, 'Tweede Waterbeleidsnota – Inclusief Waterbeheerskwesties' (2013) <www.integraal waterbeleid.be/nl/publicaties/afbeeldingen/Tweede%20Waterbeleidsnota_LR.pdf/at_download/file> accessed 13 July 2017.

CIW, 'Flemish River Basin Management Plan for the Scheldt 2016-2021' (2015) <www.integraalwaterbeleid.be/nl/stroomgebiedbeheerplannen/stroomgebiedbeheerplannen-2016-2021/documenten/Vlaams_deel_stroomgebied_Schelde.pdf>.

CIW, 'Annex III to the Flemish River Basin Management Plan for the Scheldt 2016-2021' (2015), <www.integraalwaterbeleid.be/nl/stroomgebiedbeheerplannen/overwegingsdocumenten/overwegingsdocument-bij-beheerplannen-vlaamse-delen-schelde/at_download/file> accessed 14 April 2017.

Cohen-Shacham E, Walters, G, Janzen, C and Maginnis, S, 'Nature-Based Solutions to Address Societal Challenges' (2016) <https://portals.iucn.org/library/sites/library/files/documents/2016-036.pdf> accessed 17 July 2017.

Comité de Bassin Artois-Picardie, 'Schéma Directeur d'Aménagement et de Gestion des Eaux du Bassin Artois-Picardie pour la période 2016 à 2021' (2016) <www.artois-picardie.eaufrance.fr/IMG/pdf/sdage_2016-2021.pdf> accessed 12 July 2017.

Commissie van Advies inzake de Waterstaatswetgeving, 'Samenwerking over Grenzen. Relatie tussen Europese Kaderrichtlijn Water en Internationale Verdragen' (2004) <http://cawsw.nl/documenten/Samenwerking over grenzen implementatie EKW_CAW advies_22112004.pdf> accessed 13 July 2017.

Conference of Upper Delaware Townships, 'River Management Plan for the Upper Delaware Scenic and Recreational River' (1986) <www.upperdelawarecouncil.org/wp-content/uploads/Upper_Delaware_River_Management_Plan_1986_updated_2015.pdf> accessed 14 July 2017.

DePhilip M and Moberg T, 'Ecosystem Flow Recommendations for the Delaware River Basin' (2013) <www.nj.gov/drbc/library/documents/TNC_DRBFlowRpt_dec2013.pdf> accessed 17 July 2017.

Di Vittorio S, Cole N and Cooper T, 'Accountability in Emerging Forms of Governance: A Comparison of the California Bay-Delta Process and the European Water Framework Directive' (2008) Water Resources Center, UC Berkeley <http://escholarship.org/uc/item/9cg8n96f> accessed 17 July 2017.

Decision of the Political College of the Flemish-Dutch Scheldt Commission related to the Creation and Activation of the "Scheldt Council", 15 April 2014 <www.benelux.int/files/4013/9754/6191/intellingsbesluitSchelderaad.pdf> accessed 15 April 2017.

Delaware River Basin Commission, 'Comprehensive Plan' (2001) <www.nj.gov/drbc/library/documents/comprehensive_plan.pdf> accessed 14 July 2017.

Delaware River Basin Commission, 'Water Resources Plan for the Delaware River Basin' (2004) <www.nj.gov/drbc/library/documents/BasinPlan_Sept04.pdf> accessed 4 May 2017.

Delaware River Basin Commission, 'Statement of Carol Collier before the Subcommittee on Water Resources and Environment Committee on Transportation and Infrastructure United States House of Representatives on Comprehensive Watershed Management and Planning' (2008) <http://nj.gov/drbc/library/documents/crc_testimony062408.pdf> accessed 4 May 2017.

Delaware River Basin Commission, 'Administrative Manual: Rules of Practice and Procedure' (2015) <www.nj.gov/drbc/library/documents/admin_manual.pdf> accessed 17 July 2017.

Delaware River Basin Commission, 'Flexible Flow Management Program' (2016) <https://water.usgs.gov/osw/odrm/ffmp/2016/2016_FFMP_Agreement_Signed.pdf> accessed 13 July 2017.

Delli Priscoli J, 'Case Study of River Basin Organizations' Oregon State University <www.transboundarywaters.orst.edu/research/case_studies/River_Basin_Organization_New.htm> accessed 14 July 2017.

De Wit K, Meersschaut Y and Sas M, 'Third Enlargement Program of the River Scheldt' (2000) VLIZ <www.vliz.be/imisdocs/publications/224252.pdf> accessed 13 July 2017.

Environmental Council for Abington Township, Letter to the Delaware River Basin Commission regarding the Natural Gas Development Guidelines <www.state.nj.us/drbc/library/documents/NGC/Officials/PAAbington041511.pdf> accessed 5 May 2017.

Dyson M, Bergkamp G and Scanlon J, 'Flow: The Essentials of Environmental Flows' (IUCN 2003) <https://www.iucn.org/content/flow-essentials-environmental-flows-0> accessed 17 July 2017.

EPA Office of Water, 'Watershed Approach Framework' (1996) EPA 840-S-96-001 <https://www.epa.gov/sites/production/files/2015-06/documents/watershed-approach-frame work.pdf> accessed 13 July 2017.

Flemish Parliament, debate Committee on public works, mobility and energy, meeting (15 October 2002) <https://docs.vlaamsparlement.be/website/htm-vrg/327696.html> accessed 12 July 2017.

Environmental Justice Foundation, 'No Place Like Home: Where next for Climate Change Refugees?' (2009) <https://ejfoundation.org//resources/downloads/no-place-like-home.pdf> accessed 17 July 2017.

European Parliament, 'Draft Report on the proposal for a decision of the European Parliament and of the Council on mobilisation of the European Union Solidarity Fund, in accordance with point 26 of the Interinstitutional Agreement of 17 May 2006 between the European Parliament, the Council and the Commission on budgetary discipline and sound financial management' (2010) 2010/2237 (BUD) <www.europarl.europa.eu/meetdocs/2009_2014/documents/budg/pr/835/835711/835711en.pdf> accessed 12 July 2017.

European Parliament, 'Report on European Territorial Cooperation – best practices and innovative measures' (2015/2280(INI)) (2015) <www.europarl.europa.eu/sides/

getDoc.do?pubRef=-//EP//NONSGML+REPORT+A8-2016-0202+0+DOC+PDF+V0 //EN>accessed 9 June 2016.

Flemish Government, 'Vlaams Adaptatieplan 2013-2020' (2013) <https://www.lne.be/sites/default/files/atoms/files/2013-06-28_VAP.pdf> accessed 12 July 2017.

Flemish-Dutch Scheldt Commission, 'Agenda for the Future' (2015) <www.vnsc.eu/uploads/2015/11/vnsc-agendavoordetoekomst-hr.pdf> accessed 25 April 2017.

Flemish Parliament, 'Commission meeting foreign policy of 1 April 2004' (2004) C172-BU17, Parl. St. 2013-2014 <http://docs.vlaamsparlement.be/docs/handelingen_commissies/2013-2014/c0m172bui7-01042014.pdf> accessed 13 July 2017.

Gallatin A, Report of the Secretary of the Treasury; on the Subject of Public Roads and Canals; made in pursuance of a Resolution of the Senate, of March 2, 1807 (1808) (The Gallatin Plan) <http://oll.libertyfund.org/titles/2046> accessed 13 July 2017.

Garmestani AS Allen, CR, Ruhl, JB, Holling, CS, 'The Integration of Social-Ecological Resilience and Law' (2014) Nebraska Cooperative Fish & Wildlife Research Unit – Staff Publications <http://digitalcomons.unl.edu/ncfwrustaff/144> accessed 17 July 2017.

Gilbert A, Schaafsma M, De Nocker L, Liekens I, Broekx S, 'Case Study Status Report Scheldt River Basin' (2007) Flemish Institute for Technological Research and Institute for Environmental Studies <www.ivm.vu.nl/en/Images/D26_Status_report_international_Scheldt_case_study_tcm234-188709.pdf> accessed 12 July 2017.

Gleick PH and Palaniappan M, 'Peak Water Limits to Freshwater Withdrawal and Use' (2010) 107 Proceedings of the National Academy of Sciences <www.pnas.org/cgi/doi/10.1073/pnas.1004812107 PNAS> accessed 17 July 2017.

Global Water Partnership Technical Advisory Committee, 'Integrated Water Resources Management', vol 62 (2000) TAC Background Papers n 4 <www.gwp.org/globalassets/global/gwp-cacena_files/en/pdf/tec04.pdf> accessed 17 July 2017.

Gray D and Sadoff CW, 'Water for Growth and Development: A Theme Document of the 4 Th World Water Forum Water for Growth and Development' (2006) The World Bank <http://siteresources.worldbank.org/INTWRD/Resources/FINAL_0601_SUBMITTED_Water_for_Growth_and_Development.pdf> accessed 17 July 2017.

Grieshaber K, 'Central Europe Floods Are Most Dramatic In A Decade' *The Huffington Post* (2013) <www.independent.co.uk/environment/nature/central-europe-under-water-floods-wreak-havoc-in-germany-and-hungary-8650751.html> accessed 17 July 2017.

Holmes BH, 'A History of Federal Water Resources Programs, 1800-1960' (1972) US Department of Agriculture, Economic Research Service <https://archive.org/details/historyoffederal1233holm> accessed 13 July 2017.

IDMC, 'Global Report on Internal Displacement' (IDMC 2016) 21 <www.internal-displacement.org/assets/publications/2016/2016-global-report-internal-displacement-IDMC.pdf> accessed 21 February 2017.

International Commission for the Protection of the Danube River, 'Danube River Basin Strategy for Public Participation in River Basin Management Planning 2003-2009' (ICPDR 2003) <https://www.icpdr.org/main/sites/default/files/PP%20Strategy%20-%20FINAL.pdf> accessed 10 July 2017.

Internationale Commissie ter Bescherming van de Rijn, 'Reactie van de ICBR / Het Coördineringscomité Op de Zienswijzen T.a.v. Het Concept-ORBP 2014' (2015) <www.iksr.org/fileadmin/user_upload/Dokumente_nl/Rijn_actueel/R%C3%A9actie_ORBP_2014.pdf> accessed 27 April 2017.

International Scheldt Commission, 'Report of the 21st plenary meeting' (2013) <www.isc-cie.org/images/Documents/PLEN_1301_PV.pdf> accessed 13 July 2017.

International Scheldt Commission, 'Work plan for the period 2013-2015' <www.isc-cie.org/images/Documents/PLEN_1301_WERKPLAN2013-2015_REV1.pdf> accessed 13 May 2014.

International Scheldt Commission, 'Roof Report of the River Basin Management Plan for the International River Basin District Scheldt 2016-2021' (2015) <www.isc-cie.org/images/Documents/ODB2-PFPG2_RAPPORT_ENG_V17.pdf> 10.

John H (ed), 'Tips and Tricks for Water Framework Directive Implementation' (2004) WWF and EEB <www.rivernet.org/general/docs/200403_EEB_WWF_Tips&Tricks.pdf> accessed 20 October 2014.

League of Women Voters of Pennsylvania Citizen Education Fund Common Ground Project, 'Water Use and Water Rights in Pennsylvania' (1998) <http://wren.palwv.org/documents/WaterUse_000.pdf> accessed 14 July 2017.

Nicolas Levrat, 'The European Grouping of Territorial Cooperation' (2007) <http://cor.europa.eu/en/documentation/studies/Documents/The%20European%20Grouping%20of%20Territorial%20Cooperation%20-%20EGTC/The%20European%20Grouping%20of%20Territorial%20Cooperation%20-%20EGTC%20-%20EN.pdf> accessed 13 June 2017.

Mechlem K, 'Groundwater Governance – A Global Framework for Action' (2012) GEF Thematic Paper n 6: Legal & Institutional Frameworks <www.fao.org/3/a-i6040e.pdf> accessed 17 July 2017.

Ministerie van Infrastructuur en Milieu and Ministerie van Economische Zaken, 'Nationaal Waterplan' (2009) <https://www.rijksoverheid.nl/binaries/rijksoverheid/documenten/brochures/2011/03/28/nationaal-waterplan/12dw2010g226.pdf> accessed 12 July 2017.

Ministerie van Infrastructuur en Milieu, 'Overstromingsrisicobeheerplan voor het Stroomgebied van de Schelde 2016-2020' (2015) <https://www.rijksoverheid.nl/binaries/rijksoverheid/documenten/rapporten/2014/12/12/4-4-overstromingsricicobeheerplan-schelde/4-4-overstromingsricicobeheerplan-schelde.pdf> accessed 13 July 2017.

Ministerie van Infrastructuur en Milieu and Ministerie van Economische Zaken, 'Nationaal Waterplan' (2009) <https://www.rijksoverheid.nl/binaries/rijksoverheid/documenten/brochures/2011/03/28/nationaal-waterplan/12dw2010g226.pdf> accessed 12 July 2017.

OECD 'Seine Basin, Île-de-France, 2014: Resilience to Major Floods' (2014) OECD <https://www.oecd.org/gov/risk/Flood-risk-management-seine-river-executive-summary.pdf> accessed 12 July 2017.

OECD, 'OECD Principles on Water Governance' (2015) OECD <https://www.oecd.org/cfe/regional-policy/OECD-Principles-on-Water-Governance-brochure.pdf> accessed 17 July 2017.

Partnership for the Delaware Estuary, 'Comprehensive Conservation and Management Plan for the Delaware Estuary' (1996) <https://s3.amazonaws.com/delawareestuary/pdf/CCMP.pdf> accessed 4 May 2017.

Paterson T, 'Central Europe under Water: Floods Wreak Havoc in Germany and Hungary' *The Independent* (2013) <www.independent.co.uk/environment/nature/central-europe-under-water-floods-wreak-havoc-in-germany-and-hungary-8650751.html> accessed 17 July 2017.

Philadelphia Water Department, 'The Schuylkill River Watershed Source Water Protection Plan Philadelphia Water Department' (2006) <www.phillywatersheds.org/doc/Schuylkill_SWPP_2006.pdf> accessed 4 May 2017.

Poelmans L, 'Modelling Urban Expansion and its Hydrological Impacts' (Ph.D. Thesis, KU Leuven 2010) <http://statbel.fgov.be/nl/binaries/MODELLING%20URBAN%20EXPANSION%20AND%20ITS%20HYDROLOGICAL%20IMPACTS_tcm325-116170.pdf> accessed 12 July 2017.

Préfet Coordonnateur de Bassin Artois-Picardie, 'Projet de Plan de Gestion Des Risques D'inondation Du Bassin Artois-Picardie 2016-2021' (2015) <https://www.hauts-de-france.developpement-durable.gouv.fr/IMG/pdf/1-pgri_artois-picardie-version-finale.pdf> accessed 12 July 2017.

Rosado R and Cuomo A, 'Local Government Handbook' (2011) New York State, Department of State 2011 <https://www.dos.ny.gov/lg/publications/Local_Government_Handbook.pdf> accessed 13 July 2017.

Rosillon F, 'Eau et Territoire À Travers L'expérience Des Contrats de Rivière En Wallonie (Belgique)' (2015) 25-26 Territoire en Mouvement Revue de Géographie et Aménagement 1 <http://tem.revues.org/2814> accessed 12 July 2017.

Rüttinger L, Smith D, Stang G, Tänzler D, Vivekananda J, 'A New Climate for Peace: Taking Action on Climate and Fragility Risks', (Adelphi, International Alert, Woodrow Wilson International Center for Scholars, European Union Institute for Security Studies 2015) <http://newsroom.unfccc.int/media/252731/newclimateforpeace.pdf> accessed 17 July 2017.

Secretariaat Bekken van de Gentse Kanalen, 'Stroomgebiedbeheerplan Voor de Schelde 2016-2021: Bekkenspecifiek Deel Gentse Kanalen' (2016) <www.integraalwaterbeleid.be/nl/stroomgebiedbeheerplannen/stroomgebiedbeheerplannen-2016-2021/documenten/Bekken_Gentse_Kanalen.pdf/view> accessed 13 July 2017.

Secretariaat Benedenscheldebekken, 'Stroomgebiedbeheerplan Voor de Schelde 2016-2021: Bekkenspecifiek Deel Benedenscheldebekken' (2016) <www.integraalwaterbeleid.be/nl/stroomgebiedbeheerplannen/stroomgebiedbeheerplannen-2016-2021/documenten/Benedenscheldebekken.pdf/view> accessed 13 July 2017.

Secretariaat Brugse Polders, 'Stroomgebiedbeheerplan voor de Schelde 2016-2021: Bekkenspecifiek Deel Brugse Polders (2016) <www.integraalwaterbeleid.be/nl/stroomgebiedbeheerplannen/stroomgebiedbeheerplannen-2016-2021/documenten/Bekken_Brugse_Polders.pdf/at_download/file> accessed 13 July 2017.

Secretariaat Demerbekken, 'Stroomgebiedbeheerplan Voor de Schelde 2016-2021: Bekkenspecifiek Deel Demerbekken' (2016) <www.integraalwaterbeleid.be/nl/bekkens/demerbekken/planning-2016-2021/bekkenspecifiek-deel-ijzerbekken> accessed 13 July 2017.

Secretariaat Denderbekken, 'Stroomgebiedbeheerplan Voor de Schelde 2016-2021: Bekkenspecifiek Deel Dender' (2016) <www.integraalwaterbeleid.be/nl/stroomgebiedbeheerplannen/overwegingsdocumenten/od_den/at_download/file> accessed 13 July 2017.

SERV, 'Ontwerp van decreet tot wijziging van de Wet van 28 december 1967 betreffende de onbevaarbare waterlopen, meer bepaald de wijziging van de classificatie en andere diverse wijzigingen' (12 September 2013) n 2013/039 <https://www.minaraad.be/themas/hinder/ontwerpdecreet-onbevaarbare-waterlopen-adviesvraag/13-039%20advies%20onbevaarbare%20waterlopen.pdf/download> accessed 12 July 2017.

Service Public de Wallonie, 'Plan de Gestion des Risques d'Inondations pour le District Hydrographique Internationale de l'Escaut' (2016) <http://environnement.wallonie.be/inondations/files/pgri/PGRI_ESCAUT_20160304.pdf> accessed 12 July 2017.

Sanchez R and Schmidt S, 'Environmental Flows as A Tool to Achieve the WFD Objectives' (INTECSA-INARSA 2012) 20 <www.pianc.org/downloads/euwfd/Eflows%20discussion%20paper%20draft%20NAVI%20TG%20response%20v2%2010-8-12.pdf> accessed 14 July 2017.

Speed R, Yuanyuan, L, Le Quesne T, 'River Basin Planning: Principles, Procedures and Approaches for Strategic Basin Planning' (2013) UNESCO <http://unesdoc.unesco.org/images/0022/002208/220875^{e}.pdf> accessed 17 July 2017.

Stratec, Environmental Impact Assessment of the Flood Risk Management Plan of the International River Basin District of the Scheldt (2015) 26 <http://environnement.wallonie.be/inondations/files/rie/RIE_PGRIEscaut_vf_NL.pdf>, accessed 31 March 2016.

Strategische Adviesraad internationaal Vlaanderen en de Vlaamse Adviesraad voor Bestuurszaken, 'Advies over de instemming met het derde Protocol bij de Europese Kaderovereenkomst inzake grensoverschrijdende samenwerking betreffende euroregionale samenwerkingsverbanden' (20 July 2011) Parl. St. 1393 (2011-2012) <https://www.vlaanderen.be/nl/publicaties?uitgever=Strategische%20Adviesraad%20Internationaal%20Vlaanderen%20-%20SAR%20iV%20/%20Flemish%20Foreign%20Affairs%20Council> accessed 13 July 2017.

Technische Scheldecommissie, 'Long-Term Vision Scheldt-Estuary' (2001) <www.vnsc.eu/uploads/2012/01/ltv-schelde-estuarium-18-januari-2001.pdf> accessed 14 May 2014.

Terenzio V, 'Development and Effect of the Delaware River Basin Compact' (1962) 54 American Water Works Association 1445 <www.mnawwa.org/about/councils/Letterhead20102011.pdf> accessed 13 July 2017.

UN-Water, 'Transboundary Waters: Sharing Benefits, Sharing Responsibilities' (2008) 2 <www.unwater.org/app/uploads/2017/05/UNW_TRANSBOUNDARY.pdf> accessed 16 July 2017.

Unece, 'Guide to Implementing the Water Convention' (2013) <https://www.unece.org/fileadmin/DAM/env/water/publications/WAT_Guide_to_implementing_Convention/ECE_MP.WAT_39_Guide_to_implementing_water_convention_small_size_ENG.pdf> accessed 17 July 2017.

United Nations International Strategy for Disaster Reduction, 'UNISDR Terminology on Disaster Risk Reduction' (2009) <https://www.unisdr.org/files/7817_UNISDRTerminologyEnglish.pdf> accessed 17 July 2017.

Upper Delaware Council, 'Partnering to Protect the River' (2008) <www.upperdelawarecouncil.org/wp-content/uploads/partnering_to_protect_river_supplement.pdf> 5 May 2017.

US National Climate Assessment, 'Climate Change Impacts in the United States' (2014) 71 <http://s3.amazonaws.com/nca2014/high/NCA3_Climate_Change_Impacts_in_the_United%20States_HighRes.pdf> accessed 13 July 2017.

Vlaams-Nederlandse Scheldecommissie, 'Evaluatie Verdrag Beleid en Beheer Schelde-Estuarium' (2013) <www.vnsc.eu/uploads/2015/06/evaluatie-beleid-en-beheer-schelde-estuarium-1.pdf> accessed 26 April 2017.

Watermark Initiative, 'US Water Stewardship: A Critical Assessment of Interstate Watershed Agreements' (2009) <http://watermarkinitiative.com/wp-content/uploads/2013/10/WMI_Report_09-13.pdf> accessed 10 May 2017.

Weingartner A and Raich-montiu J, 'Proposal for a Water Security Plan to Improve the Detection of Threats in the Distribution Network Affecting Drinking Water Quality' (Joint Research Centre, European Union 2015) <https://erncip-project.jrc.ec.europa.eu/sites/default/files/JRC101135_Proposal%20for%20a%20water%20security%20plan%20to%20improve%20the%20detection%20of%20threats%20in%20the%20distribution%20network%20affecting%20drinking%20water%20quality.pdf> accessed 13 June 2017.

World Business Council for Sustainable Development, 'Business Guide to Circular Water Management: Spotlight on Reduce, Reuse and Recycle Contents' (2017) <www.wbcsd.org/contentwbc/download/3437/44956> accessed 13 June 2017.

Wouters J and Vidal M, 'Grensoverschrijdende Samenwerking Tussen West-Vlaanderen En Noord-Frankrijk Na de Inwerkingtreding van Het Frans-Belgisch Akkoord van 16 September 2002' (2005) Provincie West-Vlaanderen – wvi – Leiedal <https://lirias.kuleuven.be/bitstream/123456789/154985/1/GOSA+West-Vlaanderen.pdf> accessed 17 July 2017.

Wouters P, Vinogradov S, Allan A, Jones P and Rieu-Clarke A, 'Sharing Transboundary Waters – An Integrated Assessment of Equitable Entitlement: the Legal Assessment Model' (2005) IHP Technical Documents in Hydrology <http://unesdoc.unesco.org/images/0013/001397/139794e.pdf> accessed 17 July 2017.

Wouters P, 'International Law – Facilitating Transboundary Water Cooperation' (2012) TEC Background Papers n 17 <www.gwp.org/globalassets/global/toolbox/publications/background-papers/17-international-law---facilitating-transboundary-water-cooperation-2013-english.pdf> accessed 13 July 2017.

Young C, 'Shaping Water Policy: Political Influences in the European Union and United States' (2011) Claremont-UC Undergraduate Research Conference on the European Union <http://scholarship.claremont.edu/cgi/viewcontent.cgi?article=1058&context=urceu> accessed 13 July 2017.

Zaporozhets V, 'Power Distribution in French River Basin Committees' (2015) Toulouse School of Economics, Working Paper n TSE-558 <www.tse-fr.eu/sites/default/files/TSE/documents/doc/wp/2015/wp_tse_558.pdf> accessed 13 July 2017.

Zillmer S and Toptsidou M, 'Potential and Limits of the EGTC Instrument for Enhancing Integration across Borders' (Spatial Foresight 2014) <www.spatialforesight.eu/tl_files/files/editors/dokumente/Brief-2014-5-141111.pdf> accessed 14 June 2017.

www.ingramcontent.com/pod-product-compliance
Ingram Content Group UK Ltd.
Pitfield, Milton Keynes, MK11 3LW, UK
UKHW051919100926
8556IPUK00021B/1268

9 781780 686318